Plant Pigments and Their Manipulation

Plant Pigments and Their Manipulation

Contributors

Fernanda Ventorim Pacheco, Helbert Rezende de Oliveira Silveira et al.

AURIS
Reference

www.aurisreference.com

Plant Pigments and Their Manipulation

Contributors: Fernanda Ventorim Pacheco, Helbert Rezende de Oliveira Silveira et al.

Published by Auris Reference Limited

www.aurisreference.com

United Kingdom

Copyright 2016

Printed in 2017 for Sale in the Indian Subcontinent

Plant Pigments and Their Manipulation

ISBN: 978-1-78154-867-7

British Library Cataloguing in Publication Data
A CIP record for this book is available from the British Library

Printed in the United Kingdom

Exclusively distributed by CBS Publishers & Distributors Pvt. Ltd.

Sales & Distribution Rights only for India, Pakistan, Bangladesh, Sri Lanka, Nepal and Bhutan. This book is not to be sold outside these territories.

Contents

List of Abbreviations

ARI	Anthocyanin reflectance index
CLH	Chlorophyllase
DAG	Department of Agriculture
DMSO	Dimethyl sulfoxide
DOPA	Dihydroxyphenylalanine
DSSC	Dye sensitized solar cell
DVP	Deficit vapor pressure
FM	Fresh mass
GFP	Green fluorescent protein
HPLC	High performance liquid chromatography
HPLC	High-pressure liquid chromatography
HT	Titanium paste
IPCE	Incident photon-to-current conversion efficiencies
IPCE	Incident photo-to-current conversion efficiencies
MNRE	Ministry of New and Renewable Energy
NBDVI	Narrow band indices include difference vegetation index
ORF	Open reading frames
PAR	Photosynthetically active radiation
PEC	Photoelectrochemical
PPFD	Photosynthetically active photon flux density
PRI	Photochemical reflectance index
PVI	Perpendicular vegetation index
TiO2	Titanium dioxide
TOH	Tyrosine hydroxylase
UBD	Universiti Brunei Darussalam
VPD	vapor pressure deficit

List of Contributors

Fernanda Ventorim Pacheco
Setor de Fisiologia Vegetal, Departamento de Biologia, UFLA-Universidade Federal de Lavras, Lavras, Brasil

Helbert Rezende de Oliveira Silveira
Setor de Fisiologia Vegetal, Departamento de Biologia, UFLA-Universidade Federal de Lavras, Lavras, Brasil

Amauri Alves Alvarenga
Setor de Fisiologia Vegetal, Departamento de Biologia, UFLA-Universidade Federal de Lavras, Lavras, Brasil

Ivan Caldeira Almeida Alvarenga
Departamento de Agricultura Geral, UFLA-Universidade Federal de Lavras, Lavras, Brazil

José Eduardo Brasil Pereira Pinto
Departamento de Agricultura Geral, UFLA-Universidade Federal de Lavras, Lavras, Brazil

Jean Marcel Sousa Lira
Setor de Fisiologia Vegetal, Departamento de Biologia, UFLA-Universidade Federal de Lavras, Lavras, Brasil

Jingfeng Huang
Institute of Agricultural Remote Sensing & Information Application, Zijingang Campus, Zhejiang University, Hangzhou, China

Chen Wei
Zhejiang Meteorological Service Center, Hangzhou, China

Yao Zhang
Institute of Agricultural Remote Sensing & Information Application, Zijingang Campus, Zhejiang University, Hangzhou, China

George Alan Blackburn
Lancaster Environment Centre, Lancaster University, Lancaster, United Kingdom

Xiuzhen Wang
Institute of Remote Sensing and Earth Sciences, Hangzhou Normal University, Hangzhou, China

Chuanwen Wei
Institute of Agricultural Remote Sensing & Information Application, Zijin-gang Campus, Zhejiang University, Hangzhou, China

Jing Wang
Institute of Agricultural Remote Sensing & Information Application, Zijin-gang Campus, Zhejiang University, Hangzhou, China

Arturo Solís Herrera
Human Photosynthesis Study Center, Aguascalientes, Mexico

Azamjon B. Soliev
Department of Environmental Systems Engineering, Kochi University of Technology, 185 Miyanokuchi, Tosayamada, Kami, Kochi 782-8502, Japan

Kakushi Hosokawa
Department of Environmental Systems Engineering, Kochi University of Technology, 185 Miyanokuchi, Tosayamada, Kami, Kochi 782-8502, Japan

Keiichi Enomoto
Department of Environmental Systems Engineering, Kochi University of Technology, 185 Miyanokuchi, Tosayamada, Kami, Kochi 782-8502, Japan

Aiman Yusoff
Faculty of Science & Institute for Biodiversity & Environmental Research, Universiti Brunei Darussalam, Tungku Link, Gadong, BE1410, Brunei Da-russalam

N. T. R. N. Kumara
Applied Physics Program, Faculty of Science, Universiti Brunei Darussalam, Jalan Tungku Link, Gadong BE1410, Brunei Darussalam

Andery Lim
Faculty of Science & Institute for Biodiversity & Environmental Research, Universiti Brunei Darussalam, Tungku Link, Gadong, BE1410, Brunei Da-russalam

Piyasiri Ekanayake
Applied Physics Program, Faculty of Science, Universiti Brunei Darussalam, Jalan Tungku Link, Gadong BE1410, Brunei Darussalam

Kushan U. Tennakoon
Faculty of Science & Institute for Biodiversity & Environmental Research, Universiti Brunei Darussalam, Tungku Link, Gadong, BE1410, Brunei Darussalam

Giorgio Manenti
Istituto di Scienze Botaniche, Universita di Milano, Milano, Italy

Giuliano Tedesco
Istituto di Scienze Botaniche, Universita di Milano, Milano, Italy

Kristin R. Abney
Plant Sciences Department, The University of Tennessee, 2431 Joe Johnson Drive, Knoxville, TN 37996, USA

Dean A. Kopsell
Plant Sciences Department, The University of Tennessee, 2431 Joe Johnson Drive, Knoxville, TN 37996, USA

Carl E. Sams
Plant Sciences Department, The University of Tennessee, 2431 Joe Johnson Drive, Knoxville, TN 37996, USA

Svetlana Zivanovic
Department of Food Science and Technology, The University of Tennessee, 2605 River Drive, Knoxville, TN 37996, USA

David E. Kopsell
Department of Agriculture, Illinois State University, Normal, IL 61790, USA

Reena Kushwaha
Department of Chemistry, Faculty of Science, Banaras Hindu University, Varanasi 221005, India

Pankaj Srivastava
Department of Chemistry, Faculty of Science, Banaras Hindu University, Varanasi 221005, India

Lal Bahadur
Department of Chemistry, Faculty of Science, Banaras Hindu University, Varanasi 221005, India

Alítcia Moraes Kleinowski
Laboratório de Cultura de Tecidos de Plantas; Departamento de Botânica; Instituto de Biologia;Universidade Federal de Pelotas; Capão do Leão - RS - Brasil

Isabel Rodrigues Brandão
Laboratório de Cultura de Tecidos de Plantas; Departamento de Botânica; Instituto de Biologia;Universidade Federal de Pelotas; Capão do Leão - RS - Brasil

Andersom Milech Einhardt
Laboratório de Cultura de Tecidos de Plantas; Departamento de Botânica; Instituto de Biologia;Universidade Federal de Pelotas; Capão do Leão - RS - Brasil

Márcia Vaz Ribeiro
Laboratório de Cultura de Tecidos de Plantas; Departamento de Botânica; Instituto de Biologia;Universidade Federal de Pelotas; Capão do Leão - RS - Brasil

José Antonio Peters
Laboratório de Cultura de Tecidos de Plantas; Departamento de Botânica; Instituto de Biologia;Universidade Federal de Pelotas; Capão do Leão - RS - Brasil

Eugenia Jacira Bolacel Braga
Laboratório de Cultura de Tecidos de Plantas; Departamento de Botânica; Instituto de Biologia;Universidade Federal de Pelotas; Capão do Leão - RS - Brasil

Nilangani N Harris
New Zealand Institute for Plant & Food Research Limited, Private Bag 11-600, Palmerston North, New Zealand

Commonwealth Scientific and Industrial Research Organization, Ecosystem Sciences, Urrbrea, South Australia 5064, Australia

John Javellana
New Zealand Institute for Plant & Food Research Limited, Private Bag 11-600, Palmerston North, New Zealand

Kevin M Davies
New Zealand Institute for Plant & Food Research Limited, Private Bag 11-600, Palmerston North, New Zealand

David H Lewis
New Zealand Institute for Plant & Food Research Limited, Private Bag 11-600, Palmerston North, New Zealand

Paula E Jameson
School of Biological Sciences, University of Canterbury, Private Bag 4-800, Christchurch, New Zealand

Simon C Deroles
New Zealand Institute for Plant & Food Research Limited, Private Bag 11-600, Palmerston North, New Zealand

Kate E Calcott
New Zealand Institute for Plant & Food Research Limited, Private Bag 11-600, Palmerston North, New Zealand
Victoria University of Wellington, Wellington 6140, New Zealand

Kevin S Gould
Victoria University of Wellington, Wellington 6140, New Zealand

Kathy E Schwinn
New Zealand Institute for Plant & Food Research Limited, Private Bag 11-600, Palmerston North, New Zealand

Xueyun Hu
Institute of Low Temperature Science, Hokkaido University

Ayumi Tanaka
Institute of Low Temperature Science, Hokkaido University
CREST/JST, Hokkaido University

Ryouichi Tanaka
Institute of Low Temperature Science, Hokkaido University
CREST/JST, Hokkaido University

Lilian Cristina Baldon Aizza
Departamento de Biologia Vegetal. Rua Monteiro Lobato 970, Instituto de Biologia, Universidade Estadual de Campinas, Cidade Universitária Zeferino Vaz, 13083-970 Campinas, SP, Brazil

Marcelo Carnier Dornelas
Departamento de Biologia Vegetal. Rua Monteiro Lobato 970, Instituto de Biologia, Universidade Estadual de Campinas, Cidade Universitária Zeferino Vaz, 13083-970 Campinas, SP, Brazil

Neeranuch Chairungsi
Department of Chemistry, Faculty of Science, Chiangmai University, Chiang Mai, 50200, Thailand

Kanlaya Jumpatong
Department of Chemistry, Faculty of Science, Chiangmai University, Chiang Mai, 50200, Thailand

Weerachai Phutdhawong
Department of Chemistry, Faculty of Science, Maejo University, Chiang Mai, 50292, Thailand

Duang Buddhasukh
Department of Chemistry, Faculty of Science, Chiangmai University, Chiang Mai, 50200, Thailand

Preface

Plant pigments represent a large and active field of research that impacts fundamental studies of plant function and gene expression, agriculture, the processing of foods and human health. The text *Plant Pigments and their Manipulation* provides a broad review of all the major plant pigment groups, considering the underlying biology, genetic manipulation, and applications of plant pigments within agriculture, food technology and human health. The aim of first chapter is to evaluate the effect of irradiance spectrum in gas exchange and production of photosynthetic pigments of Piper aduncum, considering their medicinal importance and lack of knowledge of the adaptive capacity in photosynthetic process at different ecoambientes. Second chapter undertakes a systematic meta-analysis to determine whether passive optical hyperspectral remote sensing techniques are sufficiently well developed to quantify individual plant pigments, which operational solutions are available for wider plant science and the areas which now require greater focus. The biological pigments in plants physiology have been discussed in third chapter. Fourth chapter gives an overview of the pigmented natural compounds isolated from bacteria of marine origin, based on accumulated data. Impacts of temperature on the stability of tropical plant pigments as sensitizers for dye sensitized solar cells have been investigated in fifth chapter. Sixth chapter describes ultrastructure and pigments of ivy (Hedera helix L.) varieties with green and variegated leaves. In seventh chapter, 16 cultigens of bunching onion (Allium fistulosum L.) are grown in a glasshouse under filtered UV radiation (control) or supplemental UV-B radiation to determine impacts on growth, physiological parameters, and nutritional quality. Eighth chapter reports the performance of four natural dyes extracted from the leaves of teak (Tectona grandis), tamarind (Tamarindus indica), eucalyptus (Eucalyptus globulus), and the flower of crimson bottle brush (Callistemon citrinus). The aim of ninth chapter is to investigate the influence of tyrosine on the *in vitro* growth and the production of the betacyanin pigment in *Alternanthera philoxeroides* and *A. tenella*. In tenth chapter, we show betalain production in species that normally produce anthocyanins, through a combination of genetic modification and substrate feeding. Eleventh chapter presents a general overview of plant pigmentation, together with some general functional and economic aspects. In twelfth chapter, we assess how much chlorophyllide is formed during pigment extraction compared to the amount that naturally occurs in leaves. A genomic approach to study anthocyanin synthesis and flower pigmentation in passionflowers has been proposed in last chapter.

Chapter 1

GAS EXCHANGE AND PRODUCTION OF PHOTOSYNTHETIC PIGMENTS OF PIPER ADUNCUM L. GROWN AT DIFFERENT IRRADIANCES

Fernanda Ventorim Pacheco[1], Helbert Rezende de Oliveira Silveira[1], Amauri Alves Alvarenga[1], Ivan Caldeira Almeida Alvarenga[2], José Eduardo Brasil Pereira Pinto[2], Jean Marcel Sousa Lira[1]

[1]Setor de Fisiologia Vegetal, Departamento de Biologia, UFLA-Universidade Federal de Lavras, Lavras, Brasil

[2]Departamento de Agricultura Geral, UFLA-Universidade Federal de Lavras, Lavras, Brazil

ABSTRACT

The species Piper aduncum is a plant with great economic potential, because the essential oil has antimicrobial activity and insecticides. Thus, as the light directly affects photosynthesis process, the present study evaluated the gas exchanges' variables and photosynthetic pigments' production of P. aduncum grown under different irradiances. Treatments consisted in environments with 100%, 70% and 50% of irradiance and environments with colored nets (red and blue). After 150 days of culture, photosynthetic activity (A), stomatal conductance (gs) transpiration rate (E), internal CO_2 **concentration** (Ci), deficit vapor pressure (DVP), leaf temperature, concentration of internal/external CO_2 (Ci/Ca), carboxylation efficiency (A/Ci) and efficiency water use (A/E) were evaluated. Pigments chlorophyll a, b, carotenoids, total and chlorophyll a/b ratio were also measured. From the observed data, it can be concluded that the species Piper aduncun developed various mechanisms for adaptation to different irradiance conditions. Moreover, it can be inferred that their photosynthetic process is more efficient and has larger spectral intensities of light, with higher efficiency photosynthetic when grown under 70% and 100% of radiation.

INTRODUCTION

Piper aduncun L. (Piperaceae), popularly known as monkey-pepper and long-pepper, is a pioneer shrub, native in Americas—Brazil, Isfoundin Acre, Pará, Mato Grosso, Ceará, Bahia, Minas Gerais, Espirito Santo, Rio de Janeiro, São Paulo and Paraná States [1,2]. Although it is not cultivated commercially, the species has a great potential because of its antimicrobial property, insecticide and essential oil of low toxicity, making them highly valuable for pharmaceutical and chemical industry [3-6]. The search for vegetables that exhibit antimicrobial characteristics has increased in recent years; however, most studies focus solely on the essential oil produced by some medicinal species. The studies of agronomic areas are scarce and limited to a very restricted number of species.

Moreover, cultivation in farming systems, with either economic or conservational purpose, requires a lot of care which depends on the previous knowledge of the physiological and ecological requirements at various stages of vegetative cycle [7]. So, the cultivation of medicinal plants in different environmental conditions is an important and attractive theme for agronomic research, which gives incipient knowledge of the responses of these species in protecting environment.

The radiation in terms of quality and quantity, is surely one of the most important factors in the physical environment, which can affect the morphology and physiology in plants [8,9]. High light intensities, in general, may promote a decrease in photosynthetic efficiency and result in degradation of photosynthetic apparatus [10]. Lima Junior et al. [11], working with Cupaniavernalis, observed changes in light intensity to promote morphological alterations, affecting CO_2 assimilation and transpiration directly, and promoting changes in net photosynthesis. Nevertheless, plants have answers diversified in relation to this environmental factor and most of the studies do not evaluate physiological variables directly, especially those who involved gas exchange, in order to check physiological responses [12].

Adjustments in photosynthetic apparatus determine the adaptive plasticity of species to different irradiance conditions. According to Alvarenga et al. [13], plants with larger efficiency in conversion of radiant energy into chemical energy, achieve more performance and productivity. In this way, studies on gas exchange may be one way to verify the best condition of radiation, which enables greater essential oil and biomass yields in medicinal plants. According to Zavala [14], there are positive correlations between CO_2 assimilation rates and growth and the secondary metabolism.

The aim of this study was to evaluate the effect of irradiance spectrum in gas exchange and production of photosynthetic pigments of Piper aduncum, considering their medicinal importance and lack of knowledge of the adaptive capacity in photosynthetic process at different ecoambientes.

Material and methods

Experiment Site

The experiment was conducted between April and August of 2012, at Gota da Esperança Farm, belonging to the Department of Agriculture (DAG), in Federal University of Lavras, with the following geographical coordinates: 21°14'07"S and 44°58'22"W, at 879 maltitude. The averages of climatic conditions observed during the experiment were provided by Climatological Station of the Department of Agricultural Engineering of UFLA, had maximum temperature of 25.06°C and a minimum of 13.54°C, precipitation of 1.28 mm and a relative humidity of 72.7%.

Plants and Cultivation

Seeds from matrix plants were pre-germinated in Petri dishes on three sheets of filter paper and kept in a growth chamber at 25°C and under a photoperiod of 12 hours, for 30 days. After this period, the seedlings were transferred to polystyrene trays with 72 cells filled with substrate Plantmax® and kept in greenhouse with 50% of shade, until they reach 2.5 cm. After reaching 2.5 cm, the plants, even in the trays, were acclimatized for 14 days before planting in their respective treatments, irrigated daily, once for the plants for cultivation in full sun (100% radiation) were previously acclimatized for 7 days at 70% of irradiance and subsequently for 7 days in full sun before the final planting.

After acclimatization, the seedlings were transplanted to plastic pots 6 liters, containing substrate composed of subsoil, sand and cattle manure in the proportion of 2:1:1, and disposed in different irradiance treatments. The physico-chemical characteristics of the soil were analyzed by the Laboratory Analysis of Soils of Federal University of Lavrasand were: pH: 5.4, P: 4.13 mg·dm^{-3}; K: 73.32 mg·dm^{-3}, Ca: 2.30 cmolc·dm^{-3}, Mg: 0.30 cmolc·dm^{-3}, Al: 0.10 cmolc·dm^{-3}, H + Al: 2.90 cmolc· dm^{-3}, V: 49.00%; organic matter: 2.10 dag·kg^{-1}, Clay: 70.00 dag·kg^{-1}; Silt: 16.00 dag·kg^{-1} and Sand: 14.00 dag·kg^{-1}.

Variables Analyzed

At five months of growth, gas exchange were evaluated, selected at random, two fully expanded leaves, located between the second and third node in five plants per treatment. The variables analyzed were: net photosynthesis (A), stomatal conductance (gs), transpiration (E), intercellular CO_2 concentration (Ci), vapor pressure deficit (VPD) and leaf temperature (TF). In addition, were calculated ratio Ci/Ca, carboxylation efficiency (A/Ci), and the instantaneous efficiency of use of water (A/E).

The measurements were conducted at 9 am, without cloudiness, with portable CO_2 infrared analyzer (LI-6400, LI-COR®), which was operated system with artificial source of photosynthetically active radiation (Blue + Red LED LI-6400-02B, LI-COR®, Lincoln, USA). The density of photosynthetically active photon flux density (PPFD) was 900 mol $m^{-2} \cdot s^{-1}$ treatment with 50% irradiance and the colored nets, 1200 µmol $m^{-2} \cdot s^{-1}$ and treating with 70% of irradiance and 2000 µmol $m^{-2} \cdot s^{-1}$ for the treatment of irradiance at 100%, which was standardized according to the ambient radiation incident on each day of analysis. The CO_2 concentration atmosphere in the chamber was 380 ± 3 µmol^{-1}mol CO_2.

The photosynthetic pigments analyzed chlorophyll a, chlorophyll b, a/b ratio, total chlorophyll and carotenoids. The extraction was performed according to the methodology reported by Lichtenthaler and Buschmann [15], being collected fully expanded leaves located at the third node, five months after transplantation. After collection, leaves were placed in aluminum foil and transported in polystyrene boxes containing ice for immediate extraction and quantification of pigments.

For extraction of the pigments weighed 200 mg of fresh leaves and homogenized with 10 mL of 80% acetone (v/v), filtered through glass wool, completing the volume for 30 mL 80% acetone. Immediately following this procedure was carried the reading of the absorbance at 663.2 nm, 646.8 nm and 470 nm. All procedures were done in the dark, to preserve the integrity of pigments. The entire procedure was performed in the dark to prevent degradation of chlorophylls. The content of chlorophyll a, b and carotenoids were calculated using the Equations (1)-(3) respectively, and expressed in mg g^{-1} fresh weight

$$\left[\left(12.25 \times A_{663.2} \right) - \left(2.79 \times A_{646.8} \right) \right] \tag{1}$$

$$\left[\left(21.5 \times A_{646.8} \right) - \left(5.1 \times A_{663.2} \right) \right] \tag{2}$$

$$\left[\left(\left(1000 \times A_{470} \right) - \left(1.82 \times A_{663.2} \right) - \left(82.02 \times A_{646.2} \right) \right) / 198 \right] \tag{3}$$

After determining the contents, were done relations chlorophyll a/b and total chlorophylls, summing the contents of chlorophyll a and b.

Treatments

The treatments were characterized by cultivation of plants under five spectral irradiance produced by Sombrite® (30% and 50%), two Chromatinet® nets in red (R) and blue (B) colors, which block 50% of incident radiation, and one treatment at full sun. Irrigation was performed daily, keeping the soil at field capacity.

With the assistance of a portable spectroradiometer (USB-650 Red Tide) coupled to source of electromagnetic radiation DT-MINI (200 to 2000 nm) and a probe reflectance R400-7-VIS-NIR (US BioSolutions Ocean Optics®) we evaluated the radiation spectrum of different environments, with a spectral resolution of 1 nm. The irradiance normalized for each environment presented higher values in terms of quantity and size of the spectrum, for the environment with 100% irradiance, followed by the environment R, 70% and 50% and the B environment (Figure 1). Note, that the blue net gave a peak of irradiance between 450 - 550 nm and red net between 490 - 690 nm.

Statistics

The experimental design was completely randomized, and the observed data submitted to the Lilliefors normality test at 5% significance and submitted to analysis of variance, the means compared by the Scott-Knott test at 5% of probability, by the program SAEG [16].

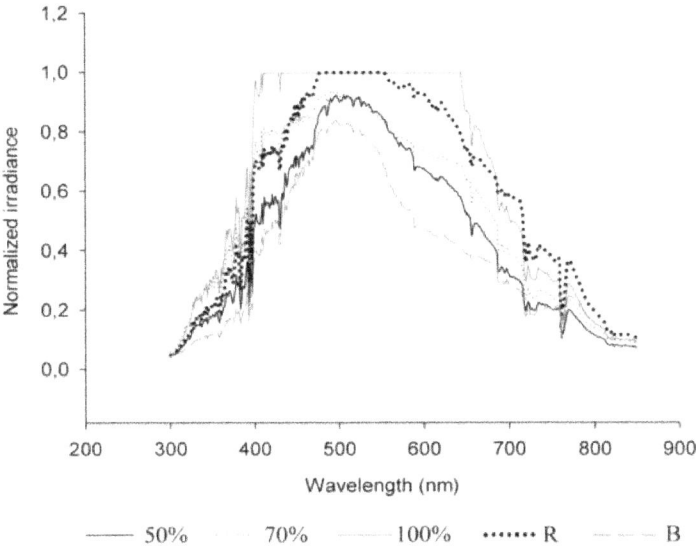

Figure 1: Normalized irradiance for wavelength radiation in five environments: 50%, 70% and 100% of irradiance, R (red net) and B (blue net).

RESULTS AND DISCUSSION

Significant differences were found among treatments in the gas exchanges' variables which were evaluated (Figure 2). The net photosynthesis was superior in the treatment with 100% irradiance, and lower in plants cultivated under colored nets. Variations in the spectral irradiance stimulate rapid responses in photosynthetic activity and the increase in brightness provides an increase in photosynthesis until they reach the saturation point [17]. Therefore, the rise in photosynthesis observed at treatments with higher irradiances might have supplied better conditions for CO_2 assimilation allowing greater photosynthetic activity. According to Souza et al. [18], plants growing under high radiation might develop mesophyll cells with more chloroplasts, enabling greater photosynthetic activity.

The low net photosynthesis observed in plants grown under red nets may be due to low chlorophyll content, low carboxylation efficiency and CO_2 assimilation showed in this treatment (Figure 3). Furthermore, the low irradiance (50%) afforded this treatment resulted in lower photosynthetic activity, as observed in in vitro studies [19].

In another study, it was observed that blue light has an important role in regulating the parameters of photosynthesis [20]. However, the balance between the red and blue is more important than the concentration of these bands in the spectrum resulting from such colored nets [21]. Similar results were observed for Capsicum annuum grown under white, yellow and red light, where despite of blocking 35% of the radiation, the higher photosynthetic rate was observed under white light [21].

The stomatic conductance presented a variable response in relation to irradiance conditions, which was higher in treatments with 100%, 50% and red net. Furthermore, plants grown under red nets showed higher values of Ci, which are lower in treatments with 70% radiation and blue net. The higher stomatal conductance observed in red net may have been answerable for the greater Ci observed in this treatment.

These results demonstrate that photosynthetic activity observed in treatment at 100% of irradiance may have been afforded thru higher stomatal conductance observed under these conditions. Furthermore, the low photosynthetic activity observed in plants grown with 50% of the radiation and red net can be attributed to other factors, not the internal CO_2 concentration, although this relationship was observed for plants grown in 70% of radiation and blue net. Studies with Vaccinium corymbosum at different shading levels (40%, 50% and 75% of shading) do not show changes in stomatal conductance [22]. Furthermore, stomatal conductance affected by different levels of irradiation has already been reported in the literature [23]. The treatments with 70% and 100% of irradiation were presented with the highest rates of transpiration, leaf temperature and vapor pressure deficit.

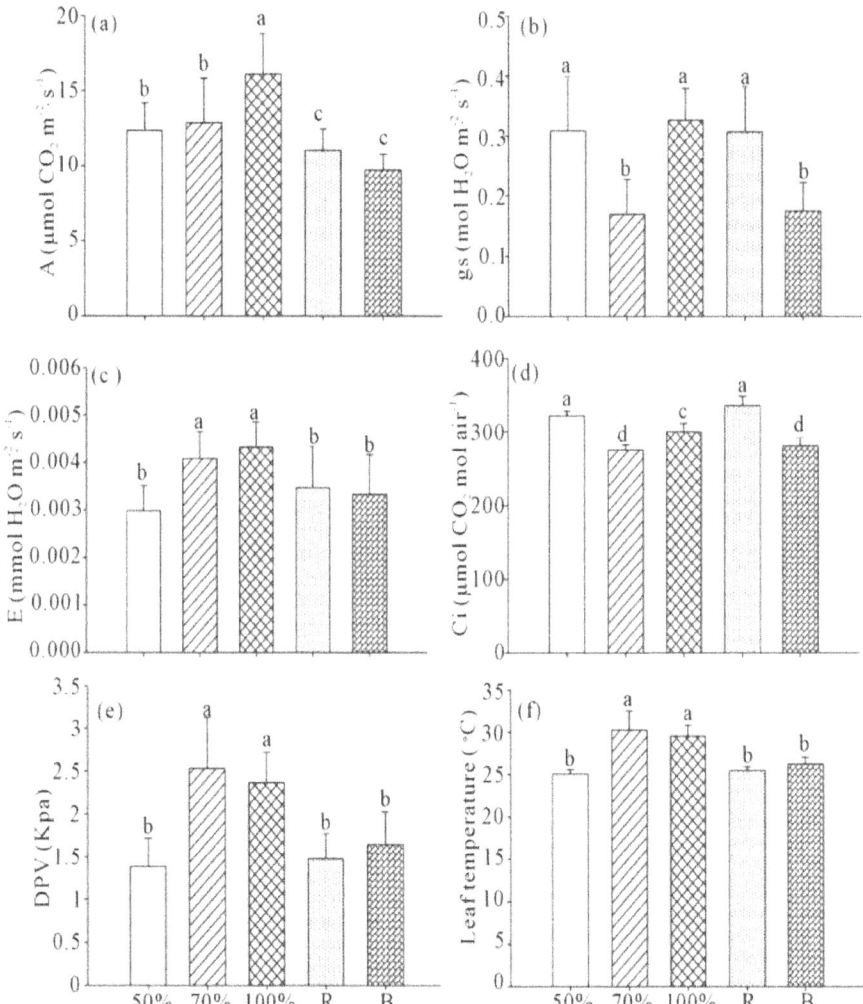

Figure 2: (a) Photosynthetic activity; (b) Stomatal conductance; (c) Rate of perspiration; (d) Concentration of intracellular of CO_2; (e) vapor pressured eficit; (f) Leaf temperature of Piper aduncun grown under different spectral irradiance—50% of irradiance, 70% of irradiance, 100% of irradiance, R (red net) and B (blue net). *Means followed by the same letter do not differ by Scott-Knot test at 5% probability.

In this way, we can infer that the high irradiance might favor the process of photorespiration. However, as the plants were grown at field capacity, i.e. not suffering water stress, this process may not have occurred, while the losses occurring in dry matter or its partition were due to the low efficiencies of water use observed this condition, which necessitates the further studies to check their productivity in such situations.

Effect of high irradiance favoring the process of photorespiration was documented in another study, where plants were not kept on field capacity of the soil [24]. The low transpiration observed in treatments with colored nets and 50% of radiation is considered as a strategy to reduce losses of carbon under this condition. As observed by Yang et al. [25], the effect of different light intensity in photosynthesis was found.

High transpiration, leaf temperature and vapor pressure deficit found in treatment with 100% of irradiance associated with a high stomatal conductance can indicate that the species does not have a strong regulatory mechanism to open and close the stomata. Furthermore, it showed a decrease in the photosynthetic process in plants grown under low-radiation and colored net conditions and gave a lower water loss and leaf temperature. Thus, nets can have promoted a reduction in photorespiration and stabilized the photosynthesis [22].

The relation of Ci/Ca indirectly shows limitations in photosynthesis caused for decrease in acquisition of a non-stomatal CO_2. Given that the external concentration of CO_2 is constant (Ca), the increase in Ci/Ca is due solely to changes in the internal concentration.

If Ci is increasing, it means that the CO_2 present in mesophyll cells is not being set during carboxilative possibly for damage sustained in its structure reducing net photosynthesis. With this manner, it can be seen that the greatest ratio Ci/Ca has been found in plants grown with 50% of brightness and in red net. Moreover, the lowest ratios were observed in treatments with 70% and 100% luminosity. These results are in accordance with the photosynthetic rates found in these treatments, indicating plants grown in treatments with 70% and 100% of brightness were those with lower relations Ci/Ca owing to CO_2 consumed in the carboxylation process.

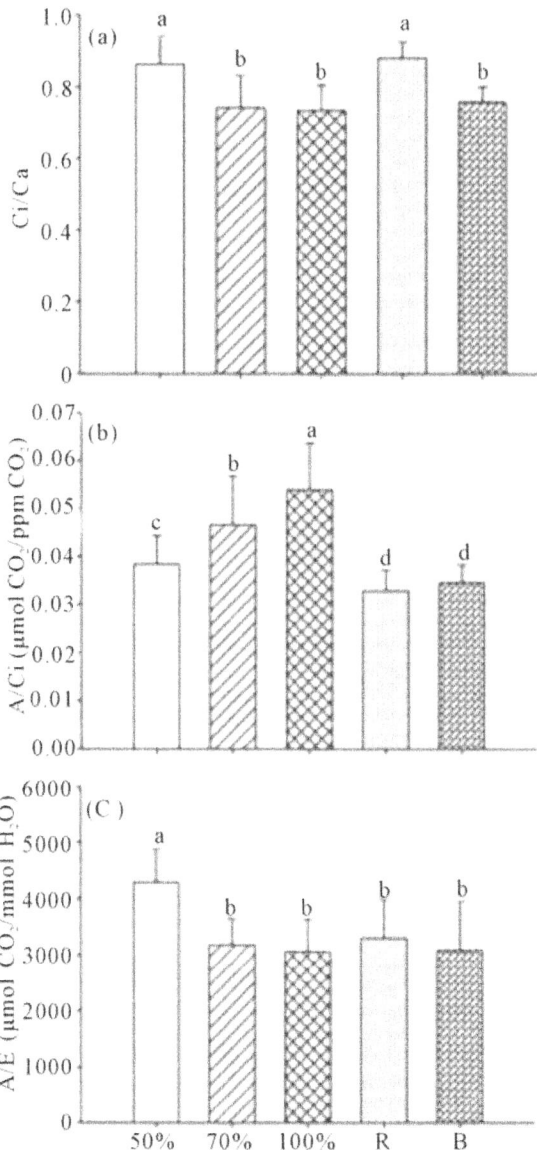

Figure 3: (a) Intra/extra celular CO_2; (b) Carboxylation eficience; (c) Water use ef-ficiency. 50% of irradiance, 70%, of irradiance, 100% of irradiance, R (red net) and B (blue net). *Means followed by the same letter do not differ by ScottKnot test at 5% probability.

The A/Ci ratios were higher in plants grown under 100% of irradiance and lower under 50% and colored nets (Figure 3). In this way, it can be suggested that photosynthetic activity and carboxylation efficiency are more related to the availability of radiation than spectral quality. It can be seen that colored nets had a blockage of radiation equivalent to 50%. The A/E ratio showing us the instantaneous efficiency of water use was higher in treatment with 50% of radiation and lower in treatments with 70% and 100%. This ratio can indicate the efficiency of plants using water, representing photosynthetic properties at a given transpiration rate [26].

Thus, it was observed that plants cultivated under 100% of irradiance though showed the highest photo-synthetic rates. They had a great loss of water through transpiration, which reduced their water use efficiency. Additionally, this parameter can be used to denote the adaptability of plants to environmental changes [27].

When compared with the plants grown in 50% of irradiance, there were lower transpiration rates and higher efficient use of water, which may indicate that the plant has the ability to maintain a certain level of photo-synthesis and carbon assimilation. The greater efficiency of water use in plants grown under partial shade was also observed in Illicium lanceolatum [27].

The plants grown under full sun have higher rates photosynthesis associated with high perspiration, high leaf temperatures and low efficiency of water use, which can promote a lower optimization in production of carbonaceous compounds in relation to water loss. The leaves of Piper aduncun had higher quantity of total and chlorophyll a in treatments under 70% of irradiance. Chlorophyll b levels were changed due to the variation in irradiance spectrum, where the quality of radiation damaged the pigment production (Figure 4).

The ratio of chlorophyll a/b was also changed for the treatment being superior in 70% of irradiance.

The increment of chlorophyll b in environments with a reduced irradiance spectrum is an important characteristic for adaptation to the shaded, because this pigment captures photons in longer wavelengths observed in this type of environment [28]. So, it can be inferred that species studies do no present adaptation in environments with low levels of irradiance because the increasing levels of chlorophyll b in fully radiation. This result is consistent with the observed gas exchange variables.

There was a greater photosynthetic activity in the treatment with 100% followed by 70% of irradiance, indicating an enlargement of absorption spectra even under conditions of high light intensities. While, these characteristics are in accordance with classification of ecological pioneer from this species, as seen sometimes considered a weed [29,30].

The content of chlorophyll a, b and total observed in treatment with 70% of irradiance indicates that it can adapt this luminous condition. In addition, the greatest amount of pigments can show a high chlorophyll synthesis and a slower degradation of chlorophyll b, relative to chlorophyll a balance between the photosystems [31]. The pigments content found in plants grown under red net might be related to smallest relationship red: far red (R/FR) [18]. Moreover, lows levels observed in leaves which have grown in blue net are at odds with the literature, given that, the blue light affects the biosynthesis of chlorophyll and other pigments by genetic regulation [32]

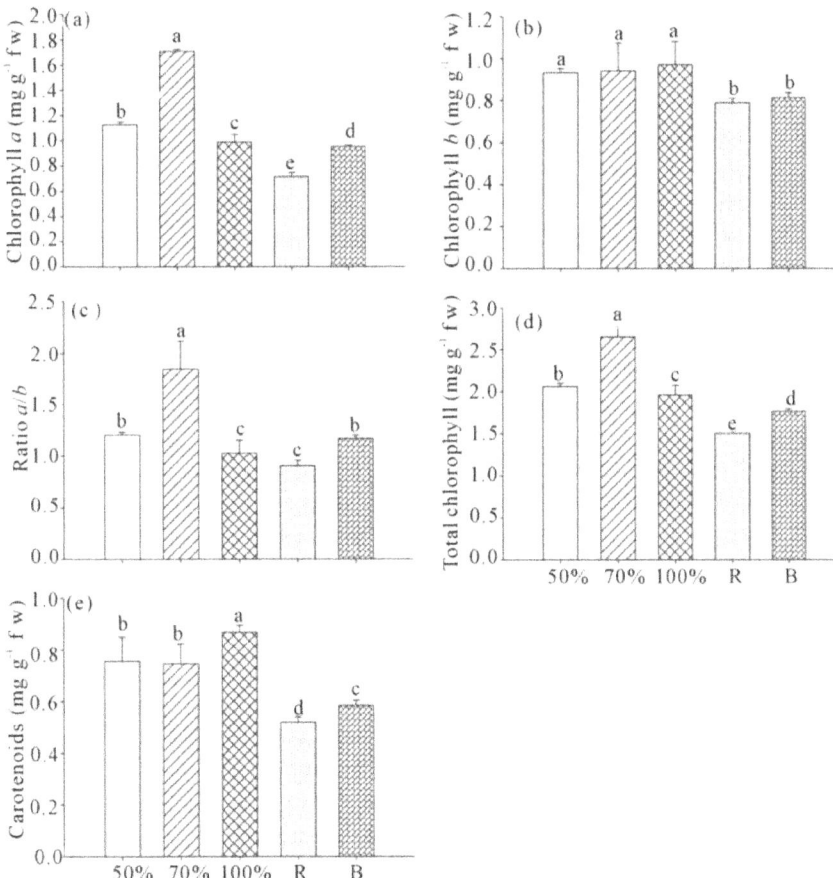

Figure 4: Content of pigments: (a) chlorophyll a; (b) chlorophyll b; (c) Ratio chlorophyll a/b; (d) Total chlorophyll; (e) Carotenoids. 50% of irradiance, 70%, of irradiance, 100% of irradiance, R (red net) and B (blue net). Means followed by the same letter do not differ by Scott-Knot test at 5% probability.

According to Larcher [33], changes in chlorophyll biosynthesis thru changes in the spectral quality could provide advantages for growth and reproduction of plants. So, these results indicate that long pepper does not adapt chromatically to progress its photosynthetic performance. On the other hand, an increased level of chlorophyll in plants grown under blue net has been observed for other medicinal plants, such as: Catharanthus roseus [34], Ocimum gratissimum [35] and Mikania laevigata [18].

The higher content of carotenoid was seen in treatment with 100% and lower irradiance in the colored net. Carotenoids are pigments that act as sunscreens by rapid quenching of excited states of chlorophyll protecting it from oxidation [36]. In this way, high levels of carotenoids in plants under 100% of radiation are associated with a sufficient protection of the receptor complexes of photosystems light, proving the high photosynthetic activity observed in this work. High levels of carotenoids in plants grown under full sunlight were observed in Artemisia vulgaris [37].

From the data obtained, it can be observed that Piper aduncun develops different mechanisms for adaptation to different irradiance conditions. Moreover, it can be inferred that the photosynthetic process and production of pigments are more efficient at light intensities and wider spectra, showing higher photosynthetic activity when grown under 100% and 70% irradiation.

However, to remain under these conditions of light, a high-water availability is necessary, since the efficiency of water usage is reduced due to the high-temperature leaf.

It is noteworthy that studies are needed to verify the production and translocation of assimilating these different conditions of radiation, aimed at increased production of essential oil.

ACKNOWLEDGEMENTS

The authors thank FAPEMIG, CNPq and CAPES for the granting of scholarships and financial aid to perform the work.

REFERENCES

1. M. Fazolin, J. L. V. Estrela, V. Catani and C. R. Costa, "Potencialidades da Pimenta de Macaco (Piper aduncun L.): Características Gerais e Resultados de Pesquisas," Embrapa Acre, Rio Branco, 2006, p. 53. http://www.infoteca.cnptia.embrapa.br/handle/doc/505568

2. S. F. R. Rocha, L. C. Ming, F. C. M. Chaves and F. M. Scarda, "Role of Light and Phytochrome on Piper aduncun L. Germination: An Adaptive

and Environmental Approach," Journal of Herbs, Spices & Medicinal Plant, Vol. 11, No. 3, 2008, pp. 85-96.http://dx.doi.org/10.1300/J044v11n03_08

3. M. Fazolin, J. L. V. Estrela, V. Catani, M. S. Lima and M. R. Alécio, "Toxicity of Piper aduncum Oil to Adults of Cerotomatingomarianu Bechyné (Coleoptera: Chrysomelidae)," Neotropical Entomology, Vol. 34, No. 3, 2005, pp. 485-489.http://dx.doi.org/10.1590/S1519-566X2005000300018

4. C. R. Lara Junior, G. L. Oliveira, B. C. F. Mota, M. F. G. Fernandes, L. S. Figueiredo, E. R. Martins, D. L. Moreira and M. A. C. Kaplan, "Antimicrobial Activity of Essential Oil of Piper aduncum L. (Piperaceae)," Journal of Medicinal Plants Research, Vol. 21, No. 6, 2012, pp. 3800- 3805.

5. N. Misni, H. Othman and S. Sulaman, "The Effect of Piper aduncum Linn. (Family: Piperaceae) Essential Oil as Aerosol Spray against Aedesaegypti (L.) and Aedesalbopictus Skuse," Tropical Biomedicine, Vol. 28, No. 2, 2011, pp. 249-258. http://www.msptm.org/files/249_-_258_Norashiqin_Misni.pdf

6. P. J. C. Sousa, C. A. L. Barros, J. C. S. Rocha, D. S. Lira, G. M. Monteiro and J. G. S. Bia, "Toxicological Evaluation of the Essential Oil of Piper aduncum L.," Revista Brasileira de Farmacognosia, Vol. 18, No. 2, 2008, pp. 217-221. http://dx.doi.org/10.1590/S0102-695X2008000200013

7. E. J. A. Santiago, J. E. B. P. Pinto, E. M. Castro, O. A. Lameira, H. E. O. Conceição and M. L. Gavilanes, "Aspects of Leaf Anatomy of Long Pepper (Piper hispidinervium C.D.C.) under Different Light Conditions," Ciência e Agrotecnologia, Vol. 25, No. 5, 2001, pp. 1035-1042. http://www.alice.cnptia.embrapa.br/handle/doc/401601

8. R. Wang and Z. H. Guo, "Photosynthetic Responses of Schimasuperba Grown in Different Light Regimes of Subtropical Evergreen Broadleaf Forest," Forest Research, Vol. 20, 2007, pp. 688-693. http://en.cnki.com.cn/Article_en/CJFDTOTAL-LYKX200705018.htm

9. L. V. Kurepin, R. J. N. Emery, R. P. Pharis and D. M. Reid, "Uncoupling Light Quality from Light Irradiance Effects in Helianthus annuus Shoots: Putative Roles for Plant Hormones in Leaf and Internode Growth," Journal of Experimental Botany, Vol. 58, No. 8, 2007, pp. 2145- 2157. http://dx.doi.org/10.1093/jxb/erm068

10. J. A. Marchese, R. S. Mattana, L. C. Ming, F. Broetto, P. F. Vendramini and R. M. Moraes, "Irradiance Stress Responses of Gas Exchange and Antioxidant Enzyme Contents in Pariparoba [Pothomorpheumbellata (L.)

Miq.] Plants," Photosynthetica, Vol. 46, No. 4, 2008, pp. 501- 505. http://dx.doi.org/10.1007/s11099-008-0085-x

11. E. C. Lima Junior, A. A. Alvarenga, E. M. Castro, C. V. Vieira and J. P. R. A. D. Barbosa, "Physioanatomy Traits of Leaves in Young Plants of Cupaniavernalis camb. Subjected to Different Shading Levels," Brazilian Journal of Forest Science, Vol. 30, No. 1, 2006, pp. 33-41.

12. V. S. R. Das, "Photosynthesis, Regulation under Varying Light Regimes," Science Publishers, Inc., Enfield, New Hampshire, 2004.

13. A. Alvarenga, E. M. Castro, E. R. Lima Junior and M. M. Bgalhães, "Effects of Different Light Levels on the Initial Growth and Photosynthesis of Croton urucurana Baill. in Southeastern Brazil," Brazilian Journal of Forest Science, Vol. 27, No. 1, 2003. pp. 53-57.

14. J. A. Zavala and D. A. Ravetta, "Allocation of Photoassimilates to bioBss, Resin and Carbohydrates in Grindelia chiloensis as Affected by Light Intensity," Field Crops Research, Vol. 69, No. 2, 2001, pp. 143-149. http://dx.doi.org/10.1016/S0378-4290(00)00136-2

15. H. K. Lichtenthaler and C. Buschmann, "Chorophylls and Carotenoids: Measurement and Characterization by UVVIS Spectroscopy," In: R. E. Wrolstad, et al., Eds., Current Protocols in Food Analytical Chemistry, John Wiley & Sons, Davis, 2001.

16. SAEG, "Sistema para Análises Estatísticas," Vers.9.1, Fundação Arthur Bernardes-UFV-Viçosa, 2007. http://www.ufv.br/saeg/

17. N. K. Boardmann, "Comparative Photosynthesis of Sun and Shade Plants," Annual Review of Plant Physiology, Vol. 28, 1977, pp. 355-377. http://dx.doi.org/10.1146/annurev.pp.28.060177.002035

18. G. S. Souza, E. M. Castro, A. M. Soares, J. E. B. P.Pinto, M. G. Resende and S. K. V. Bertolucci, "Crescimento, Teor de Óleo Essencial e Conteúdo de cuBrina de Plantas Jovens de Guaco (Mikaniaglomerata Sprengel) Cultivadas sob Blhas Coloridas," Biotemas, Vol. 24, No. 3, 2011, pp. 1-11.

19. M. C. Dias, G. Pinto, C. M. Correia, J. Moutinho-Pereira, S. Silva and C. Santos, "Photosynthetic Parameters of Ulmusminor Plantlets Affected by Irradiance during Acclimatization," Biologia Plantarum, Vol. 57, No. 1, 2013, pp. 33-40.http://dx.doi.org/10.1007/s10535-012-0234-8

20. J. Y. Yamasaki, "Is Light Quality Involved in the Regulation of the Photosynthetic Apparatus in Attached Rice Leaves?" Photosynthesis Research, Vol. 105, No. 1, 2010, pp. 63-71. http://dx.doi.org/10.1007/s11120-010-9567-3

21. Y. Kong, L. Avraham, K. Ratner and Y. Shahak, "Response of Photosynthetic Parameters of Sweet Pepper Leaves to Light Quality Bnipulation by Photoselective Shade Nets," Acta Horticulturae, Vol. 956, 2012, pp. 501- 506. http://www.actahort.org/books/956/956_59.htm

22. G. A. Lobos, J. B. Retamales, J. F. Hancock, J. A. Flore, N. Cobo and A. Pozo, "Spectral Irradiance, Gas Exchange Characteristics and Leaf Traits of Vacciniumcorymbosum L. 'Elliott' Grown under Photo-Selective Nets," Environmental and Experimental Botany, Vol. 75, 2012, pp. 142- 149. http://dx.doi.org/10.1016/j.envexpbot.2011.09.006

23. R. Pieruschka, G. Huber and J. A Berry, "Control of Transpiration by Radiation," Proceedings of the National Academy Sciences, Vol. 107, No. 30, 2010, pp. 13372- 13377.http://dx.doi.org/10.1073/pnas.0913177107

24. S. Kangasjavir, J. NeukerBns, S. Li, E. Aro and G. Noctor, "Photosynthesis, Photorespiration, and Light Signalling in Defense Responses," Journal of Experimental Botany, Vol. 63, No. 4, 2012, pp. 1619-1636. http://dx.doi.org/10.1093/jxb/err402

25. W. Yang, F. Liu, L. Zhou, S. Zhang and S. An, "Growth and Photosynthetic Responses of Canariumpimela and Nepheliumtopengii Seedlings to a Light Gradient," Agroforest System, Vol. 87, No. 3, 2013, pp. 507- 516. http://dx.doi.org/10.1007/s10457-012-9570-0

26. P. X. Su, L. X. Zhang, M. W. Du, Y. R. Bi, A. F. Zhao and X. M. Liu, "Photosynthetic Character and Water Use Efficiency of Different Leaf Shapes of Populuseuphratica and Their Response to CO_2 Enrichment," Acta PhytoecologicaSinica, Vol. 27, 2003, pp. 34-40.

27. Y. Cao, B. Zhou, S. Chen, J. Xiao and X. Wang, "The Photosynthetic Physiological Properties of Illiciumlanceolatum Plants Growing under Different Light Intensity Conditions," African Journal of Agricultural Research, Vol. 26, No. 6, 2011, pp. 5736-5741.

28. F. H. Whatley and F. R. Whatley, "A Luz e a Vida das Plantas: Temas de Biologia," EDUSP, São Paulo, 1982, Vol. 30, p. 101.

29. P. Alvarenga, S. A. Botelho and I. M. Pereira, "Evaluation of Natural Recovery of Ciliary Forests in Spring in the South Area of Minas Gerais," Cerne, Vol. 12, No. 4, 2006, pp. 360-372. http://www.redalyc.org/articulo.oa?id=74412408

30. S. F. R. Rocha, L. C. Ming, F. C. M. Chaves and F. M. Scarda, "Role of Light and Phytochrome on Piper aduncun L. Germination: An Adaptive and Environmental Approach," Journal of Herbs, Spices & Medicinal Plant, Vol. 11, No. 3, 2008, pp. 85-96.http://dx.doi.org/10.1300/J044v11n03_08

31. V. L. Engel and F. Poggiani, "Study of Foliar Chlorophyll Concentration and Its Light Absorption Spectrum as Related to Shading at the Juvenile Phase of Four Native Forest Tree Species," Brazilian Journal of Plant Physiology, Vol. 3, No. 1, 1991, pp. 39-45. http://www.cnpdia.embrapa.br/rbfv/pdfs/v3n1p39.pdf

32. Y. Tsunoyama, K. Morikawa, T. Shiina and Y. Toyoshima, "Blue Light Specific and Differential Expression of Plastid Sigma Factor, Sig5 in Arabdopsis thaliana," FEBS Letters, Vol. 516, No. 1, 2002, pp. 225-228. http://dx.doi.org/10.1016/S0014-5793(02)02538-3

33. W. Larcher, "Physiological Plant Ecology," 4th Edition, Springer, Berlin, 2004, p. 531.

34. A. M. Melo and A. A. Alvarenga, "Shading of 'Pacifica White' Catharanthusroseus (L.) G. Don Plants with Colored Nets: Vegetative Development," Ciência e Agrotecnologia, Vol. 33, No. 2, 2009, pp. 514-520.

35. R. Martins, A. A. Alvarenga, E. M. Castro, A. P. O. Silva and E. Alves, "Pigmentscontentand Alfavaca-Cravo Chloroplast Structure Cultivate Undercolored Nets," Ciência Rural, Vol. 39, No. 1, 2009, pp. 82-87. http://dx.doi.org/10.1590/S0103-84782008005000040

36. L. Taiz and E. Zeiger, "Plant Physiology," 4th Edition, Sinauer, Sunderland, Massachusetts, 2009, p. 719.

37. M. I. Oliveira, E. M. Castro, L. C. B. Costa and C. Oliveira, "Biometric, Anatomical and Physiological Aspects of Artemisia vulgaris L. Grown under Colored Screens," Revista Brasileira de Plantas Medicinais, Vol. 11, No. 1, 2009, pp. 56-62.http://dx.doi.org/10.1590/S1516-05722009000100010

Chapter 2

META-ANALYSIS OF THE DETECTION OF PLANT PIGMENT CONCENTRATIONS USING HYPERSPECTRAL REMOTELY SENSED DATA

Jingfeng Huang[1], Chen Wei1,[2,] Yao Zhang[1], George Alan Blackburn[3], Xiuzhen Wang[4], Chuanwen Wei[1], Jing Wang[1]

[1] Institute of Agricultural Remote Sensing & Information Application, Zijingang Campus, Zhejiang University, Hangzhou, China

[2] Zhejiang Meteorological Service Center, Hangzhou, China

[3] Lancaster Environment Centre, Lancaster University, Lancaster, United Kingdom

[4] Institute of Remote Sensing and Earth Sciences, Hangzhou Normal University, Hangzhou, China

ABSTRACT

Passive optical hyperspectral remote sensing of plant pigments offers potential for understanding plant ecophysiological processes across a range of spatial scales. Following a number of decades of research in this field, this paper undertakes a systematic meta-analysis of 85 articles to determine whether passive optical hyperspectral remote sensing techniques are sufficiently well developed to quantify individual plant pigments, which operational solutions are available for wider plant science and the areas which now require greater focus. The findings indicate that predictive relationships are strong for all pigments at the leaf scale but these decrease and become more variable across pigment types at the canopy and landscape scales. At leaf scale it is clear that specific sets of optimal wavelengths can be recommended for operational methodologies: total chlorophyll and chlorophyll a quantification is based on reflectance in the green (550–560nm) and red edge (680–750nm) regions; chlorophyll b on the red, (630–660nm), red edge (670–710nm) and the near-infrared (800–810nm); carotenoids on the 500–580nm region; and anthocyanins on the green (550–560nm), red edge (700–710nm) and near-infrared (780–790nm). For total chlorophyll the optimal wavelengths are valid across canopy and landscape scales and there is some evidence that the same applies for chlorophyll*a*.

INTRODUCTION

A pigment is a material that changes the spectral distribution of reflected or transmitted light as the result of wavelength-selective absorption which is determined by the physical properties of the pigment itself. Plant pigments play an important role in light capture, photosystem protection, and in various growth and development functions. The photosynthetic pigments control the amount of solar radiation absorbed by a leaf and thus determine photosynthetic potential and primary production [1,2]. Pigment concentrations are also related to plant stress (excess direct sunlight, UV–B irradiation, low temperature, water stress, nitrogen deficiencies and so on) and senescence (e.g., [3–9]). Therefore, accurate measurements of the temporal dynamics and spatial variations of pigment concentration using remotely sensed data can provide a basis for monitoring physiological and ecological processes [10,11].

The spectral absorbance properties of pigments offer the possibility of using measurements of reflected radiation as a non-destructive method for quantifying pigments. Different approaches have arisen recently to remotely estimate pigment concentrations from a wide variety of wavelengths and sensor types. These studies produced variable results, and none have been demonstrated to have satisfactory performance under all growth and environmental conditions. These inconsistencies may stem from the fact that the experimental results are influenced by a number of factors including different species, experimental conditions and analytical methods used [11].

Recent review articles have attempted to assimilate knowledge in this field of passive optical hyperspectral remote sensing with the sun as energy source. Blackburn [10] reviewed the developing technologies and analytical methods for quantitative estimation of pigment across a range of spatial scales using passive optical hyperspectral remote sensing. Ustin *et al.* [11] appraised the most widely used methodologies for retrieving pigment information with hyperspectral data at the leaf scale. However, it has been demonstrated that traditional qualitative reviewers may subjectively select their preferred studies when faced with conflicting results on a single question [12]. In contrast, it has been argued that meta-analysis can take the results from primary research articles and quantitatively analyze and synthesize these data in an attempt to arrive at more robust conclusions. As such, meta-analysis review papers make the shift from a narrative-driven to a data-driven approach [13,14].

Glass [15] published the first article to lay out the essential rationale of meta-analysis. As a fully general set of methods, meta-analysis has been widely applied to the integration of literatures in many areas of empirical science, including ecology [14]. This form of analysis has, for example, been used to determine the response of biodiversity to intensive biomass production, the effects of elevated CO_2 on plant–arthropod interactions, the influence of plant invasion on carbon and nitrogen cycles and the causes and consequences of variations in leaf mass per area [16–19]. Today, many findings and advances are being made not only by those who do primary research studies, but also by those who use meta-analysis to discover the latent meaning of existing research literatures [13]. Recently, meta-analysis has been employed in remote sensing research. Garbulsky *et al.* [20] performed a meta-analysis to assess the use of the photochemical reflectance index (PRI) as an indicator of radiation use efficiencies at the leaf, canopy and ecosystem scales for different time scales and vegetation types. Zolkos *et al.* [21] conducted a meta-analysis of publications on LiDAR remote sensing estimation of terrestrial aboveground biomass. These investigations show that meta-analysis can be used to systematically integrate the results from a collection of studies, and through statistical comparison, assess the relationships between remotely sensed measurements and variables of interest.

Here, a meta-analysis of data from a wide selection of studies reporting the passive optical hyperspectral remote sensing of pigments was used to quantify the development of this scientific field, identify optimal wavelengths for retrieval of individual pigments and evaluate the strength of the relationships between pigment concentration and remotely sensed data across pigment types and scales.

MATERIALS AND METHODS

Study Selection and Data Extraction

Databases of Elsevier, Springer and Web of Science, licensed to Zhejiang University, were used for source data from inception to August 2014. The following key words were used: pigment, chlorophyll, carotenoids, carotene, xanthophyll, anthocyanins, anthoxanthin in combination with the terms reflectance, estimation, quantification, retrieval, prediction and remote sensing. More than 4500 citations were collected as a result of this initial search.

Then the abstracts of these articles were reviewed and considered for inclusion in the meta-analysis. The following criteria were applied to ensure homogeneity in methodology. First, the studies had to include a chemical measurement of pigment concentration (total chlorophyll, chlorophyll *a*, chlorophyll *b*, carotenoids, xanthophyll, carotene or anthocyanins). Second, the article had to report the quantification of pigments using remotely sensed data. Third, the authors must have provided the following statistical information: (1) coefficient of determination for the relationships between pigment concentration and remotely sensed measurements; (2) the wavelength(s) used to estimate pigment concentration; and (3) training sample sizes.

Based on the first two decision rules, 135 articles were selected. According to the final criterion, 50 studies were excluded because of insufficient statistical information. Finally, 85 articles were used in the meta-analysis, which reported results at different spatial and temporal scales and from a wide range of vegetation types between 1977 and 2014. The number of studies selected at various stages is shown in the flow diagram in Fig 1. Some studies reported multiple results for different pigment types or vegetation types. Different types of sensors were used in these studies, from spectrophotometers and hand-held spectroradiometers to satellite sensors. All the sensors were working in reflectance mode. Within the selected articles 44 were working at the leaf scale, 21 at the canopy scale, 15 at the landscape scale, 2 at the leaf and canopy scales, 1 at the leaf and landscape scales, and 2 covered the leaf, canopy and landscape scales. The term "canopy" refers to either a single plant or a monospecific stand where the experimental results are influenced by a number of controlling factors, such as orientation of leaves (leaf angle distribution; *LAD*), variations in number of leaf layers (*LAI*), presence of non-leaf elements, multiple scattering and areas of shadow [10,22], the term "landscape" refers to a mixed-species stand where the reflectance spectrum from airborne and spaceborne sensors is subject to even more controlling factors, such as atmospheric conditions, instrucment sensitivity (signal-to-noise ratio) and spatial resolution. In total, the sample size from all the selected studies is 16100. The Preferred Reporting Items for Meta-Analyses is shown in S1 PRISMA Checklist.

Relevant information was extracted from each study in the final set: ① scales (leaf, canopy, landscape), ② pigment types, ③ species, ④ wavelengths, ⑤ coefficient of determination, ⑥ sample sizes, ⑦ sensors, ⑧ authors and ⑨ year of publication. In order to reduce human error in data extraction and coding, two sets of reviewers independently screened articles in accordance with those inclusion criteria discussed above, evaluated the quality and extracted the data from the eligible studies. The results from one group were cross-checked by the other group. Divergences of opinion about article selection and data extraction were settled by discussion. Table 1 is a summary of the studies contained in this research. This list is not exhaustive but it does cover most papers published related to quantification of pigments using remotely sensed data that met the selection criteria. Table 2 provides a statistical summary of the data extracted from the studies included in the meta-analysis.

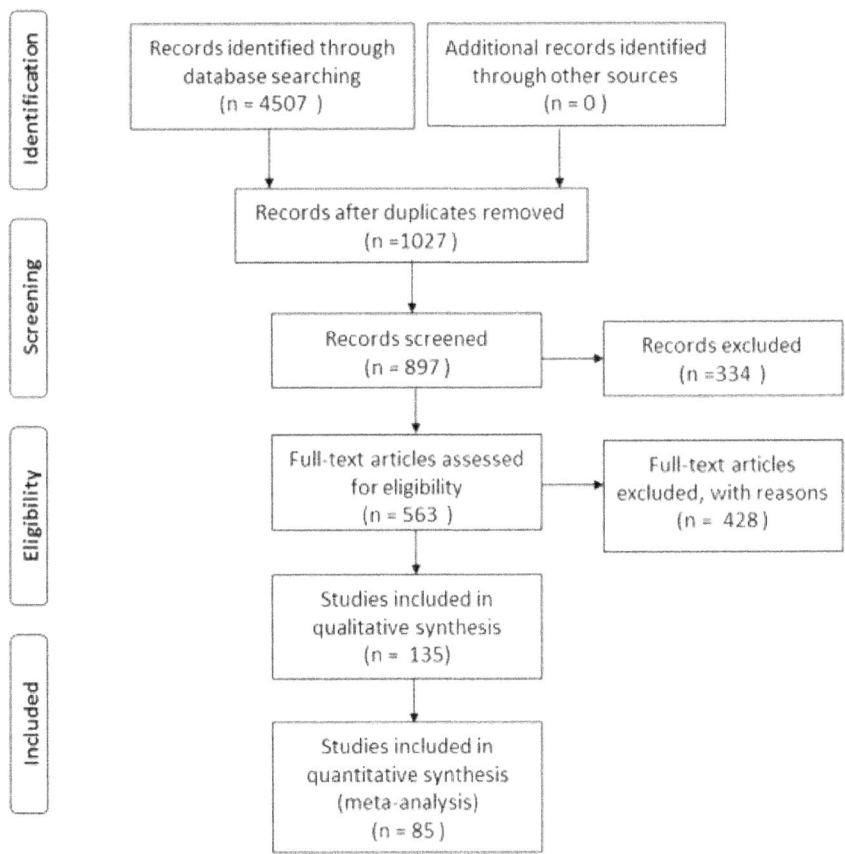

Figure 1: Selection of studies for inclusion in the meta-analysis.

Table 1: A summary of the studies contained in this research that linked remotely sensed data with pigment

Scale	Pigment Type	Year	Species	Sensor	Reference
leaves	Chl tot	1992	Amaranthus tricolor	Specpho	[23]
leaves	Chl tot	1995	Slash pine	Specrad	[24]
leaves	Chl tot	1995	Bigleaf maple	Specrad	[25]
leaves	Chl tot	1996	Horse Chestnut, Norway maple,Cotoneaster, Tobacco	Specpho	[26]
leaves	Chl tot	1996	Norway Maple, Horse Chestnut	Specpho	[27]
leaves	Chl tot	1997	Norway Maple, Horse Chestnut, Fig, Cotoneaster, Tobacco,Oleander, Hibiscus, Vine, Rose	Specpho	[28]
leaves	Chl tot	1998	Tobacco, Horse Chestnut, Cotoneaster	Specpho	[29]
leaves	Chl tot	1999	Beech tree, Elm tree,Wild vine shurb	Specpho	[30]
leaves	Chl tot	1999	Bragg Soybean	Specrad	[31]
leaves	Chl tot	2002	53 species	Specrad	[32]
leaves	Chl tot	2002	Paper birch	Specrad	[33]
leaves	Chl tot	2003	Bigleaf Maple, Horse Chestnut, Wild vine, Beech	Specpho	[34]
leaves	Chl tot	2005	Cotton	Specrad	[35]
leaves	Chl tot	2007	Winter wheat	Specpho	[36]
leaves	Chl tot	2012	15 different species(Beech, Fraxinus lanuginosa, Acer Japonicum, Magnolia obovata and so on)	Specrad	[37]
leaves	Chl tot	2014	Douglas fir	Specrad	[38]
leaves	Chl a	1994	Norway Maple, Horse Chestnut	Specpho	[39]
leaves	Chl a	1994	Norway Maple, Horse Chestnut	Specpho	[40]
leaves	Chl a	1996	Norway Maple, Horse Chestnut	Specpho	[41]
leaves	Cars/Chl tot	1977	Cantaloupe, Corn, Spinach Cotton, Cucumber, tobacco, Head lettuce, Grain sorghum	Specpho	[42]
leaves	Cars/Chl tot	1992	Sunflower	Specrad	[43]
leaves	Cars/Chl tot	1999	Norway Maple, Potato, Lemon, Apple, Coleus	Specpho	[7]
leaves	Cars/Chl tot	2006	24 species of woody trees and shurbs	Specpho	[44]
leaves	Anths/Cars/Chl tot	1999	Quercus agrifolia, Pseudotsuga menziesii	Specpho	[45]
leaves	Anths/Cars/Chl tot	2003	Apple	Specpho	[46]
leaves	Anths/Cars/Chl tot	2004	Norway maple, Maize, Dogwood,Horse chestnut, Second-flush beech, Wild vine shrub, Cotoneaster, Pelargonium zonale	Specpho	[47]
leaves	Chl tot/Anths	2014	Chilean strawberry	Specrad	[48]
leaves	Cars/Chl a/ Chl b	1992	Soybean	Specrad	[49]
leaves	Cars/Chl a/Chl b	1998	Beech, Oak, Maple, Sweet chestnut	Specrad	[50]
leaves	Cars/Chl a/Chl b	2005	Rice	Specrad	[51]
leaves	Chl tot/Chl a/Chl b	1999	Norway Maple, Horse Chestnut, Beech, Oak	Specrad	[52]
leaves	Chl tot/Chl a/Chl b	2001	Croton, Elaeagnus, Japanese pittosporum,Benjamin fig	Specrad	[53]
leaves	Chl tot/Chl a/Chl b	2010	Flowering cherry	Specrad	[54]
leaves	Chl tot/Chl a	1996	Tobacco	Specpho	[55]
leaves	Chl tot/Chl a	1999	Eucalyptus	Specrad	[56]
leaves	Cars	2002	Norway maple, Horse chestnut,Second-flush beech	Specpho	[57]
leaves	Cars	2009	Scot pine	Specpho	[58]
leaves	Cars	2011	Bur oak, Sugar maple, LOPEX database	Specrad	[59]
Scale	Pigment Type	Year	Species	Sensor	Reference

leaves	Anths	2001	Norway maple, Cotoneaster, Dogwood	Specpho	[60]
leaves	Anths	2009	Grapevine	Specrad	[61]
leaves	Anths	2009	European hazel, Siberian dogwood, Norway maple, Virginia creeper	Specpho	[62]
leaves	Anths	2011	Grapevine	Specrad	[63]
leaves	Anths	2011	Sweet cherries	Specpho	[64]
leaves	Anths	2011	Norway maple, Horse chestnut, Beech, Virginia creeper, Dogwood	Specpho&specrad	[65]
Leaves/canopy	Chl tot	2009	Maize	Specpho	[66]
Leaves/canopy	Chl tot	2013	Irrigated maize	Specrad	[67]
Leaves/landscape	Chl tot	2014	Black Spruce, Sugar maple	Specrad&MERIS	[68]
Leaves/canopy/ landscape	Chl tot	2010	Winter Wheat, Winter Rapeseed	Specrad	[69]
Leaves/canopy/ landscape	Cars/Chl tot	2000	Sugar maple	Specrad	[70]
canopy	Chl tot	1990	Slash pine	Airborne spectro	[1]
canopy	Chl tot	1994	pepper	Specrad	[71]
canopy	Chl tot	2005	Maize, Soybean	Specrad	[72]
canopy	Chl tot	2006	Rice	Specrad	[73]
canopy	Chl tot	2007	Cotton	Specrad	[74]
canopy	Chl tot	2008	Winter wheat, Corns	Specrad	[75]
canopy	Chl tot	2008	Heterogeneous grassland	Specrad	[76]
canopy	Chl tot	2008	Heterogeneous grassland	Specrad	[77]
canopy	Chl tot	2008	Corn, Cotton	Specrad	[78]
canopy	Chl tot	2010	Rice	Specrad	[79]
canopy	Chl tot	2011	Rice	Specrad	[80]
canopy	Chl tot	2012	Potato, Grassland	Specrad	[81]
canopy	Chl tot	2013	Irrigated maize	Specrad	[82]
canopy	Chl tot	2014	Winter wheat	Specrad	[83]
canopy	Chl a	2003	Rice	Specrad	[84]
canopy	Chl a	2007	Winter Wheat	Specrad	[85]
canopy	Chl a/Chl b	2004	Winter wheat	Specrad	[86]
canopy	Chl tot/Chl a	2006	Wheat	Specrad	[87]
canopy	Cars/Chl tot	2010	Tall fescue	Specrad	[88]
canopy	Cars	2008	Kermes oak	Specrad	[89]
canopy	Cars	2008	Douglas fir	Specrad	[90]
landscape	Chl tot	2002	Corn	CASI	[91]
landscape	Chl tot	2003	Eucalypt	CASI-2	[92]
landscape	Chl tot	2004	Jack pine	CASI	[93]
landscape	Chl tot	2004	Douglas fir	MERIS	[94]
landscape	Chl tot	2007	Corn, Wheat	CASI	[95]
landscape	Chl tot	2008	Rice, Cotton	EO-1	[96]
landscape	Chl tot	2008	Garlic, Alfalfa, Onion, Sunflower, Corn, Potato, Wheat, Vineyard, Sugar beet	PROBA/CHRIS	[97]
landscape	Chl tot	2010	Flax, Tea, Chestnut, Corn, Potato, Pine, Bamboo	EO-1	[98]
landscape	Chl tot	2010	Garlic, Onion, Corn, Alfalfa, Sugar beet, Sunflower, Potato, Vineyard, Wheat	PROBA/CHRIS	[99]
landscape	Chl tot	2014	London plane, Canary Island date palm, European nettle tree, White mulberry	CASI	[100]
landscape	Chl a	2004	Winter Wheat	AVIS	[101]
landscape	Cars/Chl tot	2002	Quercus petrea. Pinus sylvestris	CASI	[102]
landscape	Chl a/Cars	2005	Rice	PHI	[103]
landscape	Cars/Chl tot/Chl a/Chl b	2008	Aspen, Birch, Spruce, Balsam fir	CASI	[104]
landscape	Anths	2009	Austrocedrus chilensis forest	Hyperion	[105]

doi:10.1371/journal.pone.0137029.t001

Specrad = spectroradiometer; Specpho = spectrophotometer; Chl tot = total chlorophyll; Chl a = chlorophyll *a*; Chl b = chlorophyll *b*; Cars = carotenoids; Anths = anthocyanins.

Table 2: Summary statistics for the selected studies and extracted data for different pigment types at leaf, canopy and landscape scales

Scale	Pigment type	Number of studies	Number of effect sizes	Total sample size	Number of wavelengths
leaves	Chl tot	34	53	6431	131
	Chl a	11	23	1595	53
	Chl b	6	10	860	24
	Cars	14	15	1381	40
	Anths	10	17	1752	43
canopy	Chl tot	20	23	1146	55
	Chl a	4	4	162	6
	Chl b	1	1	35	0
	Cars	3	2	45	7
	Anths	0	0	0	0
landscape	Chl tot	15	17	1883	46
	Chl a	3	3	153	6
	Chl b	1	1	24	2
	Cars	3	3	573	4
	Anths	1	1	60	2

doi:10.1371/journal.pone.0137029.t002

Statistical Analysis of Effect Size

The Calculation of Effect Size for Each Study

The coefficient of determination (R^2) was used to evaluate the strength of relationships between spectral reflectance and pigment concentration in each article we selected. The value of R^2, however, is affected by the number of selected wavelengths. The more wavelengths included in the model, be they relevant or not, the larger would be the R^2 [106]. The increase of R^2 is not without cost. The increasing number of selected wavelengths reduces the degrees of freedom, which reduces model robustness. The adjusted coefficient of determination was applied to correct for the degrees of freedom:

$$R_A^2 = 1 - (1 - R^2)\frac{n-1}{n-k} \tag{1}$$

where n is the sample size for each study, k is the number of independent variables in the linear or nonlinear model. Eq (1) shows that R_A^2 is always smaller than R^2 when $k > 1$, which means the growth rate of R_A^2 is lower than that of R^2 as the number of parameters increase. This result is straightforward and it has been shown that when the added parameter explains a significant amount of the behavior of the dependent variable, R_A^2 will increase; otherwise, R_A^2 will decrease [107]. So R_A^2 was chosen as the effect size statistic, the variance of effect size is calculated as [108]:

$$V_i = \frac{(1 - R_A^2)^2}{n-1}, \tag{2}$$

The resulting data set was categorized by pigment type at the scales of leaf, canopy and landscape to allow comparison.

Test of Heterogeneity for Effect Sizes

It is important to assess the heterogeneity among the results from a collection of studies before computing the mean effect size [109]. Basically, there are two possible sources of heterogeneity in meta-analysis: methodological heterogeneity and statistical heterogeneity. To ensure homogeneity in methodology, we applied a series of criteria to identify the studies to be used in the meta-analysis (as described in section 2.1). Here the I^2 statistic was used to test for the statistical heterogeneity. The I^2 statistic measures the extent of true heterogeneity dividing the difference between the result of the Q test and its degrees of freedom by the Q value itself [110]:

$$I^2 = 100\% \times \frac{Q_{tot} - df}{Q_{tot}} \tag{3}$$

where $df = N_{tot} - 1$, N_{tot} is the total number of effect sizes from all the selected studies, Q_{tot} is computed as [111]:

$$Q_{tot} = \sum_{i=1}^{N_{tot}} W_i E_i^2 - \frac{\left(\sum_{i=1}^{N_{tot}} W_i E_i \right)^2}{\sum_{i=1}^{N_{tot}} W_i} \tag{4}$$

where $W_i = 1/v_i$, E_i is adjusted coefficient of determination (R_i^2).

The I^2 statistic can be interpreted as the percentage of heterogeneous component in the total variability of effect size (Q_{tot}), so the larger the I^2 statistic is, the stronger the heterogeneity is. If I^2 exceeds 50%, the null hypothesis of homogeneity is rejected. The I^2 statistic for different pigments at different scales were calculated, all the results were lower than 50%, the null hypothesis of homogeneity for this study was accepted.

The Calculation of Mean Effect Size for Different Pigments at Different Scales

In contrast to studies based on original data, the unit of meta-analysis is the individual research study. Distinctive aspects of data analysis follow from this difference. The first complication is that the studies incorporated into the meta-analysis generally use different sample sizes and this controls the statistical properties of effect sizes [112]. From a statistical perspective, larger sample studies have less sampling error than smaller sample studies, thus more weight should be assigned to larger sample studies in the computation of the mean effect size. The other complication is inter-study variability, which is caused by the influence of an indeterminate number of characteristics that vary among the studies.

Considering the two sources of variability discussed above, a random effects model was used to compute the weighted mean of R_A^2 for different pigment types. In contrast to a fixed effects model, the weight applied to each effect size in a random effects model must represent both subject-level sampling error and the additional random variance component [112]. As such, the mean effect size becomes a reasonable estimate of the true strength of the effect in the population. Because of the generality of the random effects model, it is the preferred strategy in meta-analysis [113]. The mean effect size is computed as:

$$M_{rand} = \frac{\sum_{i=1}^{N_p} W_{i(rand)} E_i}{\sum_{i=1}^{N_p} W_{i(rand)}}$$

(5)

The variance is:

$$V_{rand} = \frac{1}{\sum_{i=1}^{N_p} W_{i(rand)}}$$

(6)

where

$$W_{i(rand)} = \frac{1}{V_i + \sigma^2},$$

$$\sigma^2 = \frac{Q_p - (N_p - 1)}{\sum_{i=1}^{N_p} W_i - \frac{\sum_{i=1}^{N_p} W_i^2}{\sum_{i=1}^{N_p} W_i}}, \quad W_i = 1 \Big/ V_i,$$

$$Q_p = \sum_{i=1}^{N_p} W_i E_i^2 - \frac{\left(\sum_{i=1}^{N_p} W_i E_i\right)^2}{\sum_{i=1}^{N_p} W_i} E_i$$

Is adjusted coefficient of determination (R_4^2) and N_p is the total number of effect sizes for a specific type of pigment at each different scales (Table 2). Using this approach the mean effect size of Chl tot, Chl a, Chl b, Cars and Anths at the scales of leaf, canopy and landscape were calculated.

A confidence interval gives the range of values within which the mean effect size is likely to be, it is useful in indicating the degree of precision of the estimate of the mean effect size. A 95% confidence interval is subsequently calculated as follows:

$$Conf_{95} = M_{rand} \pm 1.96SE_{rand} \tag{7}$$

where $SE_{rand} = \sqrt{V_{rand}}$. If the confidence intervals of multiple mean effect sizes donot overlap, then there are significant differences between these mean effect sizes.

Optimal Wavelengths for Pigment Quantification

A large number of narrow-band indices were proposed to measure plant pigments in the selected articles. These narrow band indices include difference vegetation index (NBDVI), ratio vegetation index (NDRVI), normalized difference vegetation index (NBNDVI), anthocyanin reflectance index (ARI), soil-adjusted vegetation index (SAVI), perpendicular vegetation index (PVI) and so on. The wavelengths used in these studies are different and there is lack of agreement on optimal wavelengths for pigment quantification.

Histograms and quantile plots were used to identify the optimal wavelengths for individual pigment quantification at different scales. The histogram partitions the data distribution of wavelengths into subsets of 10nm width. This enabled us to provide an overview of suitable wavelengths, which is difficult to achieve if the analysis is performed at higher spectral resolutions. Also this approach avoided inaccuracies of spectral calibration associated with the use of many different instruments across the studies incorporated into the meta-analysis. In the histogram each subset is represented by a rectangle whose height is equal to the count of observations that fall into the wavelength interval.

A quantile plot is a simple and effective way to compare different wavelength distributions. Let $\lambda_i (i = 1\,\text{to}\,G)$ be the wavelengths sorted in increasing order so that λ_1 is the smallest wavelength and λ_G is the largest. Each wavelength, λ_i, is paired with a percentage, f_i, which indicates that approximately $100\,f_i\%$ of the data are below or equal to the value, λ_i.

$$f_i = \frac{i - 0.5}{G} (i = 1, \ldots, G)$$

(8)

In a quantile plot, λ_i is graphed against f_i. This allows us to compare different wavelength distributions based on their quantiles [114].

RESULTS

Quantifying the Development of Remote Sensing Of Plant Pigment Concentrations

The number of studies used in the meta-analysis published over the period from 1977 to 2014 are shown in Fig 2, along with the 5-year running mean which summarises the overall trajectory of development in this scientific field. After the first two studies were published in 1977 there were no other publications for 11 years, but then there was fast rate of growth from 1990 to 1999. The number of publications reached top in 1999 after which the publication rate stopped increasing, indicating that research in passive optical hyperspectral remote sensing of plant pigment concentrations is within a mature phase. The overall trajectory of publications shows three periods covering the origins, development and proliferation of research in this field. This trajectory corresponds to the developmental phases of hyperspectral instruments, which started with spectrophotometers and hand-held spectroradiometers enabling leaf and canopy-scale work. With the more recent advent of airborne and spaceborne imaging spectrometers, more landscape scale analyses have become possible.

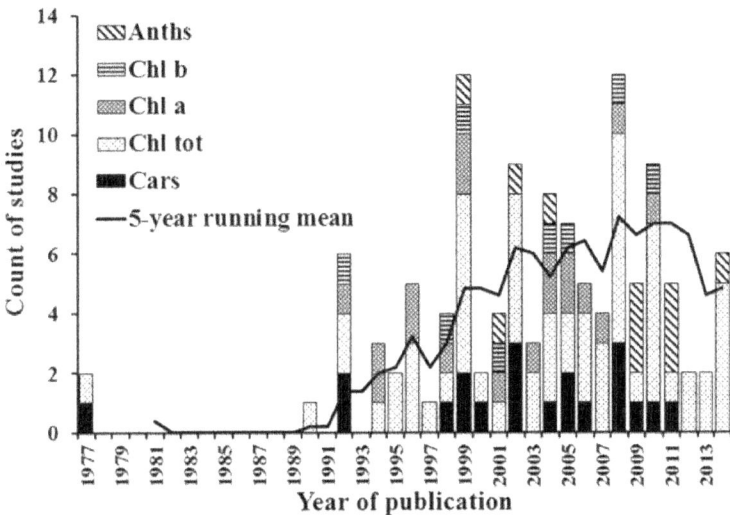

Figure 2: Histogram of numbers of selected studies published over time, showing the total in each year and the number focusing on each pigment type.

The solid line is a 5-year running mean of the total number of studies.

Despite this overall development in the field, there were substantial differences in research on different pigments. The first studies of total chlorophyll and carotenoids were published in 1977, followed by chlorophyll *a* and chlorophyll *b* in 1992 and anthocyanins in 1999. The growth rate of publications on chlorophyll *a*, chlorophyll *b*, carotenoids and anthocyanins has been significantly lower than that for total chlorophyll. These differential rates of growth are perhaps indicative of the increased difficulty in quantifying the concentrations of individual photosynthetic and protective pigments remotely.

The Relationships between Pigment Concentrations and Remotely Sensed Variables

The mean effect size for different pigments at the scales of leaf, canopy and landscape were calculated (Fig 3). At the leaf scale, the mean effect sizes were fairly consistent between different pigment types, varying from 0.87 to 0.93, while the difference in mean effect sizes between pigment types was statistically significant at the canopy and landscape scales. The mean effect size presented the highest value 0.93 (95% confidence interval, 0.92–0.95) for anthocyanins quantification at the leaf scale, far higher than the result of 0.35 (95% confidence interval, 0.18–0.51) at the landscape scale. The mean effect size for total chlorophyll quantification was 0.88 (95% confidence interval, 0.87–0.89) at the leaf scale, 0.73 (95% confidence interval, 0.69–0.77) at the canopy scale and 0.79 (95% confidence interval, 0.76–0.82) at the landscape scale. The mean effect size for carotenoids was the lowest of the various pigments at 0.87 (95% confidence interval, 0.84–0.90) at the leaf scale, still higher than the result 0.80 (95% confidence interval, 0.71–0.90) at the canopy scale and 0.85 (95% confidence interval, 0.76–0.94) at the landscape scale. The results show that these mean effect sizes varied across pigment types and scales. In general, the relationships are stronger at the leaf scale than those at the canopy and landscape scales.

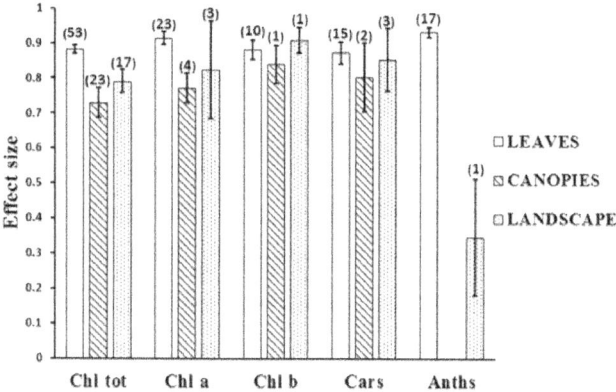

Figure 3: The mean effect size for pigment types at the scales of leaf, canopy and landscape.

(The numbers of reported relationships found in the literature are shown in brackets, error bars represent 95% confidence intervals).

Fig 3 shows that the highest number of relationships published was for pigment quantifications at the leaf scale. Pigment quantification at the canopy scale was less frequently reported in the literature and only a few studies were conducted at the landscape scale. This can be attributed to the limited availability and high costs of suitable airborne and spaceborne hyperspectral instruments [20]. For each scale, the highest number of relationships published was for total chlorophyll quantification, followed by chlorophyll *a*, carotenoids, chlorophyll *b* and anthocyanins. These findings are consistent with previous studies [10,11].

Wavelength Selection for Pigment Quantification Using Remotely Sensed Data

Optimal Wavelengths for Chlorophyll Quantification

There is a large quantity of studies on the relationships between chlorophyll concentration and remotely sensed data. The distributions of wavelengths used at the three scales are shown inFig 4. It should be noted that all of the wavelengths for pigment quantification were concentrated in the 350–950 nm region, except for total chlorophyll quantification at the canopy scale, which spread over 400–2400 nm. For comparison, wavelengths in the histograms and quantile plots were limited within the 350–950 nm region.

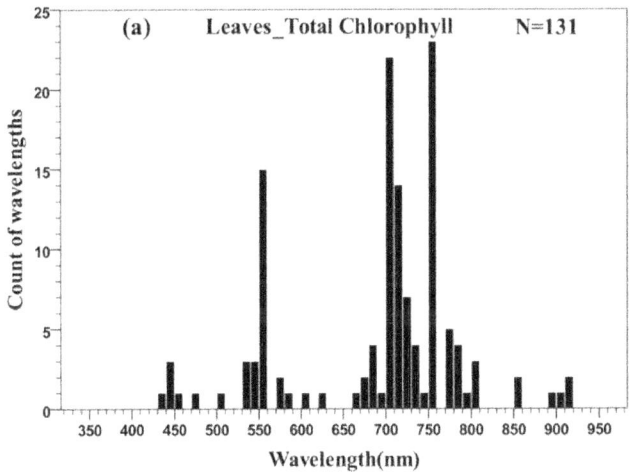

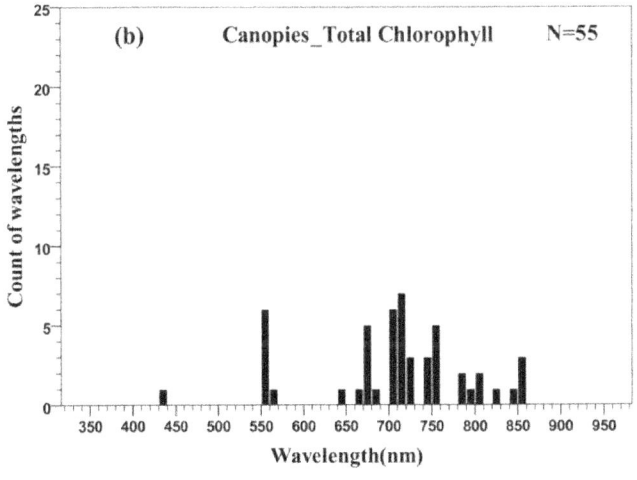

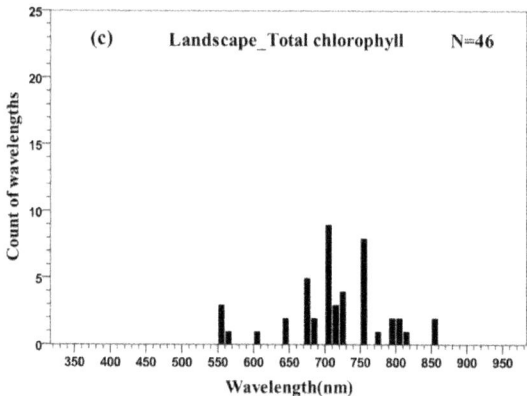

Figure 4: Histogram of wavelengths for total chlorophyll quantification using remotely sensed data at leaf (a), canopy (b) and landscape (c) scales using an interval width of 10 nm.

In general, the distribution of wavelengths displayed a double-peak feature, concentrated in the green (550–560 nm) and red edge (680–750 nm) regions rather than the main absorption wavelengths of chlorophyll (blue or red) (Fig 5). At the canopy scale, five wavelengths in the NIR to SWIR regions (1000–2400 nm) were also used for total chlorophyll quantification (not shown in Fig 4B). This is due to the major influence of canopy structure in canopy reflectance and because leaf chlorophyll concentration was relatively stable in the particular studies [76,77].

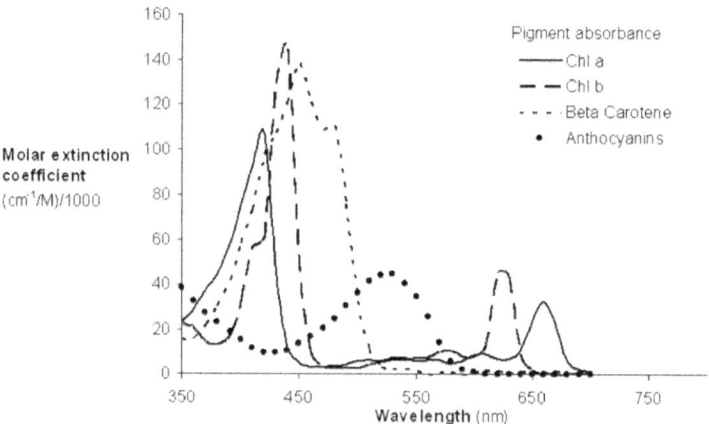

Figure 5: Absorption spectra of the major plant pigments (reproduced from Blackburn, 2007).

The distribution of wavelengths proposed for chlorophyll *a* quantification at the leaf scale was similar to that of total chlorophyll, concentrated in the green and red edge ranges (Fig 6A). At the canopy and landscape scales, the number of wavelengths is limited and is difficult to identify the central tendency of wavelength distribution (Fig 6B and Fig 6C).

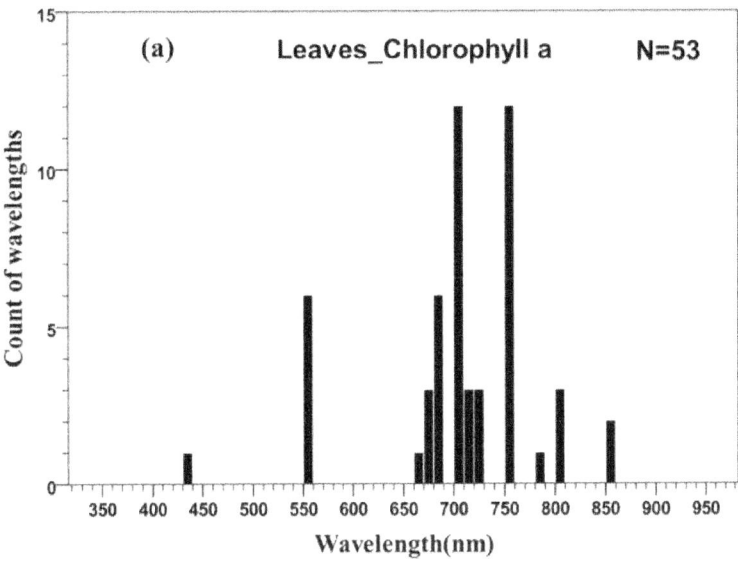

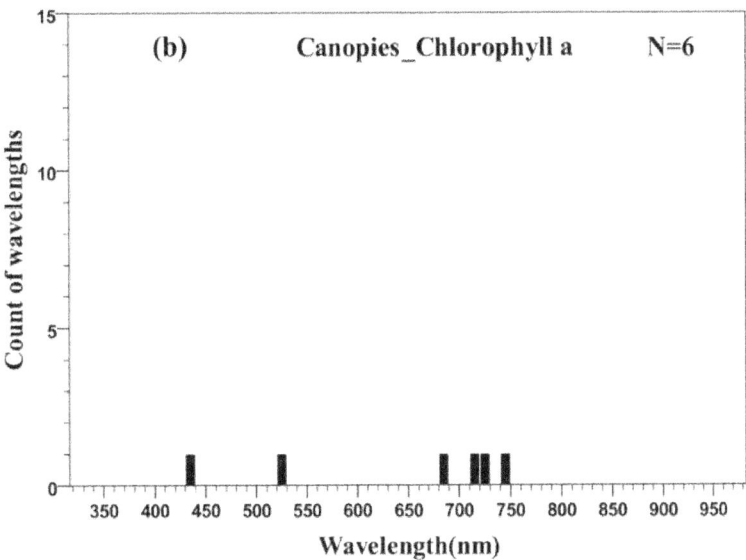

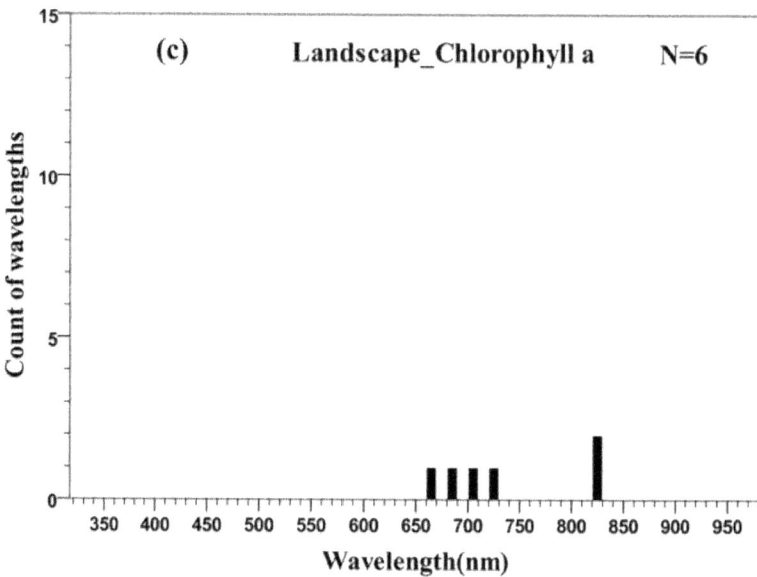

Figure 6: Histogram of wavelengths for chlorophyll *a* quantification using remotely sensed data at leaf (a), canopy (b) and landscape (c) scales by an interval width of 10 nm.

The distribution of wavelengths used for chlorophyll *b* quantification at the leaf scale were concentrated in the main absorption wavelength of chlorophyll *b* (red, 630–660 nm), the red edge (670–710 nm) and the NIR (800–810 nm) regions (Fig 7A). Only two wavelengths were selected at the landscape scale and could not be used for statistical inference (Fig 7B). The distributions of wavelengths used for quantification of different pigments at different scales can be compared in the quantile plots (Fig 8). There were similar wavelength distributions for total chlorophyll quantification at the scales of leaf, canopy and landscape (Fig 8A). For chlorophyll *a* there were similar wavelength distributions at the leaf and canopy scales, but the landscape scale differed (Fig 8B), while a comparison across scales for chlorophyll *b* was difficult due to a lack of data at scales other than the leaf (Fig 8C).

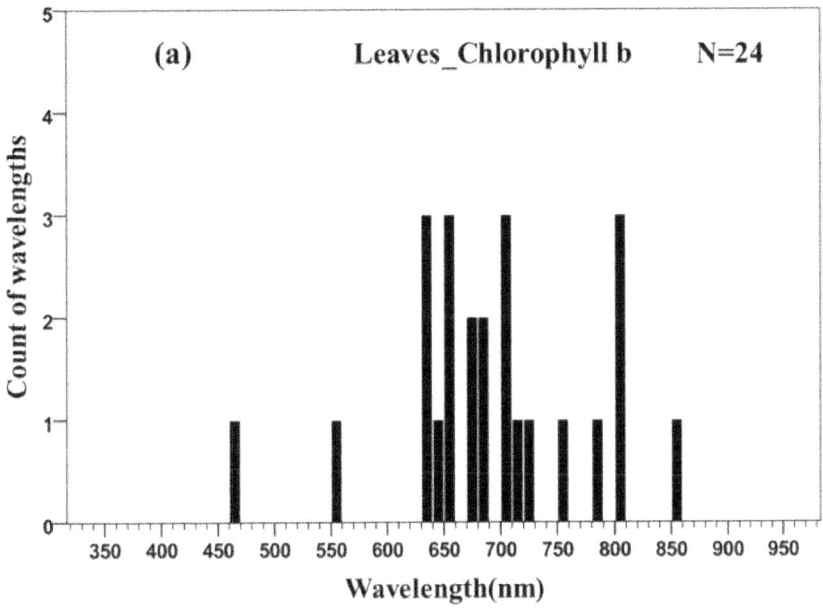

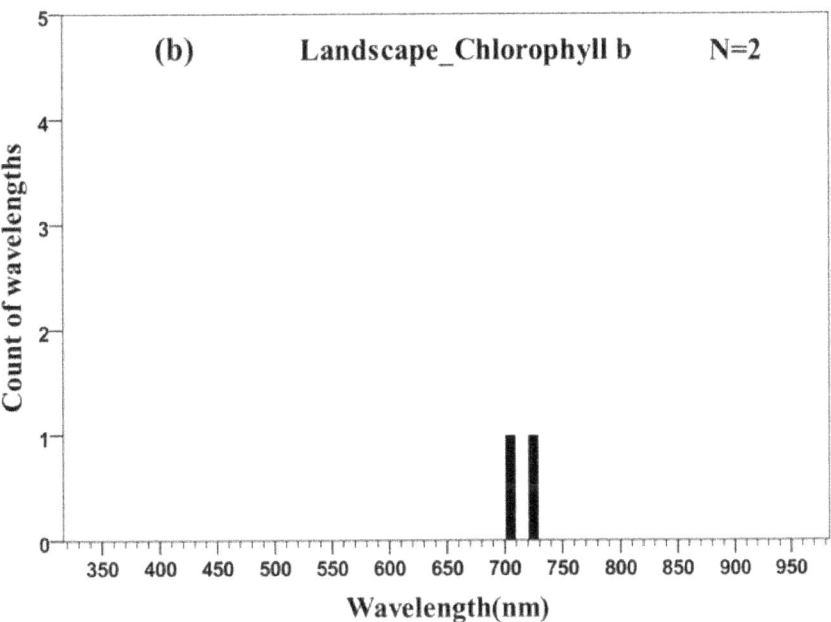

Figure 7: Histogram of wavelengths for chlorophyll *b* quantification using remotely sensed data at leaf (a) and landscape (b) scales using an interval width of 10 nm.

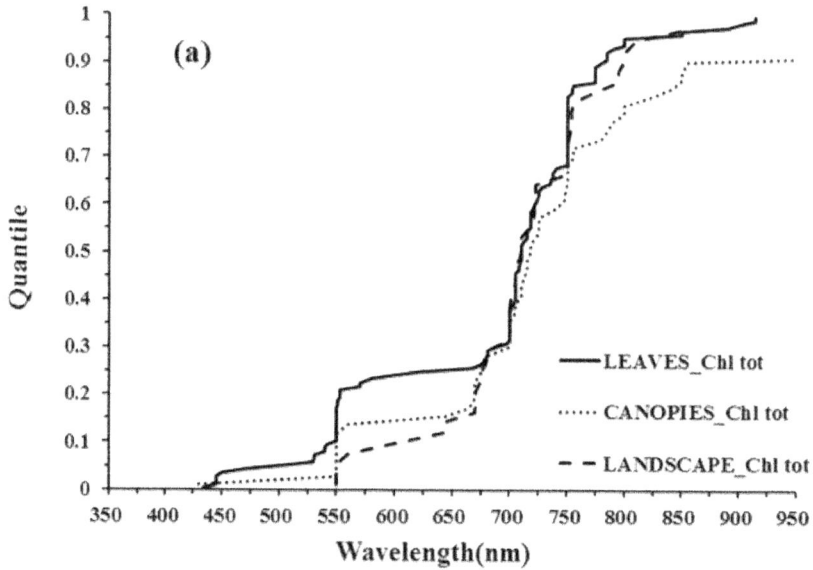

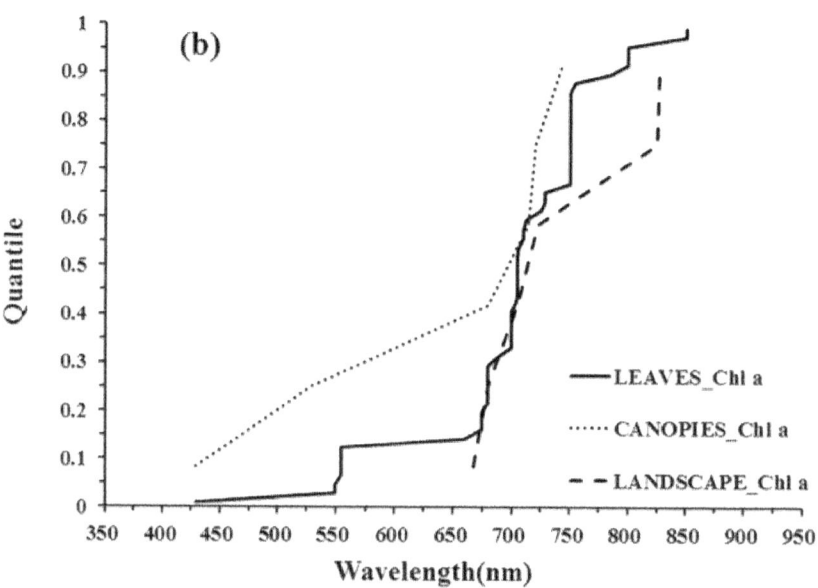

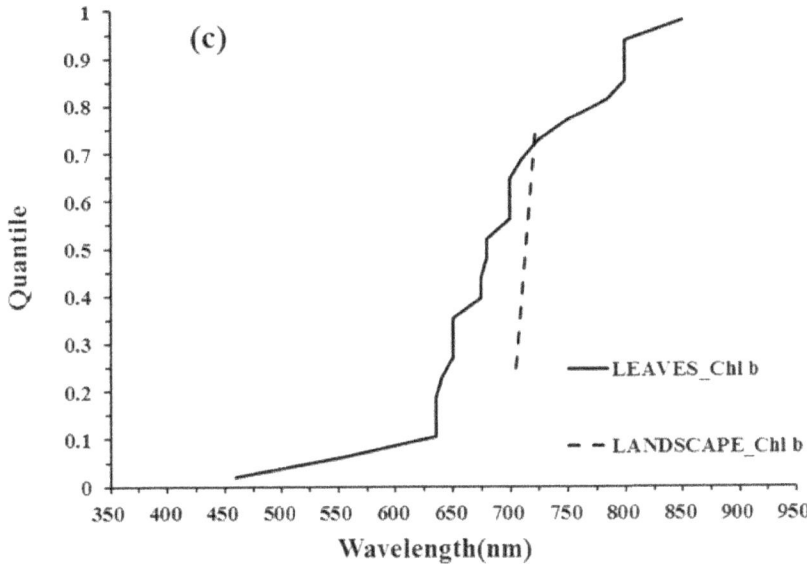

Figure 8: Quantile plots of the wavelengths used for the quantification of Chl tot (a), Chl a (b), and Chl b (c) at different scales.

At the leaf scale, the wavelength distributions for total chlorophyll and chlorophyll *a*quantification were relatively similar while there were notable differences for chlorophyll *b* (Fig 9A). In the region 425–625 nm, the wavelengths used for chlorophyll *a* quantification were concentrated in the region of the green peak in leaf reflectance (550nm), but the central tendency of wavelength distribution for chlorophyll *b* quantification was not obvious. In the red region, the wavelength distribution for chlorophyll *a* quantification was shifted to longer wavelengths than that of chlorophyll *b* (Fig 9A). The significant overlap in the absorption features of chlorophyll *a* and chlorophyll *b* (Fig 5) and the low concentrations of chlorophyll *b*with respect to chlorophyll *a* in most leaves can present difficulties in defining optimal wavelengths for chlorophyll *b* quantification. The absorption spectra of chlorophyll *a* and chlorophyll *b* both display a double-peak feature; the absorption maxima of chlorophyll *a* are at 430 and 662 nm, and chlorophyll *b* has peaks located at 453 and 642 nm (Fig 5). In the presence of carotenoids, it is difficult to separately assess chlorophyll *a* and chlorophyll *b* from reflectance data in the blue region. However, in the red region, the wavelength position of maximum absorption by chlorophyll *a* is longer than that of chlorophyll *b*, which can be exploited for chlorophyll *a* and chlorophyll *b* discrimination (as seen in Fig 9A). The capacity to use this approach to discriminate chlorophyll *a* and chlorophyll *b* is difficult to assess at the canopy and landscape scales due to the small number of studies on chlrophyll b (Fig 9B and 9C).

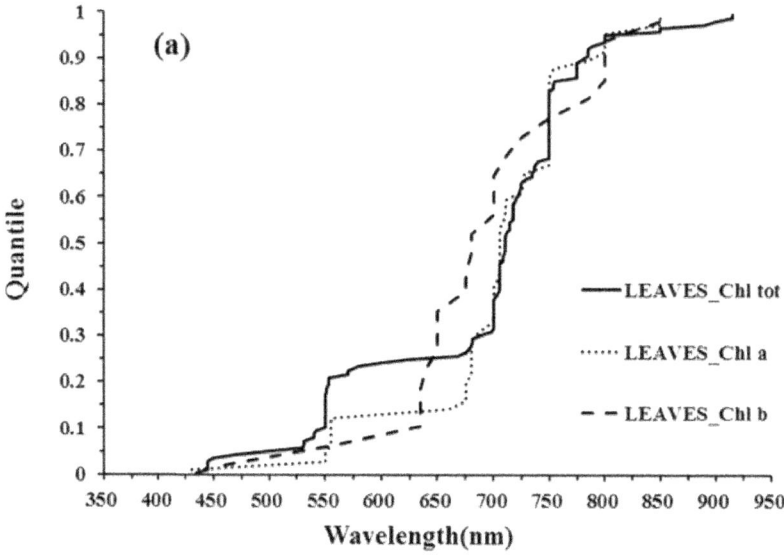

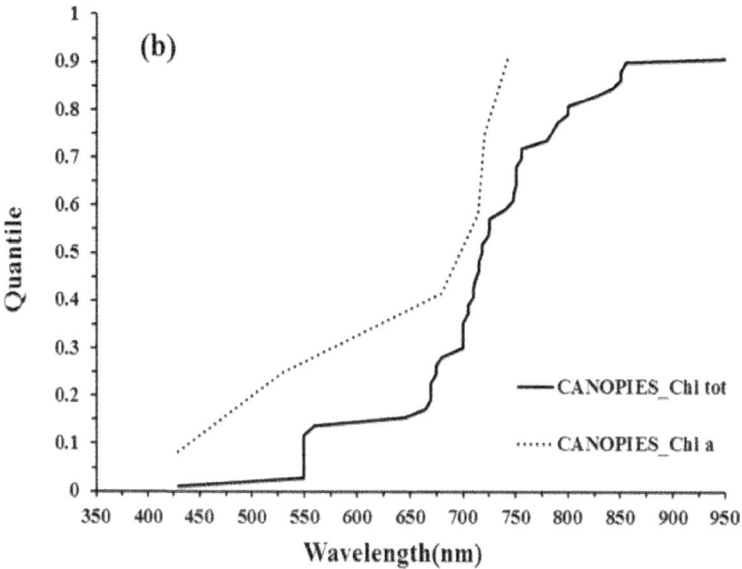

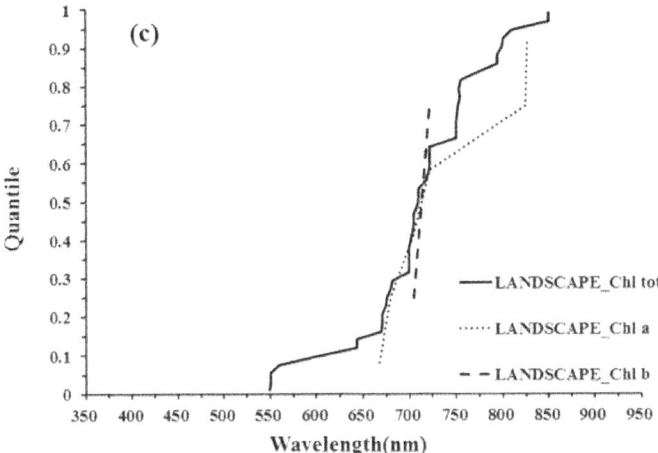

Figure 9: Quantile plot of the wavelengths used at leaf (a), canopy (b) and landscape (c) scales for the quantification of Chl tot, Chl a, and Chl b.

Optimal Wavelengths for Carotenoids Quantification

At the leaf scale, the central tendency of wavelength distribution was not obvious but was mainly concentrated in the 500–580 nm region (Fig 10A). There were similar wavelength distributions for carotenoids quantification at the leaf and canopy scales (Fig 11) but at the landscape scale, the number of wavelengths was too small for statistical inference (Fig 10C).

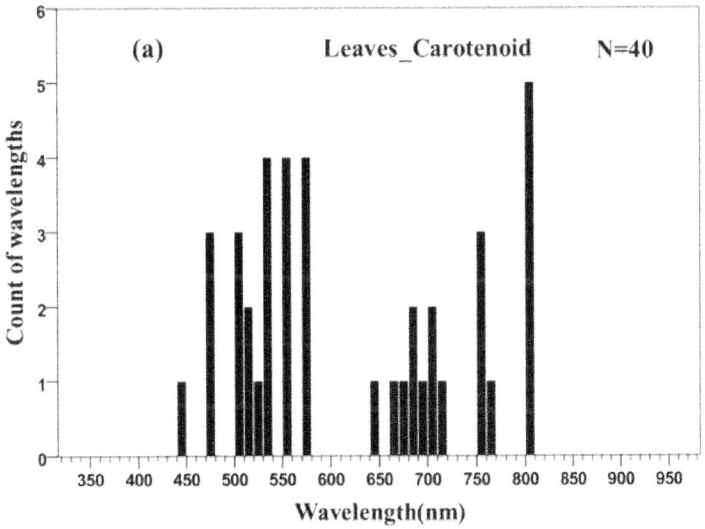

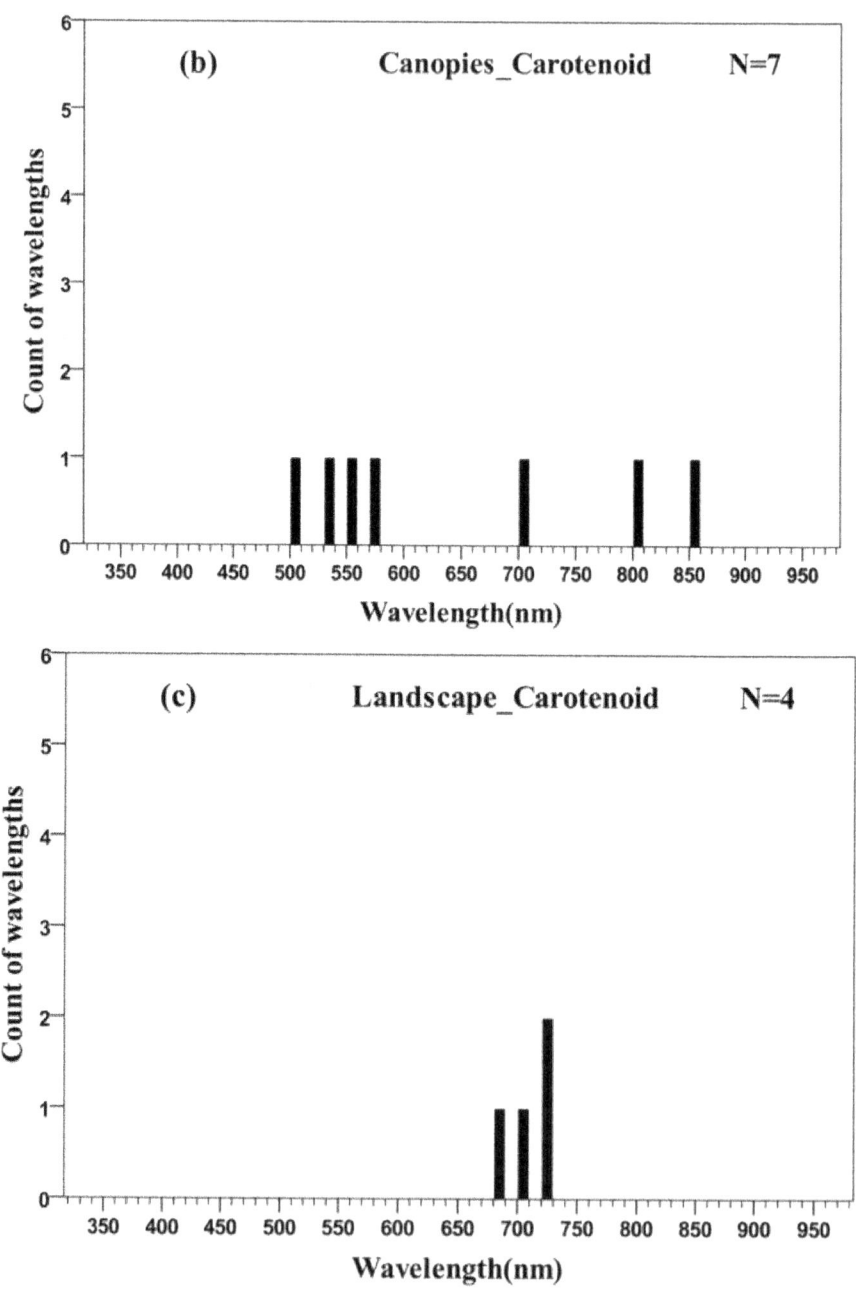

Figure 10: Histogram of wavelengths for carotenoids quantification using remotely sensed data at leaf (a), canopy (b) and landscape (c) scales using an interval width to 10 nm.

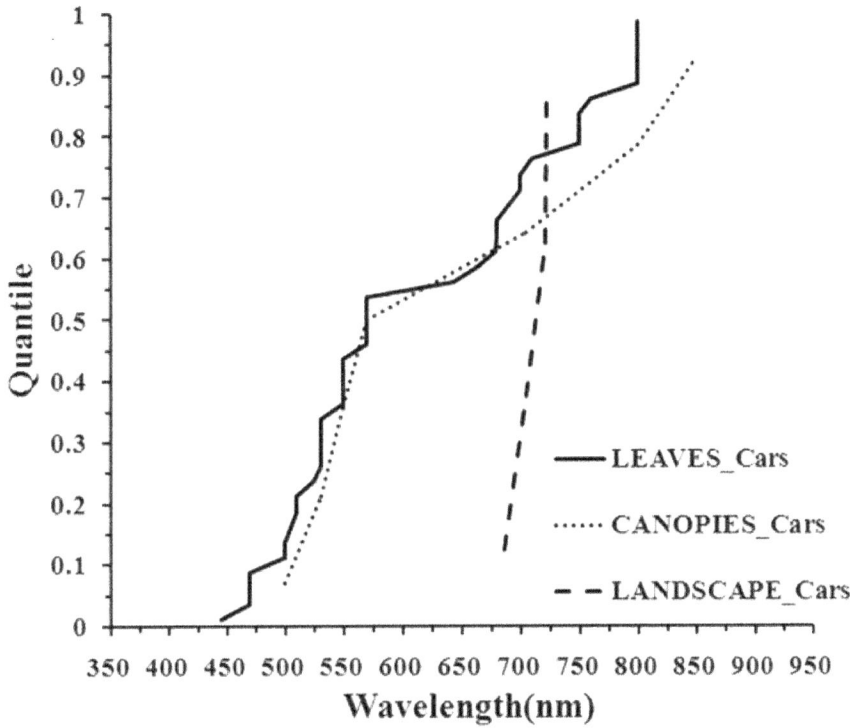

Figure 11: Quantile plot of the optimal wavelength for the quantification of Cars at different scales.

Optimal Wavelengths for Anthocyanins Quantification

Quantification of anthocyanins from reflectance data has been given less attention by the passive optical hyperspectral remote sensing community than chlorophyll and carotenoids. Most studies have concentrated on the quantification of anthocyanins at the leaf scale, with some work at the landscape scale but nothing at canopy level. At the leaf scale, the distribution of wavelengths used for quantifying anthocyanins was concentrated in the main absorption wavelength of anthocyanins (green, 550–560 nm), the red edge (700–710 nm) and the NIR (780–790 nm) ranges (Fig 12A). Similarly, the two wavelengths used to estimate anthocyanin concentration at the landscape scale were distributed in the green and red edge regions, respectively (Fig 12B and Fig 13).

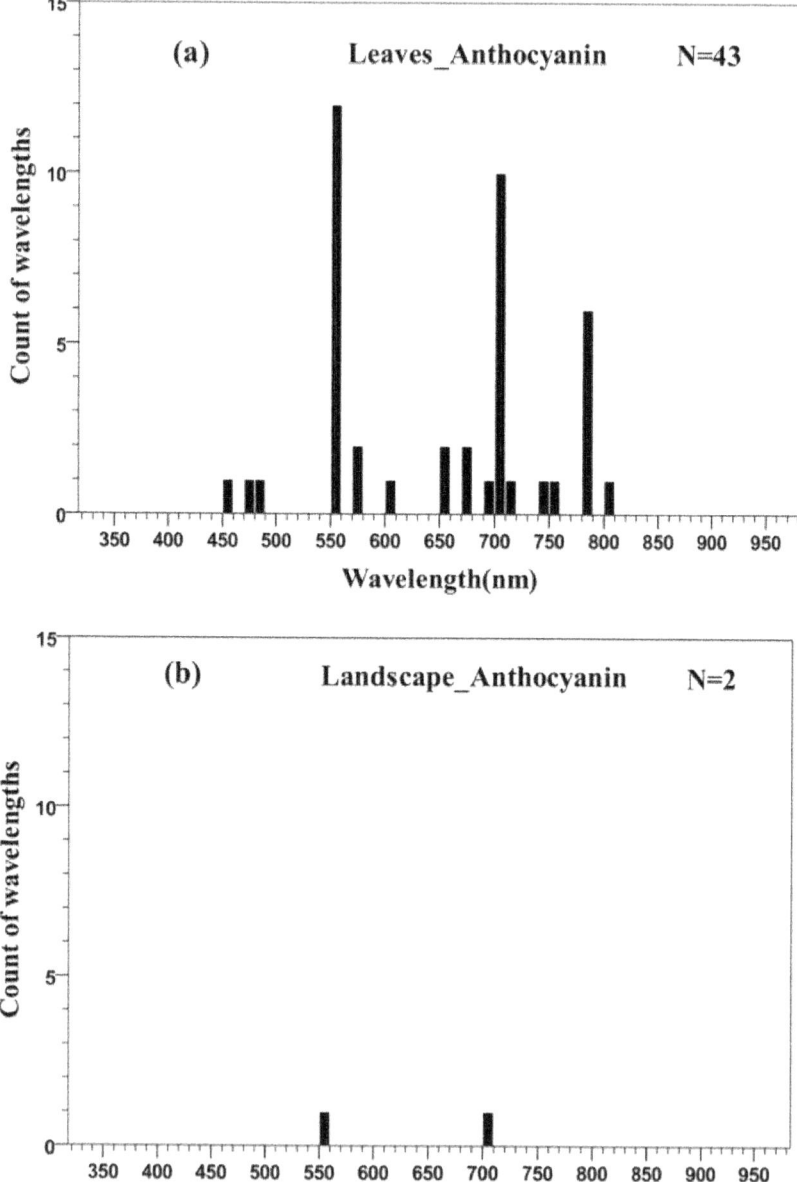

Figure 12: Histogram of wavelengths for anthocyanins quantification using remotely sensed data at leaf (a) and landscape (b) scales using an interval width to 10 nm.

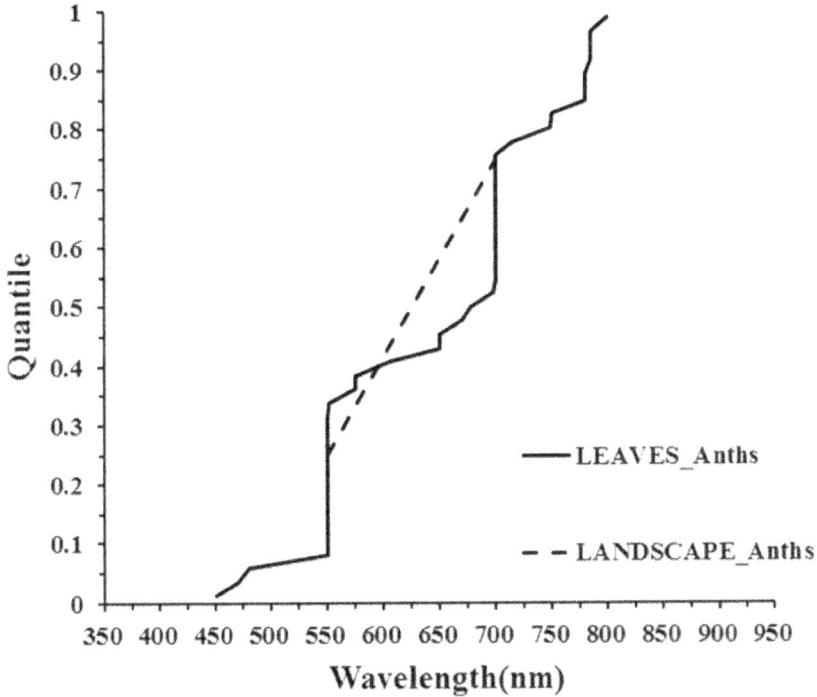

Figure 13: Quantile plot of the optimal wavelengths for the quantification of Anths at different scales.

Discussion

This meta-analysis of 85 studies has demonstrated that remotely sensed variables are good estimators of plant pigment concentration. Most of the studies were conducted at the leaf scale, while pigment quantification at the canopy and landscape scales was less frequently reported. For each scale, most of the studies were conducted for total chlorophyll quantification, followed by chlorophyll *a*, carotenoids, chlorophyll *b* and anthocyanins. These findings are consistent with previous studies [10,11].

The strength of these relationships varied across pigments types and scales. In general, the relationships are stronger at the leaf scale than those at the canopy and landscape scales. At the leaf scale, the mean effect sizes were fairly consistent across different pigment types and were all greater than 0.87, while the difference in mean effect sizes between pigment types was statistically significant at the canopy and landscape scales. This result has been widely assumed, yet a quantitative evaluation has been lacking. At the leaf scale, the methodological basis for pigment quantification has been fully explored, which provides an important basis for developing estimation models at the canopy and landscape scales. The primary goal of most leaf scale passive optical hyperspectral remote sensing studies has been to develop analytical approaches for pigment quantification that can be applied to data from airborne and spaceborne sensors [11].

At the canopy and landscape scales, the experimental results are influenced by a number of factors, which obscures the relationships between spectral reflectance and concentrations of individual pigments. The reflectance spectrum of a whole canopy is subject to canopy biophysical attributes (e.g., orientation of leaves (leaf angle distribution; *LAD*), variations in number of leaf layers (*LAI*) and foliage clumping), presence of non-leaf elements (e.g., soil reflectance and the proportions of shadowed and sunlit background), anisotropic scattering of photons to interact with multiple surfaces such as leaves, woody material and soils, viewing geometry (e.g., sun and view zenith and azimuth angles) and illumination conditions (e.g., the ratio between direct and diffuse sunlight and atmospheric condition). It is the interaction of these factors, including their potential covariance or unique behavior that drive variation in canopy and landscape reflectance characteristics in three-dimensional space [10,22].

It should be noted that part of the variability in effect sizes at the canopy scale may be entirely artifactual. These artifacts are common in experimental studies: studies vary in terms of the quality of measurement; researchers make computational errors; people make typographical errors in copying numbers from handwritten tables to computer; and sampling errors. With the advent of airborne and spaceborne imaging spectrometers, there have been opportunities to measure plant pigment concentrations at the landscape scale. The reflectance spectrum from airborne and spaceborne sensors is subject to even more controlling factors, notably, soil/litter surface reflectance, and vegetation structure. The range of controlling factors should be taken into account in subsequent analyses.

Table 2 shows that the total sample size at the leaf scale is much more than that of canopy and landscape scales. The law of large numbers correctly states that large samples are reasonable representations of the population and parameter estimation is close to the real values when the sample size is large enough. Many researchers seem to believe that the same law applies to small samples and severely underestimate the amount of variability in findings that is caused by sampling errors. As a result, they erroneously expect statistics based on small samples to be close to the real values [13]. At the canopy and landscape scales, the number of studies and total sample size is limited, which influences the robustness and accuracy of effect sizes.

Despite the significant difference in effect sizes between different scales, it was found that the wavelength distribution for total chlorophyll quantification at the scales of leaf, canopy and landscape was similar, being concentrated in the green (550–560 nm) and red edge (680–750 nm) regions rather than the main absorption wavelength of chlorophyll (blue or red). The consistency in optimal wavelengths across scales can be attributed to several factors: (1) despite the many factors influencing reflectance at the canopy and landscape scales, it is the selective absorbance properties of pigments that determines the selection of wavelengths for pigment quantification, and (2) several estimation models derived at the leaf scale were directly applied to canopy and landscape scales. This suggests that the leaf-level study has provided an important basis for developing estimation models at the canopy and landscape scales.

At the leaf scale, the distribution of wavelengths used for chlorophyll *a* quantification was similar to that of total chlorophyll; the distribution of wavelengths for chlorophyll *b* quantification was concentrated in the main absorption wavelength of chlorophyll *b* (red, 630–660 nm), the red edge (670–710 nm) and the NIR (800–810 nm) regions; the central tendency of wavelength distribution for carotenoids quantification was not obvious, but was mainly concentrated in the 500–580 nm region; for the estimation of anthocyanins, the distribution of wavelengths was concentrated in the main absorption wavelength of anthocyanins (green, 550–560 nm), the red edge (700–710 nm) and the NIR (780–790 nm) ranges. In the present meta-analysis, the lack of studies reporting the quantification of carotenoids and anthocyanins at the canopy and landscape scales has hindered cross-scale comparisons (Fig 10; Fig 12). Consequently, it is not entirely clear if the optimal wavelengths for carotenoids and anthocyanins quantification at the leaf scale are necessarily the optimal wavelengths at the canopy and landscape scales, where multiple scattering and other confounding effects may alter the spectral response of individual pigments, much in the way that pigment absorption peaks can vary depending upon their chemical and scattering medium. Therefore, more work may be needed to determine the optimal algorithms for airborne or spaceborne platforms.

It should be noted that the lack of statistical information in the studies (e.g., sample size and coefficient of determination) has hindered a more comprehensive cross-study comparison in the present research. When selecting the final set of studies, 50 studies were excluded due to the lack of statistical information. Insufficient statistical information can not only limit the research population covered by meta-analysis but also render the findings of the original study somewhat suspect. Thus, it is suggested that when conducting primary research, such information should include, but not be limited to, the sample size, the pertinent test statistic (e.g., r, t, or F), the unit of pigment concentration/content, the range of pigment concentrations/content, and estimation precision for pigment quantification (e.g. root mean squared error, RMSE).

This study has established the possibility of integrating the results of studies on the passive optical hyperspectral remote sensing of plant pigment concentrations across a range of vegetation types and scales using a meta-analysis approach. Despite the robust models for pigment prediction at the leaf scale, the continuing challenge is to properly account for the multiple factors introduced by scene components such as sunlit and shaded parts of tree crowns and gaps influencing the retrieved signal at the canopy and landscape scales. Recent work have illustrated that, in addition to other influencing factors such as illumination geometry and atmospheric conditions, canopy architecture had an important control on the applicability of models for pigment prediction. Scanning LIDAR systems have only recently become widely available which enable the estimation of the range between the sensor and a target by recording the time during which the emitted laser pulse is reflected off an object and returns to the sensor [21]. LIDAR systems have the ability to directly measure spatial variations in canopy height and other aspects of the vertical structure of canopies. Given the high degree of structural complexity at the canopy and landscape scales, it would appear that the integration of vertical canopy structural information provided by active LIDAR remote sensing with hyperspectral reflectance may has both a structural and physiological interpretation and improve the estimation of pigment concentrations over passive optical hyperspectral imagery alone [102].

ACKNOWLEDGMENTS

We acknowledge the contribution of Bao She, Weijiao Huang, Dilong Gan, Sujuan Wang and Zhewen Zhao for the literature database searches and associated support. The authors thank anonymous reviewers who provided very valuable comments also.

AUTHOR CONTRIBUTIONS

Conceived and designed the experiments: JH. Performed the experiments: Chen Wei YZ XW Chuanwen Wei JW. Analyzed the data: Chen Wei. Contributed reagents/materials/analysis tools: Chen Wei. Wrote the paper: Chen Wei GAB JH.

REFERENCES

1. Curran PJ, Dungan JL, Gholz HL. Exploring the relationship between reflectance red edge and chlorophyll content in slash pine. Tree Physiol 1990; 7: 33–48. pmid:14972904 doi: 10.1093/treephys/7.1-2-3-4.33

2. Filella I, Serrano L, Serra J, Penuelas J. Evaluating wheat nitrogen status with canopy reflectance indices and discriminant analysis. Crop Sci 1995; 35: 1400–1405. doi: 10.2135/cropsci1995.0011183x003500050023x

3. Hendry GAF, Houghton JD, Brown SB. The degradation of chlorophyll: a biological enigma. New Phytol 1987; 107: 255–302. doi: 10.1111/j.1469-8137.1987.tb00181.x

4. Merzlyak MN, Gitelson A. Why and what for the leaves are yellow in autumn? On the interpretation of optical spectra of senescing leaves (*Acerplatanoides L*). J. Plant Physiol 1995; 145: 315–320. doi: 10.1016/s0176-1617(11)81896-1

5. Demmig—Adams B, Adams WW. The role of xanthophyll cycle carotenoids in the protection of photosynthesis. Trends Plant Sci 1996; 1: 21–26. doi: 10.1016/s1360-1385(96)80019-7

6. Peñuelas J, Filella I. Visible and near-infrared reflectance techniques for diagnosing plant physiological status. Trends Plant Sci 1998; 3: 151–156. doi: 10.1016/s1360-1385(98)01213-8

7. Merzlyak MN, Gitelson AA, Chivkunova OB, Rakitin VY. Non-destructive optical detection of pigment changes during leaf senescence and fruit ripening. Physiol Plantarum 1999; 106: 135–141. doi: 10.1034/j.1399-3054.1999.106119.x

8. Chalker-Scott L. Environmental significance of anthocyanins in plant stress responses. Photochem Photobiol 1999; 70: 1–9. doi: 10.1111/j.1751-1097.1999.tb01944.x

9. Carter GA, Knapp AK. Leaf optical properties in higher plants: linking spectral characteristics to stress and chlorophyll concentration. Am J Bot 2001; 88: 677–684. pmid:11302854 doi: 10.2307/2657068

10. Blackburn GA. Hyperspectral remote sensing of plant pigments. J. Exp Bot 2007; 58: 855–867. pmid:16990372 doi: 10.1093/jxb/erl123

11. Ustin SL, Gitelson AA, Jacquemoud S, Schaepman M, Asner GP, Gamon JA, et al. Retrieval of foliar information about plant pigment systems from high resolution spectroscopy. Remote Sens Environ 2009; 113: S67–S77. doi: 10.1016/j.rse.2008.10.019

12. Hunter JE, Schmidt FL. Methods of meta-analysis: correcting error and bias in research findings. Los Angeles, USA: SAGE Publications; 2004.

p. 33–34.

13. Borenstein M, Hedges LV, Higgins JPT, Rothstein HR. Introduction to meta-analysis. West Sussex, United Kingdom: John Wiley & Sons Ltd; 2009. p. 12–13.

14. Curtis PS, Queenborough SA. Raising the standards for ecological meta–analyses. New Phytol 2012; 195: 279–281. doi: 10.1111/j.1469-8137.2012.04207.x. pmid:22702404

15. Glass GV. Primary secondary and meta-analysis of research. Educ Res 1976; 5: 3–8. doi: 10.3102/0013189x005010003

16. Verschuyl J, Riffell S, Miller D, Wigley TB. Biodiversity response to intensive biomass production from forest thinning in North American forests-A meta-analysis. Forest Ecol Manag 2011; 261: 221–232. doi: 10.1016/j.foreco.2010.10.010

17. Robinson EA Ryan GD Newman JA A meta-analytical review of the effects of elevated CO_2 on plant-arthropod interactions highlights the importance of interacting environmental and biological variables. New Phytol 2012; 194: 321–336. doi: 10.1111/j.1469-8137.2012.04074.x. pmid:22380757

18. Liao CZ Peng RH Luo YQ Zhou X H Wu X W Fang C M Chen J K Li B Altered ecosystem carbon and nitrogen cycles by plant invasion: a meta-analysis. New Phytol 2008; 177: 706–714. pmid:18042198 doi: 10.1111/j.1469-8137.2007.02290.x

19. Poorter H, Niinemets Ü, Poorter L, Wright IJ, Villar R. Causes and consequences of variation in leaf mass per area (LMA): a meta-analysis. New Phytol 2009; 182: 565–588. pmid:19434804 doi: 10.1111/j.1469-8137.2009.02830.x

20. Garbulsky MF, Peñuelas J, Gamon J, Inoue Y, Filella I. The photochemical reflectance index (PRI) and the remote sensing of leaf canopy and ecosystem radiation use efficiencies: a review and meta-analysis. Remote Sens Environ 2011; 115: 281–297. doi: 10.1016/j.rse.2010.08.023

21. Zolkos SG, Goetz SJ, Dubayah R. A meta-analysis of terrestrial aboveground biomass estimation using lidar remote sensing. Remote Sens Environ 2013; 128: 289–298. doi: 10.1016/j.rse.2012.10.017

22. Asner GP. Biophysical and biochemical sources of variability in canopy reflectance. Remote Sens Environ 1998; 64: 234–253. doi: 10.1016/s0034-4257(98)00014-5

23. Curran PJ, Dungan JL, Macler BA, Plummer SE, Peterson DL. Reflectance spectroscopy of fresh whole leaves for the estimation of

chemical concentration. Remote Sens Environ 1992; 39: 153–166. doi: 10.1016/0034-4257(92)90133-5

24. Curran PJ, Windham WR, Gholz HL. Exploring the relationship between reflectance red edge and chlorophyll concentration in slash pine leaves. Tree Physiol 1995; 15: 203–206. pmid:14965977 doi: 10.1093/treephys/15.3.203

25. Yoder BJ, Pettigrew-Crosby RE. Predicting nitrogen and chlorophyll content and concentrations from reflectance spectra (400–2500 nm) at leaf and canopy scales. Remote Sens Environ 1995; 53: 199–211. doi: 10.1016/0034-4257(95)00135-n

26. Gitelson AA, Merzlyak MN, Grits Y. Novel algorithms for remote sensing of chlorophyll content in higher plant leaves. In International Geoscience and Remote Sensing Symposium (IGARSS); Lincoln, NE, USA; May 1996. p. 2355–2357.

27. Gitelson AA, Kaufman YJ, Merzlyak MN. Use of a green channel in remote sensing of global vegetation from EOS-MODIS. Remote Sens Environ 1996; 58: 289–298. doi: 10.1016/s0034-4257(96)00072-7

28. Gitelson AA, Merzlyak MN. Remote estimation of chlorophyll content in higher plant leaves. Int J Remote Sens 1997; 18: 2691–2697. doi: 10.1080/014311697217558

29. Gitelson AA, Merzlyak MN. Remote sensing of chlorophyll concentration in higher plant leaves. Adv Space Res 1998; 22: 689–692. doi: 10.1016/s0273-1177(97)01133-2

30. Gitelson AA, Buschmann C, Lichtenthaler HK. The chlorophyll fluorescence ratio F735/F700 as an accurate measure of the chlorophyll content in plants. Remote Sens Environ 1999; 69: 296–302. doi: 10.1016/s0034-4257(99)00023-1

31. Adams ML, Philpot WD, Norvell WA. Yellowness index: an application of spectral second derivatives to estimate chlorosis of leaves in stressed vegetation. Int J Remote Sens 1999; 20: 3663–3675. doi: 10.1080/014311699211264

32. Sims DA, Gamon JA. Relationships between leaf pigment content and spectral reflectance across a wide range of species leaf structures and developmental stages. Remote Sens Environ 2002; 81: 337–354. doi: 10.1016/s0034-4257(02)00010-x

33. Richardson AD, Duigan SP, Berlyn GP. An evaluation of noninvasive methods to estimate foliar chlorophyll content. New Phytol 2002; 153: 185–194. doi: 10.1046/j.0028-646x.2001.00289.x

34. Gitelson AA, Gritz Y, Merzlyak MN. Relationships between leaf chlorophyll content and spectral reflectance and algorithms for non–destructive chlorophyll assessment in higher plant leaves. J Plant Physiol 2003; 160: 271–282. pmid:12749084 doi: 10.1078/0176-1617-00887

35. Zhao DL, Reddy KR, Kakani VG, Read JJ, Koti S. Selection of optimum reflectance ratios for estimating leaf nitrogen and chlorophyll concentrations of field-grown cotton. Agron J 2005; 97: 89–98. doi: 10.2134/agronj2005.0089

36. Kochubey SM, Kazantsev TA. Changes in the first derivatives of leaf reflectance spectra of various plants induced by variations of chlorophyll content. J. Plant Physiol 2007; 164: 1648–1655. pmid:17292510 doi: 10.1016/j.jplph.2006.11.007

37. Wang Q, Li PH. Hyperspectral indices for estimating leaf biochemical properties in temperate deciduous forests: comparison of simulated and measured reflectance data sets. Ecol Indic 2012; 14: 56–65. doi: 10.1016/j.ecolind.2011.08.021

38. Simic A, Chen JM, Leblanc SG, Dyk A, Croft H, Tian Han. Testing the top-down model inversion method of estimating leaf reflectance used to retrieve vegetation biochemical content within empirical approaches. IEEE J Sel Top Appl Earth Observ Remote Sens 2014; 7: 92–104. doi: 10.1109/jstars.2013.2271583

39. Gitelson A, Merzlyak MN. Quantitative estimation of chlorophyll-a using reflectance spectra: experiments with autumn chestnut and maple leaves. J. Photoch Photobio B 1994; 22: 247–252. doi: 10.1016/1011-1344(93)06963-4

40. Gitelson AA, Merzlyak MN. Spectral reflectance changes associated with autumn senescence of *Aesculus hippocastanum L* and *Acer platanoides L* Leaves Spectral features and relation to chlorophyll estimation. J. Plant Physiol 1994; 143: 286–292. doi: 10.1016/s0176-1617(11)81633-0

41. Gitelson AA, Merzlyak MN. Signature analysis of leaf reflectance spectra: algorithm development for remote sensing of chlorophyll. J. Plant Physiol 1996; 148: 494–500. doi: 10.1016/s0176-1617(96)80284-7

42. Thomas JR, Gausman HW. Leaf reflectance vs Leaf chlorophyll and carotenoid concentrations for eight crops. Agron J 1977; 69: 799–802. doi: 10.2134/agronj1977.00021962006900050017x

43. Gamon JA, Peñuelas J, Field CB. A narrow-waveband spectral index that tracks diurnal changes in photosynthetic efficiency. Remote Sens Environ 1992; 41: 35–44. doi: 10.1016/0034-4257(92)90059-s

44. Levizou E, Manetas Y. Photosynthetic pigment contents in twigs of 24

woody species assessed by in vivo reflectance spectroscopy indicate low chlorophyll levels but high carotenoid/chlorophyll ratios. Environ Exp Bot 2007; 59: 293–298. doi: 10.1016/j.envexpbot.2006.03.002

45. Gamon JA, Surfus JS. Assessing leaf pigment content and activity with a reflectometer. New Phytol 1999; 143: 105–117. doi: 10.1046/j.1469-8137.1999.00424.x

46. Merzlyak MN, Solovchenko AE, Gitelson AA. Reflectance spectral features and non–destructive estimation of chlorophyll carotenoid and anthocyanin content in apple fruit. Postharvest Biol Tec 2003; 27: 197–211. doi: 10.1016/s0925-5214(02)00066-2

47. Gitelson AA, Merzlyak MN. Non-destructive assessment of chlorophyll carotenoid and anthocyanin content in higher plant leaves: principles and algorithms. In Remote Sensing for Agriculture and the Environment. Stamatiadis S, LynchJ JM, Schepers JS, Eds. Ella, Greece: OECD; 2004. p. 78–94.

48. Garriga M, Retamales JB, Romero-Bravo S, Caligari PD, Lobos GA. Chlorophyll anthocyanin and gas exchange changes assessed by spectroradiometry in Fragaria chiloensis under salt stress. J. Integr Plant Biol 2014; 56: 505–15. doi: 10.1111/jipb.12193. pmid:24618024

49. Chappelle EW, Kim MS, McMurtrey JE III. Ratio analysis of reflectance spectra (RARS): an algorithm for the remote estimation of the concentrations of chlorophyll a chlorophyll b and carotenoids in soybean leaves. Remote Sens Environ 1992; 39: 239–247. doi: 10.1016/0034-4257(92)90089-3

50. Blackburn GA. Spectral indices for estimating photosynthetic pigment concentrations: a test using senescent tree leaves. Int J Remote Sens 1998; 19 657–675. doi: 10.1080/014311698215919

51. Chen L, Huang JF, Wang FM. Retrieval of pigment contents in rice leaves and panicles using hyperspectral data by artificial neuron network models. In International Geoscience and Remote Sensing Symposium (IGARSS); Seoul, Korea; July 2005. p. 1416–1419.

52. 52. Blackburn GA. Relationships between spectral reflectance and pigment concentrations in stacks of deciduous broadleaves. Remote Sens Environ 1999; 70: 224–237. doi: 10.1016/s0034-4257(99)00048-6

53. Maccioni A, Agati G, Mazzinghi P. New vegetation indices for remote measurement of chlorophylls based on leaf directional reflectance spectra. J. Photoch Photobio B 2001; 61: 52–61. doi: 10.1016/s1011-1344(01)00145-2

54. Imanishi J, Nakayama A, Suzuki Y, Imanishi A, Ueda N, Morimoto Y,

et al. Nondestructive determination of leaf chlorophyll content in two flowering cherries using reflectance and absorptance spectra. Landsc Ecol Eng 2010; 6: 219–234. doi: 10.1007/s11355-009-0101-8

55. Lichtenthaler HK, Gitelson A, Lang M. Non-destructive determination of chlorophyll content of leaves of a green and an aurea mutant of tobacco by reflectance measurements. J. Plant Physiol 1996; 148: 483–493. doi: 10.1016/s0176-1617(96)80283-5

56. Datt B. Visible/near infrared reflectance and chlorophyll content in eucalyptus leaves. Int J Remote Sens 1999; 20: 2741–2759. doi: 10.1080/014311699211778

57. Gitelson AA, Zur Y, Chivkunova OB, Merzlyak MN. Assessing carotenoid content in plant leaves with reflectance spectroscopy. Photochem Photobiol 2002; 75: 272–281. pmid:11950093 doi: 10.1562/0031-8655(2002)0750272accipl2.0.co2

58. Filella I, Porcar-Castell A, Munne-Bosch S, Back J, Garbulsky MF, Penuelas J. PRI assessment of long-term changes in carotenoids/ chlorophyll ratio and short–term changes in de-epoxidation state of the xanthophyll cycle. Int J Remote Sens 2009; 30: 4443–4455. doi: 10.1080/01431160802575661

59. Garrity SR, Eitel JUH, Vierling LA. Disentangling the relationships between plant pigments and the photochemical reflectance index reveals a new approach for remote estimation of carotenoid content. Remote Sens Environ 2011; 115: 628–635. doi: 10.1016/j.rse.2010.10.007

60. Gitelson AA, Merzlyak MN, Chivkunova OB. Optical properties and nondestructive estimation of anthocyanin content in plant leaves. Photochem Photobiol 2001; 74: 38–45. pmid:11460535 doi: 10.1562/0031-8655(2001)074<0038:opaneo>2.0.co;2

61. Steele MR, Gitelson AA, Rundquist DC, Merzlyak MN. Nondestructive estimation of anthocyanin content in grapevine leaves. Am J Enol Viticult 2009; 60: 87–92.

62. Gitelson AA, Chivkunova OB, Merzlyak MN. Nondestructive estimation of anthocyanins and chlorophylls in anthocyanic leaves. Am J Bot 2009; 96: 1861–1868. doi: 10.3732/ajb.0800395. pmid:21622307

63. Qin JL, Rundquist D, Gitelson A, Tan Z, Steele M. A non-linear model of nondestructive estimation of anthocyanin content in grapevine leaves with Visible/Red-infrared hyperspectral. In International Conference on Computer and Computing Technologies in Agriculture; Beijing, China; October 2011. p. 47–62.

64. Pappas CS, Takidelli C, Tsantili E, Tarantilis PA, Polissiou MG.

Quantitative determination of anthocyanins in three sweet cherry varieties using diffuse reflectance infrared fourier transform spectroscopy. J. Food Compos Anal 2011; 24: 17–21. doi: 10.1016/j.jfca.2010.07.001

65. Vina A, Gitelson AA. Sensitivity to foliar anthocyanin content of vegetation indices using green reflectance. IEEE Geosci Remote Sens Lett 2011; 8: 464–468. doi: 10.1109/lgrs.2010.2086430

66. Ciganda V, Gitelson A, Schepers J. Non-destructive determination of maize leaf and canopy chlorophyll content. J. Plant Physiol 2009; 166: 157–167. doi: 10.1016/j.jplph.2008.03.004. pmid:18541334

67. Schlemmer M, Gitelson A, Schepers J, Ferguson R, Peng Y, Shanahan J, et al. Remote estimation of nitrogen and chlorophyll contents in maize at leaf and canopy levels. Int J Appl Earth Obs 2013; 25: 47–54. doi: 10.1016/j.jag.2013.04.003

68. Croft H, Chen JM, Zhang Y. The applicability of empirical vegetation indices for determining leaf chlorophyll content over different leaf and canopy structures. Ecol Complex 2014; 17: 119–130. doi: 10.1016/j.ecocom.2013.11.005

69. Ju CH, Tian YC, Yao X, Cao WX, Zhu Y, Hannaway D. Estimating leaf chlorophyll content using red edge parameters. Pedosphere 2010; 20: 633–644. doi: 10.1016/s1002-0160(10)60053-7

70. ZarcoTejada PJ. Hyperspectral remote sensing of closed forest canopies: estimation of chlorophyll fluorescence and pigment content. PhD thesis, York University, Toronto, Canada 2000.

71. Filella I, Penuelas J. The red edge position and shape as indicators of plant chlorophyll content biomass and hydric status. Int J Remote Sens 1994; 15: 1459–1470. doi: 10.1080/01431169408954177

72. Gitelson AA, Vina A, Ciganda V, Rundquist DC, Arkebauer TJ. Remote estimation of canopy chlorophyll content in crops. Geophys Res Lett 2005; 32: 1–4. doi: 10.1029/2005gl022688

73. Yang XH, Huang JF, Wang FM, Wang XZ, Yi QX, Wang Y. Science letters: a modified chlorophyll absorption continuum index for chlorophyll estimation. J. Zhejiang Univ 2006; 7: 2002–2006. doi: 10.1631/jzus.2006.a2002

74. Zhao DH, Huang LM, Li JL, Qi JG. A comparative analysis of broadband and narrowband derived vegetation indices in predicting LAI and CCD of a cotton canopy. Isprs J Photogramm 2007; 62: 25–33. doi: 10.1016/j.isprsjprs.2007.01.003

75. Wu CY, Niu Z, Tang Q, Huang WJ. Estimating chlorophyll content from

hyperspectral vegetation indices: modeling and validation. Agr Forest Meteorol 2008; 148: 1230–1241. doi: 10.1016/j.agrformet.2008.03.005

76. Darvishzadeh R, Skidmore A, Schlerf M, Atzberger C, Corsi F, Cho M. LAI and chlorophyll estimation for a heterogeneous grassland using hyperspectral measurements. ISPRS J Photogramm 2008; 63: 409–426. doi: 10.1016/j.isprsjprs.2008.01.001

77. Darvishzadeh R, Skidmore A, Schlerf M, Atzberger C. Inversion of a radiative transfer model for estimating vegetation LAI and chlorophyll in a heterogeneous grassland. Remote Sens Environ 2008; 112: 2592–2604. doi: 10.1016/j.rse.2007.12.003

78. Haboudane D, Tremblay N, Miller JR, Vigneault P. Remote estimation of crop chlorophyll content using spectral indices derived from hyperspectral data. IEEE T Geosci Remote 2008; 46: 423–437. doi: 10.1109/tgrs.2007.904836

79. Liu ML, Liu XN, Li M, Fang MH, Chi WX. Neural-network model for estimating leaf chlorophyll concentration in rice under stress from heavy metals using four spectral indices. Biosystems Eng 2010; 106: 223–233. doi: 10.1016/j.biosystemseng.2009.12.008

80. Xu X, Gu X, Song X, Li C, Huang W. Assessing rice chlorophyll content with vegetation indices from hyperspectral data. In International Conference on Computer and Computing Technologies in Agriculture; Beijing, China; October 2011. p. 296–303.

81. Clevers JGPW, Kooistra L. Using Hyperspectral Remote Sensing Data for Retrieving Canopy Chlorophyll and Nitrogen Content. IEEE J Sel Top Appl Earth Observ Remote Sens 2012; 5: 574–583. doi: 10.1109/jstars.2011.2176468

82. Clevers JGPW, Gitelson AA. Remote estimation of crop and grass chlorophyll and nitrogen content using red-edge bands on Sentinel-2 and -3. Int J Appl Earth Obs 2013; 23: 344–351. doi: 10.1016/j.jag.2012.10.008

83. Vincini M, Amaducci S, Frazzi E. Empirical Estimation of Leaf Chlorophyll Density in Winter Wheat Canopies Using Sentinel – 2 Spectral Resolution. IEEE T Geosci Remote 2014; 52: 3220–3235. doi: 10.1109/tgrs.2013.2271813

84. Cheng Q, Huang JF, Wang XZ, Wang RC. In situ hyperspectral data analysis for pigment content estimation of rice leaves. J Zhejiang Univ 2003; 4: 727–733. doi: 10.1631/jzus.2003.0727

85. Li J, Jiang JB, Chen YH, Wang YY, Su W, Huang WJ. Using hyperspectral indices to estimate foliar chlorophyll a concentrations of winter wheat

under yellow rust stress. New Zeal J Agr Res 2007; 50: 1031–1036. doi: 10.1080/00288230709510382

86. Zhao X, Liu SH, Wang JD, Tian ZK. A method for estimating chlorophyll content of wheat from reflectance spectra. In International Geoscience and Remote Sensing Symposium (IGARSS); Anchorage, AK, USA; September 2004. p. 4504–4507.

87. Bannari A, Khurshid KS, Staenz K, Schwarz J. Wheat crop chlorophyll content estimation from ground–based reflectance using chlorophyll indices. In International Geoscience and Remote Sensing Symposium (IGARSS); Denver, CO, USA; July 2006. p. 112–115.

88. Yang F, Li JL, Gan XY, Qian YR, Wu XL, Yang Q. Assessing nutritional status of*Festuca arundinacea* by monitoring photosynthetic pigments from hyperspectral data. Comput Electron Agr 2010; 70: 52–59. doi: 10.1016/j.compag.2009.08.010

89. Peguero-Pina JJ, Morales F, Flexas J, Gil-Pelegrin E, Moya I. Photochemistry remotely sensed physiological reflectance index and de-epoxidation state of the xanthophyll cycle in *Quercus coccifera* under intense drought. Oecologia 2008 156: 1–11. doi: 10.1007/s00442-007-0957-y. pmid:18224338

90. Hall FG, Hilker T, Coops NC, Lyapustin A, Huemmrich KF, Middleton E, et al. Multi–angle remote sensing of forest light use efficiency by observing PRI variation with canopy shadow fraction. Remote Sens. Environ 2008; 112: 3201–3211. doi: 10.1016/j.rse.2008.03.015

91. Haboudane D, Miller JR, Tremblay N, Zarco-Tejada PJ, Dextraze L. Integrated narrow–band vegetation indices for prediction of crop chlorophyll content for application to precision agriculture. Remote Sens Environ 2002; 81: 416–426. doi: 10.1016/s0034-4257(02)00018-4

92. Coops NC, Stone C, Culvenor DS, Chisholm LA, Merton RN. Chlorophyll content in eucalypt vegetation at the leaf and canopy scales as derived from high resolution spectral data. Tree Physiol 2003; 23: 23–31. pmid:12511301 doi: 10.1093/treephys/23.1.23

93. Zarco-Tejada PJ, Miller JR, Harron J, Hu BX, Noland TL, Goel N, et al. Needle chlorophyll content estimation through model inversion using hyperspectral data from boreal conifer forest canopies. Remote Sens Environ 2004; 89: 189–199. doi: 10.1016/j.rse.2002.06.002

94. Dash J, Curran PJ. The MERIS terrestrial chlorophyll index. Int J Remote Sens 2004; 25: 5403–5413. doi: 10.1080/0143116042000274015

95. Haboudane D, Tremblay N, Vigneault P, Miller JR. Indices-based approach for crop chlorophyll content retrieval from hyperspectral data.

In International Geoscience and Remote Sensing Symposium (IGARSS); Barcelona, Spain; July 2007. p.3297–3300.

96. Rao NR, Garg PK, Ghosh SK, Dadhwal VK. Estimation of leaf total chlorophyll and nitrogen concentrations using hyperspectral satellite imagery. J Agr Sci 2008; 146: 65–75. doi: 10.1017/s0021859607007514

97. Delegido J, Fernandez G, Gandia S, Moreno J. Retrieval of chlorophyll content and LAI of crops using hyperspectral techniques: application to PROBA/CHRIS data. Int J Remote Sens 2008; 29: 7107–7127. doi: 10.1080/01431160802238401

98. Wu CY, Han XZ, Niu Z, Dong JJ. An evaluation of EO-1 hyperspectral hyperion data for chlorophyll content and leaf area index estimation. Int J Remote Sens 2010; 31: 1079–1086. doi: 10.1080/01431160903252335

99. Delegido J, Alonso L, González G, Moreno J. Estimating chlorophyll content of crops from hyperspectral data using a normalized area over reflectance curve (NAOC). Int J Appl Earth Obs 2010; 12: 165–174. doi: 10.1016/j.jag.2010.02.003

100. Delegido J, Van Wittenberghe S, Verrelst J, Ortiz V, Veroustraete F, Valcke R, et al. Chlorophyll content mapping of urban vegetation in the city of Valencia based on the hyperspectral NAOC index. Ecol Indic 2014; 40: 34–42. doi: 10.1016/j.ecolind.2014.01.002

101. Oppelt N, Mauser W. Hyperspectral monitoring of physiological parameters of wheat during a vegetation period using AVIS data. Int J Remote Sens 2004; 25: 145–159. doi: 10.1080/0143116031000115300

102. Blackburn GA. Remote sensing of forest pigments using airborne imaging spectrometer and LIDAR imagery. Remote Sens Environ 2002; 82: 311–321. doi: 10.1016/s0034-4257(02)00049-4

103. Guan YN, Guo S, Liu JG, Zhang X. Algorithms for the estimation of the concentrations of chlorophyll a and carotenoids in rice leaves from airborne hyperspectral data. In Computational Science-ICCS 2005; Atlanta, GA, USA; May 2005. p. 908–915.

104. Thomas V, Treitz P, McCaughey JH, Noland T, Rich L. Canopy chlorophyll concentration estimation using hyperspectral and lidar data for a boreal mixedwood forest in northern Ontario Canada. Int J Remote Sens 2008; 29: 1029–1052. doi: 10.1080/01431160701281023

105. Pena MA, Altmann SH. Use of satellite-derived hyperspectral indices to identify stress symptoms in an *Austrocedrus chilensis* forest infested by the aphid *Cinara cupressi*. Int J Pest Manage 2009; 55: 197–206. doi: 10.1080/09670870902725809

106. Jacquemoud S, Verdebout J, Schmuck G, Andreoli G. Hosgood B. Investigation of leaf biochemistry by statistics. Remote Sens Environ 1995; 54: 180–188. doi: 10.1016/0034-4257(95)00170-0

107. Cornell JA, Berger RD. Factors that influence the value of the coefficient of determination in simple linear and nonlinear regression models. Phytopathology 1987; 77: 63–70. doi: 10.1094/phyto-77-63

108. Gurevitch J, Curtis PS, Jones MH. Meta-analysis in ecology. Adv Ecol Res 2001; 32: 199–247. doi: 10.1016/s0065-2504(01)32013-5

109. Hedges LV. Estimation of effect size from a series of independent experiments. Psychol Bull 1982; 92: 490–499. doi: 10.1037//0033-2909.92.2.490

110. Higgins J, Thompson SG. Quantifying heterogeneity in a meta-analysis. Stat Med 2002; 21: 1539–1558. pmid:12111919 doi: 10.1002/sim.1186

111. Hedges LV, Olkin I. Statistical methods for meta-analysis. Orlando, FL, USA: Academic Press; 1985. p. 31–34.

112. Lipsey MW, Wilson DB. Practical meta-analysis. London, UK: SAGE Publications; 2000. p. 112–116.

113. Mosteller F, Colditz GA. Understanding research synthesis (meta-analysis). Annu Rev Publ Health 1996; 17 1–23. doi: 10.1146/annurev. pu.17.050196.000245

114. Han JW, Kamber M. Data mining:concepts and techniques second ed. San Francisco, CA, USA: Morgan Kaufmann; 2006. p. 53–54.

Chapter 3

THE BIOLOGICAL PIGMENTS IN PLANTS PHYSIOLOGY

Arturo Solís Herrera

Human Photosynthesis Study Center, Aguascalientes, Mexico

ABSTRACT

The physiology of the plants has come to be a formidable challenge. Despite best efforts, the mysteries remain, most of the processes identified cannot be played in vitro, and to date it cannot be said that knowledge about the complex biochemical processes of the plant is understood properly. The need to produce food is an emergency in the world, so any knowledge that will allow us to advance in this sense is important. Our discovery about the intrinsic property of melanin transform light energy into chemical energy by means of the dissociation of the molecule of water, such as chlorophyll in plants, represents a turning point in relation to the chemical reaction so far considered the most important in the world: photosynthesis.

INTRODUCTION

A pigment is defined as any substance capable of absorbing light [1] , so it could be considered pigments almost all substances. Usually, substances absorb very specific wavelengths, and emit a characteristic electromagnetic signal that allows us to identify them. Color is the property of electromagnetic radiation with a wavelength between 300 - 400 and 700 - 800 nm. The color of individual chemical substances derived from their interaction with white light. If the total radiation falling on a chemical compound passes through or is reflected from it, the substance has a white color. If white light on contact with a substance is completely absorbed, appears black. Most often chemical compounds selectively absorb within certain radiation ranges and the color of a given compound is complementary to the absorbed radiation.

This allows us to infer that the energy inherent in the light produced a change in substance or molecule that it absorbed, so far in accordance with the law of conservation of energy, whose first postulate says that energy, defined as everything that produces a change is not created or destroyed, only is thus transformed.

Therefore, the light energy absorbed by the compound or molecule, brought about a change in it, and the new molecule issued part of the absorbed energy, in the form of electromagnetic signal, for example, light with different wavelength to which was originally absorbed.

Nature is full of examples in this regard, and the plants are no exception. And also in them it happens that the wavelengths of light can be, wholly or partly, absorbed, transmitted or reflected.

And many compounds containing a plant that they can classified as pigments, for example: porphyrins, carotenoids, anthocyanins, etc. However, there are two that have powerfully drawn attention: the chlorophyll and recently the melanin. And the reason is that both molecules have very peculiar behaviors, we could call them unique in nature. This is: both molecules are able to dissociate the water molecule.

And it is the attention, because we are against the way in which nature transforms light energy into chemical energy. It is possible that there are other molecules that can do so, but the two examples most known or at least shown are chlorophyll, which until a decade ago was considered the only molecule with such capacity, and recently, melanin.

Chlorophyll, absorbs the ends of the visible spectrum, and with that power separates the liquid water in its gaseous components, hydrogen and molecular oxygen; and it carried out at room temperature and in an irreversible manner, then ejects the oxygen into the atmosphere. The chemical reaction can be written in the following way.

$$2H_2O_{(liquid)} \rightarrow 2H_{2\,(gas)} + O_{2\,(gas)}$$

It is the attention that the reaction happens so quickly and at room temperature inside the chlorophyll's leaves, as in the laboratory we can dissociate water, but requires warm water at 2000°C and the products that we get, they are not so ordered as in chlorophyll and melanin. This reaction is impelled by the extremes of visible light (violet and red)

Another interesting detail is that part of the energy that is released when water dissociates, is transported by molecular hydrogen, which was to be expected since it is the main carrier of energy in the entire universe.

On the other hand, molecular oxygen, which is very stable and toxic, is expelled into the atmosphere by the plant, and since it has always been present in the equation, the plant optimizes it, given its high toxicity.

In melanin (Greek: melanos; dark), the reaction is similar, but quicker and thousands of times more efficient, because the melanin absorbs electromagnetic full spectrum and in addition is capable not only of separate water into its components but also re-shape the molecule, this is: gaseous components takes her to the liquid state again [2] . This reaction happens inside melanin, and can be located both intra and extracellularly. The chemical reaction is written thus:

$$2H_2O_{(liquid)} \rightarrow 2H_{2\,(gas)} + O_{2\,(gas)} \rightarrow 2H_2O_{(liquid)} + 4e^-$$

Or so:

$$2H_2O_{(liquid)} \leftrightarrow 2H_{2\,(gas)} + O_{2\,(gas)} + 4e^-$$

For every two molecules of water re-formed, are generated 4 high energy electrons, which are electrons whose probability cloud is located farthest from the atomic nucleus in question as a low-energy electron or stable state. High energy electrons or excited are exchanged easily, making it almost impossible to identify the atom or molecule to which they belong.

MELANIN, THE META-CHLOROPHYLL

Melanin is what lies behind the chlorophyll, which explains the misunderstood chlorophyll, which remedied the deficiencies; in few words, is that even explains the origin of life.

Chemicals Processes are known in plants, which the mere dissociation of water by chlorophyll cannot explain, could tell that the physiology of the plant is full of mysteries until now misunderstood. The energy that emanates from the chlorophyll does not provide a satisfactory explanation for so many and so many observed phenomena, both in the root, stem and leaves. Only the circulation of the SAP not has been explained to date.

The so-called cohesion-tension theory that try to explain SAP transport by the xylem, from the roots of the plant to the leaves through the wood tubules, is to date; a theory.

The mysterious forces that drive the viscous solutions through the trunk, from the root to the leaves, in order, in a sequence that is not understood, no doubt require energy, cannot be explained by the chlorophyll, glucose or ATP, but the energy that emanates from the melanin comes to fill a hole that will allow us to advance in the understanding of the intricate biochemical processes that affect the life of the plant.

Melanin in humans had passed as unnoticed as the lignin in plants, the main interest at present was to modify the composition of lignin, which is the melanin in the plants; as a basic strategy to increase the digestibility of cell walls in order to produce biodiesel [3] .

It is expected that as you go knowing the unsuspected intrinsic property of melanin transform light energy into chemical energy by means of the water molecule dissociation, the interest becomes more reasonable, because the omnipresence of melanin explains the germination of the seeds, as it was a mystery as the seed detected level appropriate water to flourish (Figures 1-3).

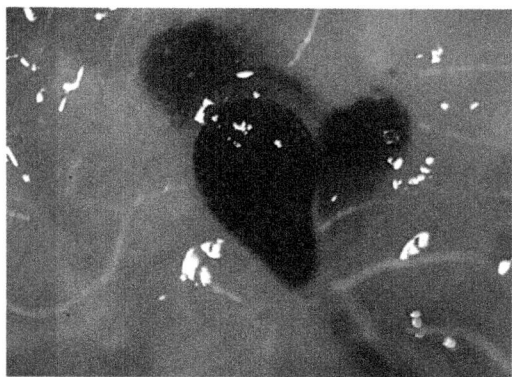

Figure 1: Melanin is present in practically all of the seeds, and the pitaya is no exception. When the amount of water is scarce, power that can produce melanin also is, by what the mechanisms leading to the emergence of seed not can be driven, when the amount of water is adequate, then the energy that emanates from the melanin will be adequate to promote each and every one of the mechanisms that lead to the germination of the seeds (16×).

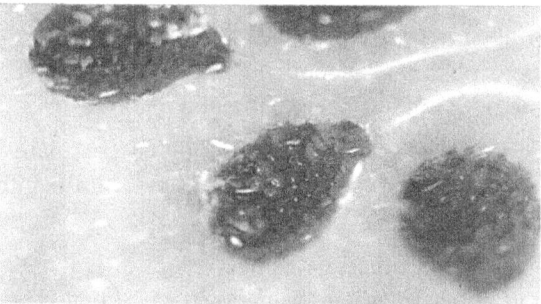

Figure 2: Seed color depends on the amount of melanin, granule size, the orientation of the molecule, the combination with other compounds, as well as the nature of the structures that are in, but, either way, its main function is the same: production of chemical energy (16×).

Figure 3: The same principle applies in all living organisms, apparently is the only or at least the best system to convert light energy into chemical energy. In Sunflower seed we can see melanin. The magnification is 25×.

The similarity of the surface of the iris of the human eye (Figure 4, Figure 5) with the surface of some seeds, for example, the peach, are given by the notable content of melanin in both structures.

The concept that photosynthetic membrane is the microscopic powerhouse of our biosphere, now is broken into small pieces.

Molecular hydrogen and high energy electrons coming from the dissociation of water, due to the melanin molecule, constitute the true universal biological engine. The oxidation of water as the basis of molecular biology is more true than ever, but as the melanindone, chlorophyll only is an additional mechanism that optimizes some biochemical processes, as does the hemoglobin in human body, as it is a molecule that also has the amazing ability to dissociate the water molecule irreversibly, and was to be expected, because the biochemical differences between chlorophyll and hemoglobin are really minimal.

Because of the huge differences between melanin and chlorophyll, it is that, based on this last, it has not been possible to explain the metabolism of plants. To date, the questions are many, despite the best efforts of researchers and Governments that support research.

BRIEF HISTORY OF THE DISCOVERY

Since melanin absorbs all kinds of energy and dissipates it separating the water molecule, their behavior is unique in nature, because it does not emit any electromagnetic, visible or invisible signal that will allow us to characterize it. As a result, the study of melanin in the laboratory has been the date; a formidable challenge [4] . So it is understandable that the finding of its intrinsic property of transform light energy into chemical energy, by means of the water molecule dissociation, came from another type of study, in this case, of a clinical trial; because we observed incidentally melanin acting in one of its multiple location that is in nature. Why is that we consider important to a brief description of the how it came to knowledge.

In 1990, we started an observational, descriptive, study about the three main causes of blindness in Mexico, and that is no coincidence that is the same around the world: Glaucoma, diabetic retinopathy, and macular degeneration related to age.

Our working hypothesis was trying to characterize the morphology of the blood vessels of the optic nerve, very small structure in humans, whose average diameter is equivalent to twelve human hair together. Our aim was to try to find vascular anatomical changes that could eventually serve as indicators of early disease and therefore allowed us early treatments.

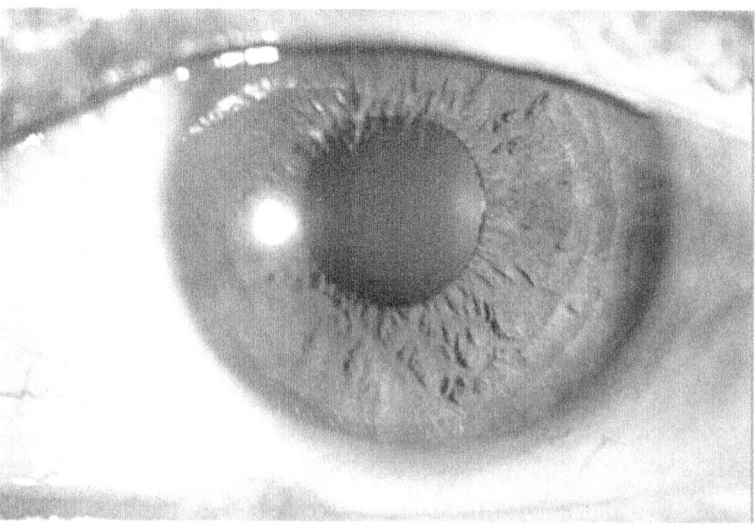

Figure 4: Melanin, wherever you are or whatever that he is called, has the same function, namely the transformation of light, visible and invisible power; in energy chemistry through dissociation and further re-formating of the water molecule. The effect of melanin on the nearest atoms and molecules is surprisingly consistent, as evidenced the huge resemblance between the surface of the iris of the eye and the surface of some seeds, bark of trees, etc.

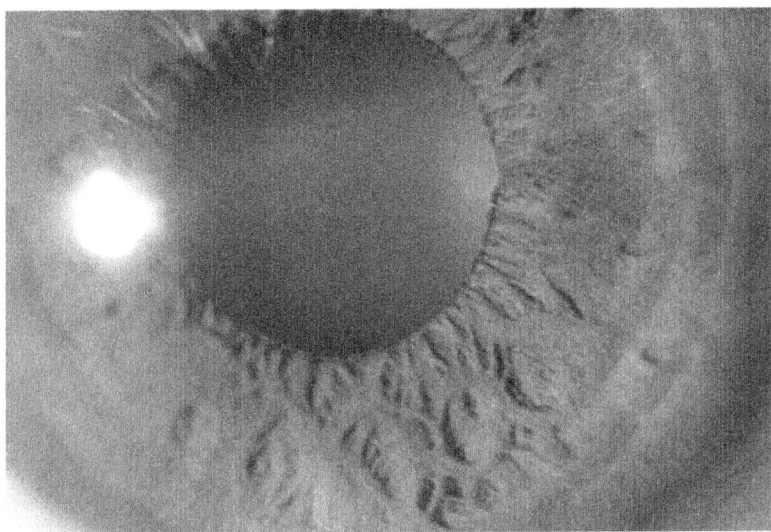

Figure 5: To greater amplification, the resemblance is even greater between the surface of the iris and some seeds, including with the bark of some trees. Is conceivable that at higher magnifications, come a time that would be difficult to discern which is which.

However, the magnifications required to carry out properly the characterization of vascular anatomical pattern, are in the order of 20× to 40×. And to these amplifications, we started to notice the omnipresence of the melanin in the vicinity of the optic disc (Figure 6).

And the insistence of nature in place melanin in the entirety of the almost 6000 patients who checked in the span of 12 years which lasted the study (1990-2002), struck us mightily, because nature just insists on important things.

So a few months of starting the study, instead of one, there were two main variables in study: blood vessels and melanin.

We were sure that melanin should play an important biological role, but we didn't understand that. Eventually we began to detect an antagonism between the blood vessels and the melanin that was becoming more and more evident. Some time later, we had a conclusion: more melanin, fewer blood vessels and vice versa (Figure 7).

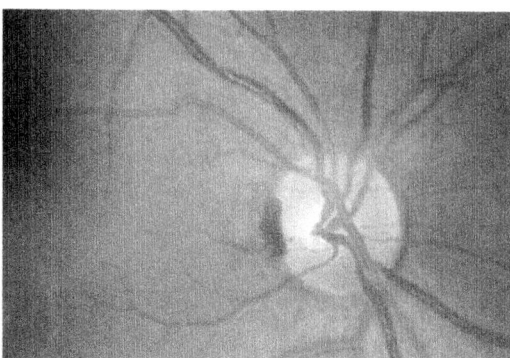

Figure 6: The picture shows the optic nerve of the right eye, with its characteristic orange colour, its oval shape, vessels entering and leaving; and in the meridian of the 9 there is a patch of dark color which corresponds to melanin.

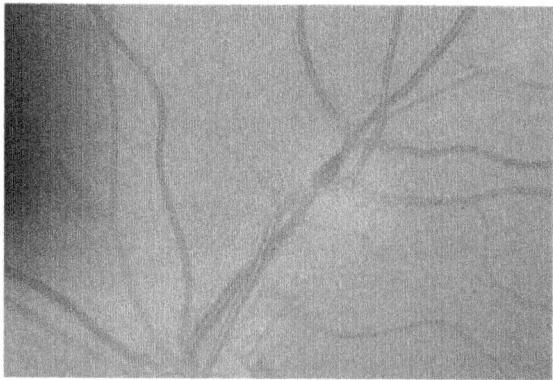

Figure 7: PICTURED are the retinal vessels, and a dark spot that surrounds them partially and which corresponds to the melanin. The effect on the vessels is somewhat subtle, because in this case only observed a light pale around the melanin. In people with fair skin angiogenesis inhibitor effect of melanin is still more subtle, and in dark-skinned people is almost impossible to detect by the amount of melanin.

What remained was to try to explain it. In principle, we did not know which predominated on which. On the one hand, melanin does not possess something like organelles that could produce some kind of transmitters, e.g. peptides, neither nor has something that could be interpreted as receivers. In relation to the blood vessels, these have rough endoplasmic reticulum, so it can synthesize factors that function as transmitters of information, but melanin has no receptors.

And something that was surprising was that the melanin antiangiogenic effect is very powerful, because it is present all the time, which was at odds with some kind of factor, for example: Pigment epithelium derived factor or Vascular endothelial growth factor; because the biological variability in the same factor and the corresponding receiver significantly reduces its effect, perhaps in 60%.

And the effect of melanin is surprisingly consistent, then it had to be something different from the known factors. By then, we realized that molecular oxygen levels were different between tissues, according to the amount of pigment. A greater amount of melanin, higher levels of molecular oxygen and vice versa. So the oxygen was the answer, because it is the best anti-angiogenic agent referred to.

We have been solving the unknowns, but now appeared a new: where does oxygen came from? The difference in oxygen levels between the pigmented tissues and the less pigmented, in our laboratory conditions, in the human eye, it goes from 34% to 54%, a huge difference.

We review several molecules, various tissues, looking for changes consistent with a constant oxygen donation, and did not find anything, it seemed that not passing anything, but a molecule or tissue that had to submit so much oxygen, , sooner or later, will has any perceptible change, and nothing.

Finally, we had only one option: the water containing the eye and that is something more than 95% of the body. And it was the perfect candidate, because in water is hardly notice the changes. But the water not releasing oxygen for free, have to pay it, i.e. energy is required.

So that left us with an unknown: where does the energy so the water yield their oxygen atoms? And after several unsuccessful attempts, the answer appeared in front of a Blackboard full of cluttered biochemical equations, was February 2002; "He is dissociating water" I thought, and to write the equation on the Blackboard, suddenly appeared the order in the rest of the equations.

$$2H_2O \rightarrow 2H_2 + O_2$$

She was the equation written, but as we did laboratory tests, we found that apart from the elevated oxygen levels, also had flow of electrons; which was easy to explain because the reaction is reversible in appropriate conditions. So, eventually we perfect the reaction in the following manner:

$$2H_2O \rightarrow 2H_2 + O_2 \rightarrow 2H_2O + 4e^-$$

Melanin releases energy in the form of symmetrical in all directions. Something like increased energy spheres (Figure 8).

The position of the granules of melanin into the cell eucarionte is strategic since it is located mainly in the perinuclear space, allowing him to be the source of energy of the nucleus, which does not possess neither mitochondria nor ATP; and at the same time it floods the cell cytoplasm with these growing spheres of energy, which are distributed throughout the entire cell, following the laws of simple diffusion (Figure 9).

The process of transformation of light energy into chemical energy seems to be the same in all living organisms, including insects (Figure 10).

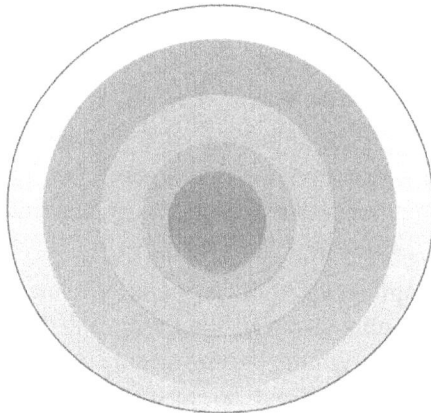

Figure 8: Melanin drawn to the Center, and the areas surrounding it, mean the different concentrations of H_2 and O_2 resulting of the dissociation of water, also the different concentrations of water re- formed, and at the same time, also with their corresponding 4 high- energy electrons for every two molecules of re-formed water.

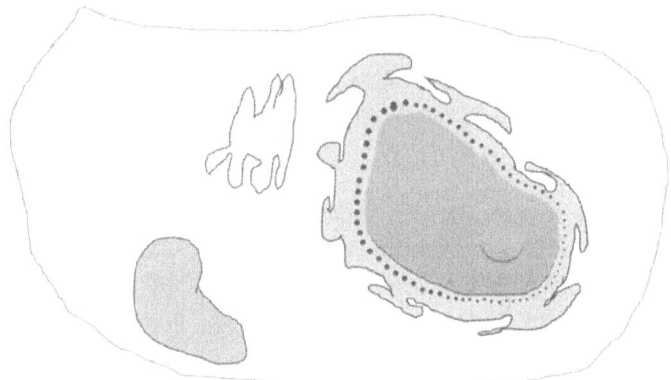

Figure 9: The scheme represents the cell membrane, mitochondria, Golgi apparatus, rough endoplasmic reticulum, which, like the granules of melanin (melanosomes) surrounds completely to the cell nucleus.

Figure 10: Melanin is also present in the common moth, and its main function is also transform light energy into chemical energy through dissociation and re-forming of the water molecule.

MELANIN AND PLANTS

The downward movement of sugary plant fluids in phloem requieres energy, indeed, and melanin is the explanation. Mistery about how plants move fluids long distances in their phloem is easiest of understand taking in account the energy that comes from melanin.

That sucrose is either used as an energy source for its own metabolism is an idea very difficult to explain. So far, there are only hypothesis, in example, hypothesis of how water rises up in the xylem to the tops of tall trees, how sucrose and a proton (H^*) are pumped into the cell; how water comes and build a pressure, how the sap can move a very significant distance; how sucrose is pumped out, how water is pumped out, how sucrose can be splitted to its monomers and thence to other hexoses.

Let us remember that any chemical reaction involves exchange of energy. The metabolism of plants as animals is poorly understood, since the power source was not known. Glucose is a special molecule that organisms use as a source of biomass as a source of carbon chains, but even there, and both animals and plants take the necessary energy from water. Energy can be defined as everything that produces a change, and the word metabolism means continuous change.

The metabolic needs of the plant are continuous, incessant, both day and night; and it is the way in which melanin delivered energy, in the form of incessant, continuous, both day and night. Therefore, the mysteries about life in the plants will have to look in the trunk, not so much in the leaves (Figure 11).

Figure 11: The chlorophyll in plants, as well as the hemoglobin in humans, possess the ability to dissociate the water molecule in an irreversible manner, using the light energy it contains. But both molecules are only an extension of the master molecule that is melanin [5] .

CONCLUSIONS

Both plants and humans are enveloped in a dense cloud of mysteries, at least in regards to your metabolism. It is no coincidence that of the 7000 cellular reactions described in the literature, only about 200 are described in the same way in the different sources of information, there are significant disputes in the remaining 6800. And it is that cell biology has puzzled trying to concatenate the glucose as a source of energy and at the same time of carbon chains; which is to say that a wonderful molecule not only brings carbon atoms to organisms but also carries the necessary energy for its own metabolism.

This wrong concept is deeply rooted in both the scientist and public in general. But glucose just provides the building-blocks of the 99% of the biomolecules that mold our organism. Thereby the body knows glucose very well, so is able to split it, to combine it with other elements or molecules, and in a very precise way; even more, the cell is able to metabolize glucose to CO_2 and water; but the energy that supposedly glucose contains is a myth.

Therefore, that biology at the molecular level has been stagnant. It is not possible to extract more energy from glucose containing that.

However, if from now forward we take into account the unexpected role of melanin as an important source bioenergetics, the role of the glucose will refine and allows us to re-order our concepts in a manner more consistent with nature, where everything depends on the generation and distribution of energy.

The discovery that the main source of energy comes from lignin [6] and chlorophyll, represents a substantial change in the study of photosynthesis in plants. We will have to reconsider many metabolic pathways so far accepted dogmatically but remain without plausible explanation, whereas to date theories. A significant advance in the knowledge will be able to be here, and it is a sharp change of direction to redirect the attention toward the trunk of the plants, rather than the leaves; but this will allow us to better understand the mysteries of the physiology of plants.

REFERENCES

1. Hoffman, J. and Puszynski, A. (2009) Chemical Engineeering and Chemical Process Technology. Vol. V. Pigments and Dyestuffs. Encyclopedia of Life Support Systems (EOLSS).

2. Solis Herrera, A., Arias Esparza, M.C., Solís Arias, R.I., et al. (2010) The Unexpected Capacity of Melanin to Dissociate the Water Molecule Fill the Gap between the Life before and after ATP. Biomedical Research, 21, 224-227.

3. Anderson, A.N., Tobimatsu, Y., Ciesielki, P.N., Ximenes, E., Ralph, J., Donohoe, B.S., Ladisch, M. and Chapple, C. (2015) Manipulation of Guaiacil and Syringyl Monomer Byosynthesis in an Arabidopsis Cinnamyl Alcohol Dehydrogenase Mutant Results in Atypical Lignin Byosynthesis and Modified Cell Wall Structure. The Plant Cell, 27, 2195-2209. http://dx.doi.org/10.1105/tpc.15.00373

4. Hill, H.Z. (1992) The Function of Melanin or Six Blind People Examine an Elephant. BioEssays, 14, 49-56. http://dx.doi.org/10.1002/bies.950140111

5. Solís-Herrera, A. And Solís-Arias, M.P. (2013) The Odyssey of Atmospheric Oxygen in Their Futile Attempt to Reach the Interior of the Cell. International Journal of Health Research and Innovation, 1, 37-52.

6. Solís-Herrera, A. And Arias-Esparza, M.C. and Solís Arias, M.P. (2015) Photosynthesis in Humans, Chapter 12 of the Book: Photosynthesis, Functional Genomics, Physiological Processes and Environmental Issues. Edited by Nafees Khan. Nova Publishers, New York, 257-279.

Chapter 4

BIOACTIVE PIGMENTS FROM MARINE BACTERIA: APPLICATIONS AND PHYSIOLOGICAL ROLES

Azamjon B. Soliev, Kakushi Hosokawa, and Keiichi Enomoto

Department of Environmental Systems Engineering, Kochi University of Technology, 185 Miyanokuchi, Tosayamada, Kami, Kochi 782-8502, Japan

ABSTRACT

Research into natural products from the marine environment, including microorganisms, has rapidly increased over the past two decades. Despite the enormous difficulty in isolating and harvesting marine bacteria, microbial metabolites are increasingly attractive to science because of their broad-ranging pharmacological activities, especially those with unique color pigments. This current review paper gives an overview of the pigmented natural compounds isolated from bacteria of marine origin, based on accumulated data in the literature. We review the biological activities of marine compounds, including recent advances in the study of pharmacological effects and other commercial applications, in addition to the biosynthesis and physiological roles of associated pigments. Chemical structures of the bioactive compounds discussed are also presented.

INTRODUCTION

Marine Bacteria and Its Role in Life Sciences

A wide variety of diseases and medical problems represent a challenging threat to humans, who since ancient times have searched for natural compounds from plants, animals, and other sources to treat them. Although the process of finding effective treatments against fatal diseases is difficult, extensive searches for natural bioactive compounds have previously yielded some successful results. The isolation and identification of specific natural compounds led to the development of folk medicine, and humans learned to separate the isolates into medicinal drugs, which could be used to treat different diseases, and poisonous substances, which could be used for nonmedicinal purposes (i.e., during tribal wars, hunting, etc.). Statistically, at least 50% of the existing drugs that are used to treat human illnesses are derived from natural products, most of which are obtained from terrestrial organisms [1]. However, due to continuous and exhaustive research, land-based natural bioactive compounds have become increasingly difficult to find. Instead, water-based natural compounds have become a more promising source, not only from a pharmacological view, but also for industrial and commercial applications.

Theoretically, life is considered to have originated in the sea and, as a result of evolutionary changes, developed into a wide variety of diverse biological systems. The Earth's surface consists of 70% water, which is inhabited by 80% of all life forms [1], and consequently aquatic organisms have a greater diversity than their terrestrial counterparts. As research into the marine environment is still in its early phases, many mysteries associated with aquatic fauna and flora have yet to be discovered. Therefore, the marine environment has recently become an attractive research subject for many investigations, because of its rich biodiversity. Despite being comprised of a diverse ecosystem, the search for marine metabolites is difficult because of the inaccessibility and nonculturability of the majority of organisms [2]. Nevertheless, the existing technologies like deep seawater pumping facilities, scuba diving, and other available equipments, have facilitated investigation of the sea environment. As a result, scientific research has increasingly focused on marine biochemistry, microbiology, and biotechnology.

Microorganisms and their isolates represent a major source of undiscovered scientific potential. It should be noted that the number of microbial organisms isolated from the vast ocean territories continues to increase each year. Consequently, natural products isolated from microorganisms inhabiting environments other than soil are an attractive research tool, not only for biochemists and microbiologists, but also for pharmacologists and clinicians. Laatsch [3] described the isolation and description of nearly 250 marine bacterial metabolites versus 150 isolated from terrestrial bacteria between 2000 and 2005. Research into marine microorganisms and their metabolites has therefore become a major task in the search for novel pharmaceuticals.

Although many compounds show promising biological activities, it is difficult to point out any particular bioactive agent that has readily been commercialized as a medicine. Currently, 13 natural products isolated from marine microorganisms are being tested in different phases of clinical trials, and a large number of others are in preclinical investigations [4], thus highlighting the potential of marine natural compounds.

Despite thousands of marine bioactive compounds having been isolated and identified, in this paper, we will focus on the pharmacologically active pigmented compounds produced by marine microorganisms exhibitingin vitro or in vivo biological activities. Although pigmented compounds produced by terrestrial bacteria are beyond the scope of this review, specific examples will still be mentioned for comparative purposes, to outline common biological activities or because identical pigments were isolated from both types of microorganisms.

Marine Microorganisms and Their Bioactive Isolates

Marine and terrestrial microfloras differ from each other due to the influence of their respective environmental conditions. Microorganisms living in the sea must be able to survive and grow in the water environment with low nutrition, high salinity, and high pressure. That is why most bacteria isolated from seawater are Gram-negative rods, as it is postulated that their outer membrane structure is evolutionarily adapted to aquatic environmental factors. Marine microorganisms can be divided on the basis of habitat into psychrophiles (living at low temperatures), halophiles (living at high salinity), and barophiles (living under high pressure). Although these characteristics highlight the differences between marine and terrestrial microorganisms, it remains difficult to separate bacterial genera on the basis of habitat due to the ubiquitous presence of similar species in both environments. As such, most bioactive compounds have been isolated from bacteria in both environments.

Marine bacteria, however, are attractive to researchers because they can potentially produce compounds with unique biological properties [5]. Until now, marine Streptomyces, Pseudomonas, Pseudoalteromonas, Bacillus, Vibrio, and Cytophaga isolated from seawater, sediments, algae, and marine invertebrates are known to produce bioactive agents. They are able to produce indole derivatives (quinones and violacein), alkaloids (prodiginines and tambjamines), polyenes, macrolides, peptides, and terpenoids. Examples of bioactive-pigmented compounds isolated from marine (and some terrestrial) bacteria are discussed below.

PIGMENTS FROM MARINE BACTERIA

Bioactive pigments from marine bacteria are summarized in Table 1.

Table 1: Biologically active pigmented compounds isolated from marine bacteria

Pigment	Activity	Bacterial strains	References
(1) Undecylprodigiosin	Anticancer	*Streptomyces rubber*	[8]
(2) Cycloprodigiosin	Immunosuppressant; Anticancer; Antimalarial	*Pseudoalteromonas denitrificans*	[18, 21, 22]
(3) Heptyl prodigiosin	Antiplasmodial	α-*Proteobacteria*	[23]
(4) Prodigiosin	Antibacterial; Anticancer; Algicidal	*Pseudoalteromonas rubra*	[14]
		Hahella chejuensis	[27]
(5) Astaxanthin (carotene)	Antioxidation	*Agrobacterium aurantiacum*	[34]
(6) Violacein	Antibiotic; Antiprotozoan; Anticancer	*Pseudoalteromonas luteoviolacea*	[48, 52, 53]
		Pseudoalteromonas tunicata	[43]
		Pseudoalteromonas sp. 520P1	[50]
		Collimonas CT	[51]
(7) Methyl saphenate (phenazine derivative)	Antibiotic	*Pseudonocardia* sp. B6273	[63]
(8) Phenazine derivatives	Cytotoxic	*Bacillus* sp.	[64]
(9) Pyocyanin and pyorubrin	Antibacterial	*Pseudomonas aeruginosa*	[58]
(10) Phenazine-1-carboxylic acid	Antibiotic	*Pseudomonas aeruginosa*	[59]
(11) 5,10-dihydrophencomycin methyl ester	Antibiotic	*Streptomycete* sp.	[65]
(12) Fridamycin D, Himalomycin A, Himalomycin B	Antibacterial	*Streptomycete* sp. B6921	[68]
(13) Chinikomycin A and Chinikomycin B, Manumycin A	Anticancer	*Streptomycete* sp. M045	[71]
(14) Tambjamines (BE-18591, pyrrole and their synthetic analogs)	Antibiotic, Anticancer	*Pseudoalteromonas tunicata*	[76, 80]
(15) Melanins	Protection from UV irradiation	*Vibrio cholerae*	[83, 84]
		Shewanella colwelliana	[83, 86]
		Alteromonas nigrifaciens	[85]
		Cellulophaga tyrosinoxydans	[88]
(16) Scytonemin	Protection from UV irradiation Anti-inflammatory, Antiproliferative	Cyanobacteria	[93]
(17) Tryptanthrin	Antibiotic	*Cytophaga/Flexibacteria* AM13,1 strain	[95]

Prodiginines

Red-pigmented prodigiosin compounds were first isolated from the ubiquitous bacterium Serratia marcescensand identified as secondary metabolites. The common aromatic chemical structure of these pigmented compounds was first named prodiginine by Gerber [6] (Figure 1). Prodigiosin was the first prodiginine for which the chemical structure was determined [7]. The name "prodigiosin" has been attributed to the isolation of prodigiosin from Bacillus prodigiosus bacterium (later renamed Serratia marcescens) [8], which was historically famed for the mysterious "bleeding bread" report [9, 10]. Prodiginines share a common pyrrolyldipyrromethene core structure and have a wide variety of biological properties, including antibacterial, antifungal, antimalarial, antibiotic, immunosuppressive, and anticancer activities [9, 11]. Such properties potentially make them one of the most powerful research tools in the past decade.

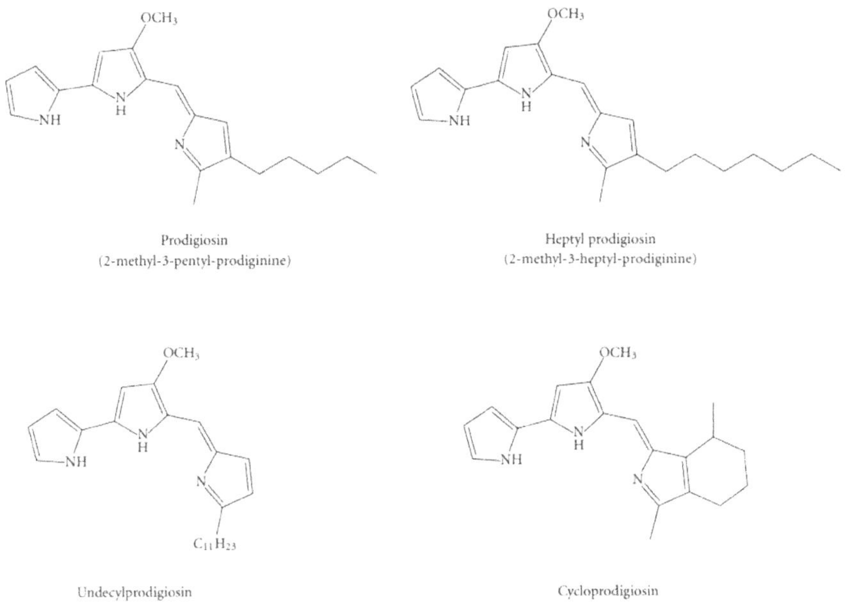

Prodigiosin
(2-methyl-3-pentyl-prodiginine)

Heptyl prodigiosin
(2-methyl-3-heptyl-prodiginine)

Undecylprodigiosin

Cycloprodigiosin

Figure 1: Prodiginine derivatives.

There are many research reports and reviews regarding prodiginines and their biological activity investigations. In addition to the Serratia, several species of marine bacteria of the genera Streptomyces [8], Actinomadura [8],Pseudomonas [12], Pseudoalteromonas [13–18], and others [19] have also been reported to produce prodigiosin and related compounds. In particular, Alteromonas denitrificans, which was isolated from the fjord systems off the west coast of Norway [16] and later reclassified as Pseudoalteromonas denitrificans [20], has been reported to produce cycloprodigiosin. This compound has immunosuppressive, antimalarial, and apoptosis-inducing activities [18, 21, 22]. Pseudoalteromonas rubra, found in the Mediterranean coastal waters [13], also produces cycloprodigiosin, in addition to prodigiosins [14, 15]. α-Proteobacteria isolated from a marine tunicate collected in Zamboanga, Philippines, was reported to produce heptyl prodigiosin. In vitro antimalarial activity against Plasmodium falciparum 3D7 (IC_{50} = 0.068 mM and SI = 20) was about 20 times the in vitro cytotoxic activity against L5178Y mouse lymphocytes [23]. In vivo experiments using Plasmodium berghei-infected mice, at concentrations of 5 mg/kg and 20 mg/kg, significantly increased their survival, while also causing sclerotic lesions at the site of injection.

Other bacteria reported to produce red pigments include Hahella [24], Vibrio [25], Zooshikella [26], andPseudoalteromonas [17], isolated from the coasts of Korea, Taiwan, and Japan. Kim et al. [27] identified red-pigmented prodiginines from Hahella chejuensis. Nakashima et al. also evaluated the biological activity of similar prodiginines from a bacterium assumed to belong to the genus Hahella [28]. Red pigment-producing bacterial species have further been isolated from river water [29, 30] and even from a swimming pool [31]. The most active prodiginine derivatives have already entered clinical trials as potential drugs against different cancer types [9].

Japan is surrounded by sea and has a bordering coastline of the Pacific Ocean in the South and the Sea of Japan in the North and West, and is consequently rich in marine resources. Therefore, one of the main tasks of our research group is to investigate the marine environment and its biodiversity, especially marine microorganisms and their respective metabolites.

Previously, a total of 85 strains of bacteria were isolated by our research group from the Pacific Ocean at a depth of 320 m off Cape Muroto in the Kochi Prefecture of Japan. Among them, 13 strains were found to produce a purple pigment and one a red pigment. The red pigment-producing bacterium was later named strain 1020R [32]. Detailed investigations have revealed that this strain is closely related to the prodigiosin-producing bacterium Pseudoalteromonas rubra and is Gram-negative with rod-shaped morphology. Physicochemical investigations have revealed that the pigment produced by this strain contains at least seven structurally similar prodiginine compounds. Chemical structures for four of these were successfully determined, and each only differed by the length of the alkyl chain attached to the C-3 position of the C-ring. These compounds were further identified as prodigiosin and its analogues 2-methyl-3-butyl-prodiginine, 2-methyl-3-pentyl-prodiginine (prodigiosin), 2-methyl-3-hexyl-prodiginine, and 2-methyl-3-heptyl-prodiginine. Compound cytotoxicity to U937 leukemia cells was strongly dependent on the length of these alkyl side chains, which decreased with an increase in chain length. 2-methyl-3-butyl-prodiginine was the most potent cytotoxic pigment among them. Molecular investigations into the cytotoxic mechanisms of these prodiginine derivatives demonstrated effects on caspase-3 activation and DNA fragmentation, indicating the potential to induce apoptosis in leukemia cells.

Carotenes

Carotenes are polyunsaturated hydrocarbons that contain 40 carbon atoms per molecule and are exclusively synthesized by plants. They are orange photosynthetic pigments important for plant photosynthesis. Recently, an unusual halophilic bacterium, which requires 15–25% salt for its normal growth, was found in Santa Pola near Alicante and on the Balearic island of Mallorca, Spain. It appeared to be red or pink due to a wide variety of isoprenoid compounds (phytoene, phytofluene, lycopene, and β-carotene) produced by this prokaryote. Oren and Rodríguez-Valera [33] investigated red-pigmented saltern crystallizer ponds in these areas of Spain and demonstrated that the pigments were carotenoid or carotenoid-like compounds produced by halophilic bacteria related to the Cytophaga-Flavobacterium-Bacteroides group. Thus, it has been shown that Salinibacter is an important component of the microbial community that contributes to the red coloration of Spanish saltern ponds.

Astaxanthin is one of the carotenoids that have commercial value as a food supplement for humans and as food additives for animals and fish (Figure 2). A carotenoid biosynthesis gene cluster for the production of astaxanthin has been isolated from the marine bacterium Agrobacterium aurantiacum [34]. Recently, another astaxanthin-producing marine bacterium was isolated and identified as Paracoccus haeundaensis [35].

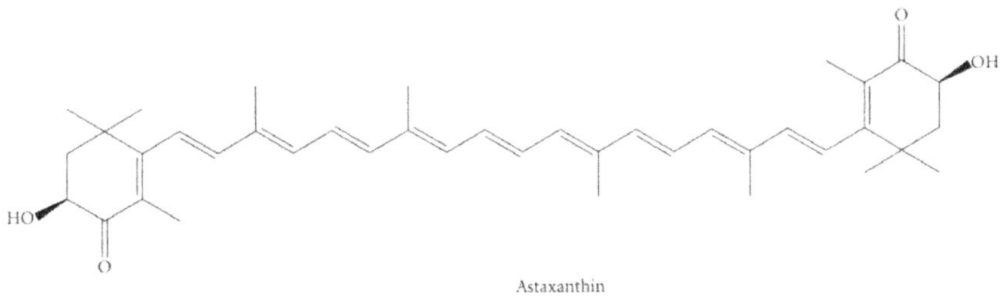

Astaxanthin

Figure 2: Astaxanthin.

Violacein

The violet pigment violacein is an indole derivative, predominantly isolated from bacteria of the genusChromobacterium that inhabit the soil and water of tropical and subtropical areas [36]. Over the past decade, the biosynthesis and biological activities of violacein have been extensively studied, and many scientific papers and reviews have been published [37–41]. Violacein has a variety of biological activities, including antiviral, antibacterial, antiulcerogenic, antileishmanial, and anticancer properties [36, 37, 41, 42] (Figure 3). Use of violacein as a chemical defense against eukaryotic predators has also been investigated [43–46].

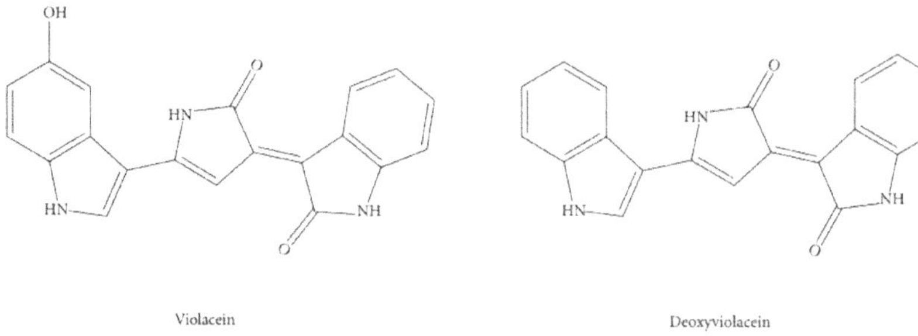

Violacein   Deoxyviolacein

Figure 3: Violacein and deoxyviolacein.

One of the first published reports on violacein production by marine bacteria was by Hamilton and Austin [47]. This bacterial strain, Chromobacterium marinum, was isolated from open ocean waters and produced a blue pigment that was identified as violacein on the basis of physicochemical characteristics [47]. Later, Gauthier [48] described 16 violet-pigmented heterotrophic bacilli isolated from Mediterranean coastal waters and proposed the name Alteromonas luteo-violaceus for these strains. Another six bacterial species were also isolated by Gauthier et al. [49] from neritic waters on the French Mediterranean coast and were very similar toAlteromonas species. These species produced characteristic pigmentations ranging from pinkish-beige with reddish-brown diffusible pigment, lemon yellow, bright red turning carmine in old cultures, and orange to greenish-brown. Light violet, dark violet, or almost black pigments were also produced and later identified as violacein. The strains showed antibiotic activity against Staphylococcus aureus [49]. Subsequently, many other reports on violacein production have been published [50, 51].

Several purple pigment-producing Alteromonas species were also isolated from Kinko Bay in Kagoshima Prefecture, Japan. One of these, Alteromonas luteoviolacea (reclassified as Pseudoalteromonas luteoviolacea), is the only extensively characterized marine bacterium ever reported that produces violacein [48, 52, 53]. Previously, we have also reported 13 strains of Gram-negative, rod-shaped bacteria that produce a violacein-like purple pigment, which were isolated from the Pacific Ocean at a depth of 320 m off the coast of Cape Muroto, Kochi Prefecture, Japan [32]. Among them, two groups of novel violacein and deoxyviolacein producing marine bacteria were isolated and characterized in detail [50]. Biological investigations of violacein produced by these strains revealed potent cytotoxic effects against U937 and HL60 leukemia cell lines, with an IC_{50} value of 0.5–1 M. The molecular mechanisms currently known to be involved in violacein cytotoxicity include caspases activation, chromatin condensation, and DNA fragmentation, which all contribute to cell apoptosis. Recently, we also demonstrated that the protein kinases actively involved in the signal transduction pathway are also targeted by violacein.

Phenazine Compounds

Phenazines are redox-active, small nitrogen-containing aromatic compounds produced by a diverse range of bacterial genera, including Streptomyces (terrestrial), Pseudomonas (ubiquitous), Actinomycetes (terrestrial and aquatic), Pelagibacter (aquatic), and Vibrio (aquatic), under the control of quorum sensing [54, 55] (Figure 4). These compounds were subjected to extensive studies due to their broad spectrum of antibiotic activities against other bacteria, fungi, or plant/animal tissues [56–62]. Phenazine color intensity may vary among the derivatives and range from blue, green, purple, yellow, red to even brown [58, 63]. More than 6,000 phenazine derivatives have been identified and described during the last two centuries [59].

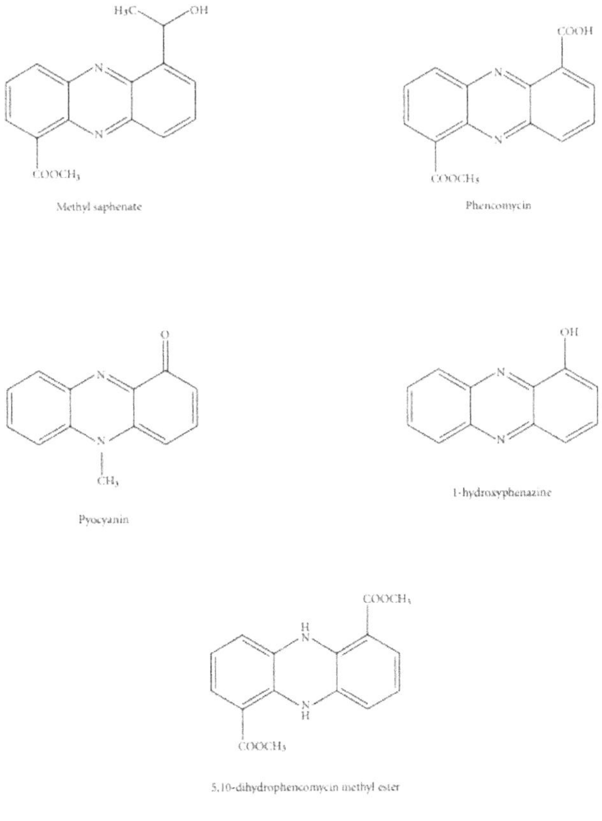

Figure 4: Phenazine derivatives.

Maskey et al. [63] reported the isolation of two yellow pigments from the marine Pseudonocardia sp. B6273, a member of the Actinomycetes. Structural investigations identified the two pigments as novel phenazostatin D, inactive against the tested microorganisms, and methyl saphenate, a known phenazine antibiotic. Li et al. [64] also reported the isolation of a novel phenazine derivative with cytotoxic effects against P388 cells, together with six previously identified compounds from the marine Bacillus sp., collected from a Pacific deep-sea sediment sample at a depth of 5059 m. A novel phenazine derivative with antibiotic activity, identified as 5,10-dihydrophencomycin methyl ester, along with (2-hydroxyphenyl)-acetamide, menaquinone MK9 (II, III, VIII, IX-H8), and phencomycin, was isolated from an unidentified marine Streptomyces sp. by Pusecker et al. [65].

Pyocyanin and 1-hydroxyphenazine also downregulate the ciliary beat frequency of respiratory epithelial cells by reducing cAMP and ATP, alter the calcium concentration by inhibition of plasma membrane Ca^{2+}-ATPase, and induce death in human neutrophils [60, 61, 66]. Due to the abundance and biotechnological application ofPseudomonas aeruginosa phenazines, pyocyanin and pyorubin have also been suggested as food colorant pigments [58].

Quinones

Quinones are additional colored compounds with an aromatic ring structure that have been isolated from marine environment [67, 68] (Figure 5). Quinone derivatives range in color from yellow to red, exhibit antiviral, anti-infective, antimicrobial, insecticidal, and anticancer activities, and have many commercial applications as natural and artificial dyes and pigments [69, 70].

Streptomyces sp. B6921 strain produced glycosylated pigmented anthracycline antibiotics, including fridamycin D and two new compounds, named himalomycin A and B, each of which displayed similar levels of strong antibacterial activity against Bacillus subtilis, Streptomyces viridochromogenes (Tü 57), S. aureus, and Escherichia coli. This strain also produced rabelomycin, N-benzylacetamide, and N-(2'-phenylethyl) acetamide [68]. Two novel pigmented antitumor antibiotics, chinikomycin A and B, together with manumycin A, were isolated from a marine Streptomyces sp. strain M045 [71]. The two chlorine containing quinone derivatives were shown not to have antiviral, antimicrobial, and phytotoxic activities; however, they exhibited antitumor activity against different human cancer cell lines. Chinikomycin A selectively inhibited the proliferation of mammary cancer, melanoma, and renal cancer cell lines, while chinikomycin B showed selective antitumor activity against a mammary cancer cell line [71].

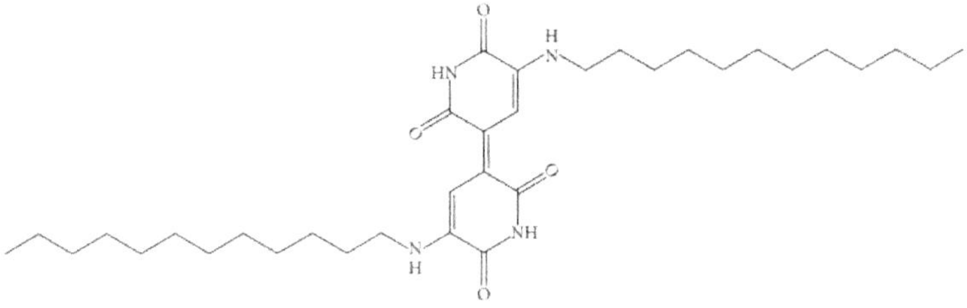

5,5'-didodecylamino-4,4'-dihydroxy-3,3'-diazodiphenoquinone-(2,2')

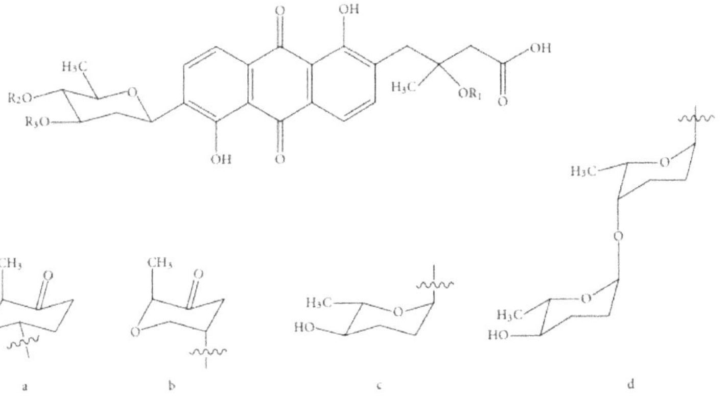

Fridamycin D: R1 = H, R2 = a, R3 = b;
Hymalomycin A: R1 = d, R2 = a, R3 = b;
Hymalomycin B: R1 = d, R2 = c, R3 = H

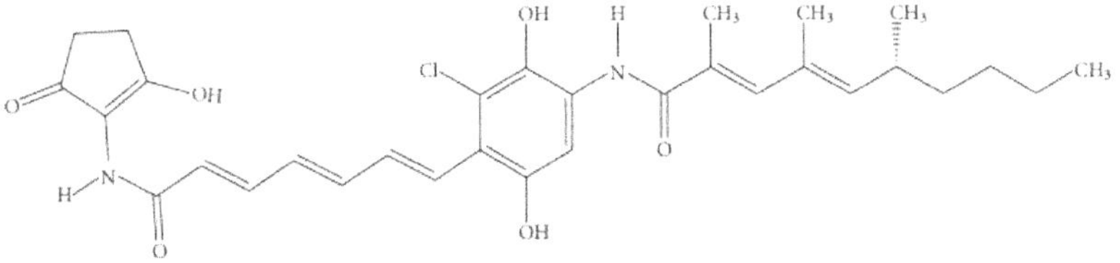

Chinikomycin A

Figure 5: Quinones.

Other bacteria, including a marine isolate Pseudomonas nigrifaciens (later reclassified as Alteromonas nigrifaciens), produce the blue pigment indigoidine [72]. Kobayashi et al. [73] isolated a new violet pigment with an alkylated indigoidine structure from Shewanella violacea, a deep-sea bacterium from sediments of Ryukyu Trench at a depth of 5110 m. This pigment was established as 5,5′-didodecylamino-4,4′-dihydroxy-3,3′-diazodiphenoquinone-(2,2′) based on X-ray diffraction analysis of single crystals. It does not have antibiotic activity against E. coli; however, it could potentially be used as a dye because of its high stability and low solubility. Thus, it could be suitable for industrial applications.

Tambjamines

It has long been noticed that marine bacteria have the ability to prevent biofouling. Holmström et al. [74] found that, amongst the marine Pseudoalteromonas species, P. tunicata has the widest range of antibiofouling activities against microorganisms, including bacteria, invertebrate larvae, algal spores, protozoan, and fungi, and provides protection for host marine organisms. These activities were linked to the production of unidentified yellow and purple pigments [75]. Recently, this yellow pigment was isolated from P. tunicata and was identified as a new member of the tambjamine class of compounds [76].

Tambjamines (Figure 6) are alkaloids isolated from various marine organisms like bryozoans, nudibranchs, and ascidians [77–79]. This yellow pigment has also been isolated from marine bacteria [76]. The tambjamines also exhibit antibiotic activity against E. coli, Staphylococcus, Vibrio anguillarum [77], B. subtilis, and Candida albicans [80, 81] and displayed cytotoxic activity against several tumor cell lines [80]. Recently, Pinkerton et al. [80, 82] reported the first total synthesis of nine tambjamines and their antimicrobial and cytotoxic activities. All of the tested tambjamines showed antibacterial, antifungal, and cytotoxic effects that contributed to cell death through apoptosis, but not necrosis. These activities were, however, lesser than the positive control (doxorubicin) [80].

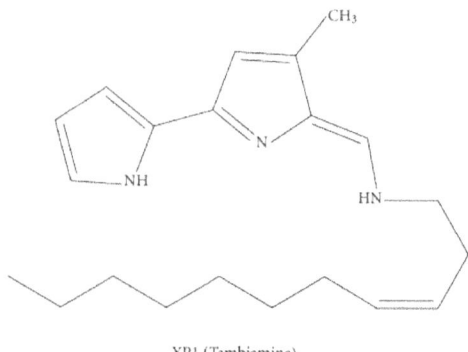

YP1 (Tambjamine)

Figure 6: Tambjamine.

Melanins

Vibrio cholerae, Shewanella colwelliana, and Alteromonas nigrifaciens were some of the first marine bacterial strains described to produce melanin or melanin-like pigments [83–86]. The pigment synthesized by Vibrio cholerae was reported to be a type of allomelanin derived from homogentisic acid [87]. Melanin formation in V. cholerae is a consequence of alterations in tyrosine catabolism and not from the tyrosinase-catalyzed melanin synthetic pathway. Cellulophaga tyrosinoxydans was reported to have tyrosinase activity and produce a yellow pigment suggested to be a pheomelanin [88].

The most illustrative example of melanin-producing marine bacteria is the actinomycetes. This is particularly the case for the genus Streptomyces, from which most compounds with known biological activity have been isolated [89]. All Streptomyces strains are reported to use tyrosinases in the synthesis of melanin pigments [90]. Another important melanin-synthesizing bacterium is Marinomonas mediterranea, which produces black eumelanin from L-tyrosine [91].

Other Pigmented Compounds

Scytonemin, a yellow-green pigment isolated from aquatic cyanobacteria, forms when the bacteria are exposed to sunlight (Figure 7). It protects bacteria by preventing about 85–90% of all UV-light from entering through the cell membrane [92]. High UV-A irradiation inhibited photosynthesis and delayed cellular growth until sufficient amounts of scytonemin had been produced by the cyanobacteria. Scytonemin may also have anti-inflammatory and antiproliferative activities by inhibiting protein kinase Cβ (PKCβ), a well-known mediator of the inflammatory process, and polo-like protein kinase 1 (PLK1), a regulator of cell cycle progression [93]. In addition, scytonemin inhibited phorbol-induced mouse ear edema and the proliferation of human umbilical vein endothelial cells.

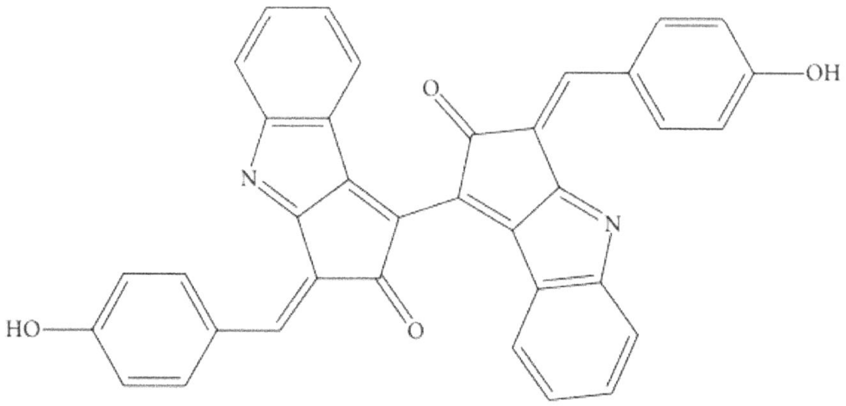

Scytonemin

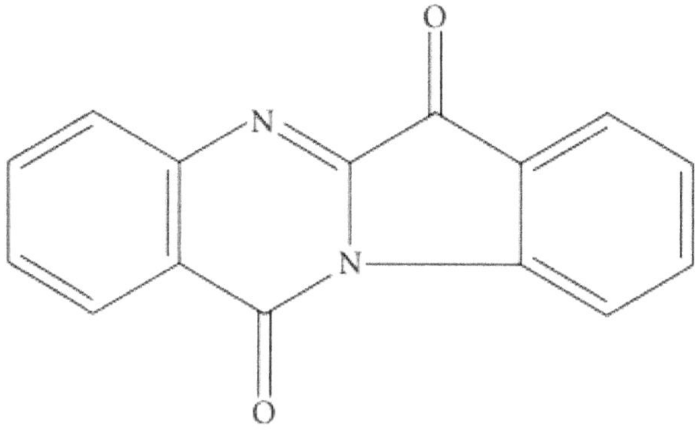

Tryptanthrin

Figure 7: Other pigmented compounds.

Recently, two γ-Proteobacteria strains of the genus Rheinheimera were isolated from the German Wadden Sea and from Øresund, Denmark that produced a deep blue pigment [94]. Structural analysis of the pigment revealed that this new compound has no similarity with any known blue pigments, like violacein and its derivatives. Due to its blue color and marine origin, the new pigment was named glaukothalin (from Greekglaukos "blue" and thalatta "sea"). The ecological role and biological activities of glaukothalin are currently under investigation.

AM13,1 strain, which was identified to belong to the Cytophaga/ Flexibacteria cluster of North Sea bacteria, was found to produce yellow tryptanthrin, a rare compound that had never before been found in bacteria [95]. This compound was suggested to be a biocondensation product of anthranilic acid and isatin and exhibited a broad yet moderate antibiotic activity. Thus, the yellow color of the AM13,1 colonies was potentially due to their tryptanthrin content. In another yellow cultured Hel21 strain, pigment color may be a consequence of carotenoid zeaxanthin or one of the many vitamin K derivatives (e.g., menaquinone MK6) [95].

BIOSYNTHESIS OF PIGMENTS

Numerous reports detail the regulation and biosynthesis of bacterial secondary metabolites. Increased research and verification of specific bacterial pathways has predominantly been due to the antibiotic, immunosuppressive, and anticancer potential of these compounds. A brief discussion of this topic is given next, as detailed information is further provided in the cited references.

Biosynthesis of bacterial prodiginines has extensively been studied and reviewed [96, 97]. Prodigiosin biosynthesis was proposed to originate during the enzymatic condensation of 2-methyl-3-n-amyl-pyrrole (MAP) and 4-methoxy-2,2'-bipyrrole-5-carbaldehyde (MBC) precursors. Prodiginine biosynthetic gene clusters for Serratia sp. ATCC 39006 [98], Serratia marcescens ATCC 274 [98], Hahella chejuensis KCTC 2396 [27, 99], and Streptomyces coelicolor A3(2) [100] have been identified, sequenced, and expressed. Several gene clusters are involved in the biosynthetic pathway, depicted as pig in Serratia strains, red in S. coelicolor A3(2), and hap (numbered) in H. chejuensis KCTC 2396, with each encoding several proteins responsible for synthesis. The largest gene cluster found in S. coelicolor A3(2) consists of four transcriptional units, whereas the other three clusters are strongly homologous to each other and are arranged uni-directionally.

In Serratia strains, pigB–pigE genes were identified to encode proteins responsible for the biosynthesis of MAP and condensation with MBC to form prodigiosin [96, 97]. A common pathway of MBC biosynthesis is proposed for all strains, in which proline, acetate, serine, and S-adenosylmethionine are incorporated into the bipyrrole at the initial stage [97]. PigA, PigF, PigG, PigH, PigI, PigJ, PigM, and PigN in Serratia strains and RedE, RedI, RedM, RedN, RedO, RedW, RedV, and RedX proteins in S. coelicolor A3(2) have been determined to participate in MBC biosynthesis [97]. PigB, PigD, and PigE enzymes in Serratia strains were proposed to be involved in the

MAP biosynthesis, which requires 2-octenal as the initial precursor [97]. Monopyrroles condense with MBC during the final step of prodigiosin and/ or undecylprodigiosin biosynthesis. PigC and its homologues catalyze this condensation in bacteria.

Some prodiginines can also be produced when monopyrroles are supplied to colorless S. marcescens mutants [8]. Addition of monopyrroles directly to a culture medium or as a vapor across the culture surface of a colorless mutant of S. marcescens resulted in the strain becoming initially pink and later red, indicating prodiginine formation [8]. Similar prodiginine biosynthesis produced by exogenously adding MAP and MBC was observed in white strains of Serratia marcescens isolated from patients [101].

The violacein biosynthesis pathway and associated biosynthetic enzymes have been extensively studied [38, 40,102], although certain reactions and intermediates are yet to be elucidated. Currently, this proposed system involves an operon of five genes, vioA–vioE, which are transcriptionally regulated by a quorum-sensing mechanism that uses acyl-homoserine lactones as autoinducers. At the early stationary phase of bacterial growth, acylhomoserine lactones accumulate in the culture medium, inducing the transcription of the viogenes. Therefore, violacein is considered a typical secondary metabolite in bacteria. The first enzyme encoded by the vio gene operon, VioA, converts L-tryptophan to indole-3-pyruvic acid imine (IPA imine), and the second enzyme, VioB, catalyzes the reaction to convert IPA imine into an unidentified compound X (possibly an IPA imine dimer) [103, 104]. Compound X then undergoes successive reactions, catalyzed by the enzymes VioE, VioD, and VioC, to produce violacein.

Phenazine pigment biosynthesis reportedly involves shikimic acid as a precursor and forms chorismic acid as an intermediate product. Two molecules of chorismic acid then form phenazine-1,6-dicarboxylic acid, which is sequentially modified to create a variety of phenazine derivatives with different biological activities [105].Pseudomonas aeruginosa PAO1 has two gene clusters (phzA1B1C1D1E1F1G1 and phzA2B2C2D2E2F2G2), with each cluster capable of producing phenazine-1-carboxylic acid (PCA) from chorismic acid [106]. It is proposed that PhzM and PhzS catalyze the subsequent conversion of PCA to pyocyanin. In addition, PhzH is responsible for producing phenazine-1-carboxamide from PCA.

Fridamycin, hymalomycin, and chinikomycin are typical bacterial compounds that share a quinone skeleton. However, little information regarding the biosynthesis of these compounds has been accumulated.

Detection and identification of the entire P. tunicata gene cluster involved in the biosynthetic pathway production of the tambjamine YP1 using recombinant

E. coli was conducted by Australian researchers Burke et al. [107]. In total, 19 proteins encoded the Tam cluster participate in the postulated biosynthetic pathway. Among them, 12 were found to have high sequence similarity to the red proteins responsible for undecylprodigiosin synthesis in S. coelicolor A3(2) and the pig proteins involved in prodigiosin biosynthesis inSerratia sp. [107]. Such similarity in the chemical structures of these two classes of compounds results in tambjamines having two pyrrole rings while the prodiginines have three. As is the case for the prodiginines, 4-methoxy-2,2-bipyrrole-5-carbaldehyde (MBC) is initially formed from proline, serine, and malonyl CoA in the tambjamine biosynthetic pathway. A double bond is inserted by TamT and an amino group is transferred by TamH to dodecenoic acid activated by AfaA, which is predicted to be an acyl-CoA synthase. The resulting dodec-3-en-1-amine is condensed with MBC by TamQ to form tambjamine YP1 [107].

In addition to V. cholera, S. colwelliana, A. nigrifaciens, and C. tyrosinoxydans, melanin syntheses have also been reported in M. mediterranea, which contains the tyrosinase gene operon [108], and in an epiphyticSaccharophagus degradans 2-40 bacterium [109]. While the specific details of melanin formation continue to be debated, well-defined biosynthetic schemes have now been proposed. Two different biosynthetic pathways synthesize the eumelanins and pheomelanins. Both pathways are initiated by the oxidation of L-tyrosine to 3,4-dihydroxyphenylalanine (DOPA) and the subsequent creation of dopaquinone by tyrosinase. The latter product is transformed either to pheomelanin by combining with cystein and forming an intermediate S-cysteinyldopa and benzothiazine or to eumelanin with intermediate leucodopachrome, dopachrome (red), 5,6-dihydroxyindole, 5,6-indolequinone (yellow) formation [69].

Nostoc punctiforme ATCC 29133 is the only scytonemin-producing organism whose genome has been fully sequenced [110]. This scytonemin biosynthesis potentially involves a gene cluster consisting of 18 open reading frames (ORFs) (NpR1276 to NpR1259). Although, the functional roles of all these ORFs are not yet fully determined, some intriguing hypotheses have been proposed. In particular, both tyrosine and tryptophan are implicated as biosynthetic precursors for scytonemin in the pigment formation pathway. NpR1275, which functionally resembles leucine dehydrogenase, is utilized in the early stages of scytonemin synthesis in N. punctiforme, thereby oxidizing tryptophan and/or tyrosine to their corresponding pyruvic acid derivative.

Alternatively, it is suggested that NpR1269, a putative prephenate dehydrogenase, generates p-hydroxyphenylpyruvic acid, which is a derivative of tyrosine in the early pathway stages. NpR1276 uses two pyruvic acid derivatives from tryptophan and tyrosine for the synthesis of a labile β-ketoacid product, which is homologous to the thiamin diphosphate- (ThDP-) dependent enzyme acetolactate synthase. NpR1274 possibly catalyzes the intermediate cyclization and decarboxylation of the β-ketoacid product to form the indole-fused cyclopentane moiety of the pigment [111]. Monomer precursors that are formed then undergo dimerization to produce scytonemin. NpR1263, which was found to be similar to a tyrosinase in melanin biosynthesis, participates in these later oxidative dimerization steps, thereby forming scytonemin [112]. Functional roles of other ORFs and their putative intermediate products for the pigment production are still under investigation.

CONCERNS REGARDING THE PHYSIOLOGICAL ROLE OF PIGMENTED COMPOUNDS

A number of bacterial species, including those inhabiting the vast marine environment, produce a wide variety of pigments that are important to cellular physiology and survival. Many of these natural metabolites were found to have antibiotic, anticancer, and immunosuppressive activities. These secondary metabolites, produced by microorganisms mostly via the quorum sensing mechanism, have the ability to inhibit the growth of or even kill bacteria and other microorganisms at very low concentrations. Due to such diverse and promising activities against different kinds of diseases, these compounds can play an important role in both pharmaceutical and agricultural research.

It still remains uncertain why these pigmented secondary metabolites from bacteria have antibiotic and/or cytotoxic activities. Although, their true physiological role is yet to be fully discovered, there are a few reports that provide reasonable explanations by making comparisons with nonpigmented bacteria. In particular, the relationships between pigment production and toxicity have been studied by Holmström et al. [113], who found that 90% of all dark-pigmented compounds taken from marine living surfaces showed inhibitory activity towards invertebrate larvae. Two fractions isolated after column chromatography, one colorless and the other a yellowish-green color, were identified as phenazine derivatives from unidentified marine Streptomycete sp. by Pusecker et al. [65]. The colorless fraction was biologically inactive, while the pigmented phenazine derivative showed highly active antibiotic properties. Previous studies have also demonstrated that marine bacterial metabolites with antibiotic properties were always pigmented [114]. Screening of 38 antibiotic-producing bacterial strains revealed that all pigmented bacteria belonging to the Pseudomonas-Alteromonas group displayed antibiotic activity, while nonpigmented bacteria were inactive.

Considering data from all reported literature, a number of reasonable biological functions for pigment production in bacteria have been established.

In general, the pigmented marine isolates seem to play two important roles: firstly, they provide an adaption to environmental conditions, and, secondly, they provide defense against predators [115]. For instance, it has been shown that the brown colored melanin pigments produced by a variety of species, as well as a yellow-green colored scytonemin pigment isolated from cyanobacteria, protect cells from UV irradiation and desiccation [69, 93]. Therefore, in order to adapt to the excessive sunlight and survive under harmful UV irradiation, bacteria must produce these indispensable compounds. Griffiths et al. [116] found that carotenoids, which were later suggested to be a substitute for sterols, are an important structural component of microbial membranes [117] and may protect bacterial cells from photooxidation or damage caused by visible light irradiation.

Several bacterial pigments that act as antagonists by exhibiting antibiotic activity against other organisms can be considered as potent weapons for survival and effective chemical defenses against eukaryotic predators. This class of bioactive agents includes almost all pigmented compounds commonly produced by Pseudoalteromonas, Pseudomonas, and Streptomyces species. These compounds inhibit the settlement of marine invertebrate larvae [118], the germination of algal spores [119] and protect the host surface by interfering with bacterial colonization and biofilm formation [74]. They may also inhibit other organisms that compete for space and nutrients.

Such hypotheses are also supported by a number of studies that found that these bacterial compounds were active against other prokaryotes and even eukaryotes [120–128]. In many studies, pigmented bacterial strains demonstrated a strong and broad range of antibiotic activities against other organisms, while nonpigmented strains did not [74, 129]. A clear correlation between pigment production and antibacterial activities of the twoSilicibacter sp. strain TM1040 and Phaeobacter strain 27-4 grown under static conditions was further reported by Bruhn et al. [129]. Mutant strains, which lacked pigment production, also lost their biological activities. Holmström et al. have also shown a close relationship between pigmentation and inhibitory activity, whereby 20 out of 22 dark pigmented bacterial strains tested displayed inhibitory activity against the settlement of two invertebrate larvae and algal spores [113].

Amongst other bacterial strains, Pseudoalteromonas has the most diverse antibiotic activities against alga biofouling, and the dark green pigmented P. tunicata exhibits the most active and broadest range of inhibitory activity when compared to other strains from this genus [74]. Two nonpigmented P. nigrifaciens and P. haloplanktis strains were also found not to display any antibiotic activities using various bioassays [74].

Blue-pigmented pyocyanin production in P. aeruginosa (Pup14B) was observed by Angell et al. to be induced by Enterobacter species (Pup14A and KM1), and this pyocyanin displayed moderate antibiotic activity againstE. coli and yeast [130]. It was experimentally demonstrated that metabolites produced by Pup14A strain are necessary for the production of this pigment in Pup14B strain [130]. Many other reports describe synergism between bacteria and higher organisms; however, this is a rare example between two bacterial species [131]. Such an unusual case contrasts with the hypothesis of the regulated biodiversity of marine bacteria, in which surface-associated microorganisms produce antimicrobial agents [74] to prevent competing microorganisms. The symbiosis of the two bacterial species is not yet fully understood, although both species appear to benefit from the pigment production.

One of the promising biological activities of marine bacteria isolates is their cytotoxic effect against cancer cells. Despite many investigations, the exact molecular mechanism of this pigmented compound cytotoxicity remains undetermined and requires further study. For example, violacein is known to cause apoptosis in tumorous cells [41]. However, the pathways leading to cell death have not yet been linked to the possible effects of the pigment, which was also shown to affect signal transduction agents, such as protein kinase and protein phosphatase family enzymes that play crucial role in cell differentiation and proliferation.

In a study by Bromberg et al., violacein showed inhibitory activity against protein phosphatases isolated from human lymphocytes [132]. A similar study was also conducted by Fürstner et al. to assess the inhibitory activity of prodigiosin derivatives [133]. Other targets of these compounds, including ion channels, are further being investigated [134–137].

Unexpected problems have also arisen when investigating marine environments. While the marine environment is a promising source for identifying microorganisms that can produce important biologically active pigments, yields of these compounds remain variable and are sometimes too low to provide enough material for drug development [138] or commercial applications. The main reason for such low yields is that these compounds are secondary metabolites and production depends on the quorum sensing mechanism.

Despite marine bacteria being capable of growing in the extremely low concentrations of nutrients that often exist in seawater, most species still require seawater or its equivalent as a growth medium for artificial culturing. Seawater is therefore used for the growth of marine bacteria, or similar levels of sodium, potassium, and magnesium chloride are supplemented in cultures. Optimal growth and the production of pigments are only sustained for most bacteria when appropriate salt mixtures are used for culturing, as is the case for the prodigiosin-producing marine Pseudomonas magnesiorubra and Vibrio psychroerythrus, among other marine species. These bacteria grew optimally and produced red pigment when cultured in seawater or its equivalent, while pigment production by the terrestrial Serratia marcescens was inhibited in 3% sea salts [8].

Enhancing low pigment productivity is one of the main issues facing researchers, and some solutions have already been reported. It is well established that antibiotic production by bacteria might be regulated both qualitatively and quantitatively by the nature of the culture medium. In particular, the addition of individual natural compounds to nutrient media or the use of gene expression methods was found to increase the pigment production far beyond expectations. For example, saturated fatty acids, especially peanut broth, was found to be a better choice in increasing prodigiosin production by 40-fold (approximately ~39 mg/mL) in S. marcescens[139].

Undecylprodigiosin synthesis by S. marcescens was also markedly enhanced by the addition of vegetable (soybean, olive, and sunflower) oils (2–6% [v/v]) and amino acids to the fermentation broth [140, 141]. Violacein production by the recombinant Citrobacter freundii strain, the genes of which were reconstructed from Duganella sp. B2, reached up to 1.68 g/L, making it fourfold higher than the highest production previously reported [142]. It is anticipated that these methods will facilitate the production of sufficient quantities of many bioactive and pharmacologically important compounds obtained from bacteria of marine origin. These compounds, including prodiginine and violacein, are now considered as potential drug candidates for potentially fatal diseases such as cancer and malaria. Although further improvement of culture methods and technologies for pigment production including recombinant technology is necessary, bioactive compounds from marine bacteria may potentially replace the existing drugs that have lower therapeutic actions.

CONCLUSIONS

Recently, a number of review papers have appeared in the literature, and they give an overview of all investigations of the marine environment and its isolates. While previous reviews have covered the biological activities of natural products isolated from marine microorganisms [115, 143] and other living organisms [144,145], our paper is the first to review the importance of pigmented compounds from marine origin and their potential pharmacological applications.

Most studies investigating marine microorganisms have shown the efficacy and the potential clinical applications of pigmented secondary metabolites in treating several diseases. These studies have also emphasized the effects of microbial metabolites as antibiotic, anticancer, and immunosuppressive compounds. Despite the enormous difficulty in isolating and harvesting marine bacteria, significant progress has been achieved in this field, and investigations of bioactive compounds produced by these species are rapidly increasing. As such, the number of compounds isolated from marine microorganisms is increasing faster when compared with terrestrial species [95].

Overall, this review of pigmented marine bioactive compounds and their pharmacological applications highlights the importance of discovering novel marine bacterial metabolites. Such compounds have a wide variety of biologically active properties and continue to provide promising avenues for both fundamental sciences and applied biomedical research.

REFERENCES

1. W. Bruckner, "Life-saving products from coral reefs," Issues in Science and Technology, vol. 18, no. 3, p. 35, 2002.

2. P. Hugenholtz and N. R. Pace, "Identifying microbial diversity in the natural environment: a molecular phylogenetic approach," Trends in Biotechnology, vol. 14, no. 6, pp. 190–197, 1996.

3. H. Laatsch, "Marine bacterial metabolites," 2005,http://wwwuser. gwdg.de/~ucoc/laatsch/Reviews__Books__Patents/R30_Marine_ BacterialMetabolites.pdf.

4. M. S. Mayer, K. B. Glaser, C. Cuevas et al., "The odyssey of marine pharmaceuticals: a current pipeline perspective," Trends in Pharmacological Sciences, vol. 31, no. 6, pp. 255–265, 2010.·

5. W. Fenical, "Chemical studies of marine bacteria: developing a new resource," Chemical Reviews, vol. 93, no. 5, pp. 1673–1683, 1993.

6. N. N. Gerber, "Prodigiosin-like pigments from Actinomadura (Nocardia) pelletieri and Actinomadura madurae," Applied Microbiology, vol. 18, no. 1, pp. 1–3, 1969.

7. H. Rapoport and K. G. Holden, "The synthesis of prodigiosin," Journal of the American Chemical Society, vol. 84, no. 4, pp. 635–642, 1962.

8. N. N. Gerber, "Prodigiosin-like pigments," CRC Critical Reviews in Microbiology, vol. 3, no. 4, pp. 469–485, 1975.

9. N. R. Williamson, P. C. Fineran, T. Gristwood, S. R. Chawrai, F. J. Leeper, and G. P. C. Salmond, "Anticancer and immunosuppressive properties of

bacterial prodiginines," Future Microbiology, vol. 2, no. 6, pp. 605–618, 2007.

10. J. W. Bennett and R. Bentley, "Seeing red: the story of prodigiosin," Advances in Applied Microbiology, vol. 47, pp. 1–32, 2000.

11. Montaner and R. Pérez-Tomás, "The prodigiosins: a new family of anticancer drugs," Current Cancer Drug Targets, vol. 3, no. 1, pp. 57–65, 2003.

12. N. M. Gandhi, J. R. Patell, J. Gandhi, N. J. De Souza, and H. Kohl, "Prodigiosin metabolites of a marinePseudomonas species," Marine Biology, vol. 34, no. 3, pp. 223–227, 1976.

13. M. J. Gauthier, "Alteromonas rubra sp. nov., a new marine antibiotic producing bacterium," International Journal of Systematic Bacteriology, vol. 26, no. 4, pp. 459–466, 1976.

14. N. N. Gerber and M. J. Gauthier, "New prodigiosin-like pigment from Alteromonas rubra," Applied and Environmental Microbiology, vol. 37, no. 6, pp. 1176–1179, 1979.

15. Fehér, R. S. Barlow, P. S. Lorenzo, and T. K. Hemscheidt, "A 2-substituted prodiginine, 2-(p-hydroxybenzyl)prodigiosin, from Pseudoalteromonas rubra," Journal of Natural Products, vol. 71, no. 11, pp. 1970–1972, 2008.

16. Ø. Enger, H. Nygaard, M. Solberg, G. Schei, J. Nielsen, and I. Dundas, "Characterization of Alteromonas denitrificans sp. nov.," International Journal of Systematic Bacteriology, vol. 37, no. 4, pp. 416–421, 1987.

17. T. Sawabe, H. Makino, M. Tatsumi et al., "Pseudoalteromonas bacteriolytica sp. nov., a marine bacterium that is the causative agent of red spot disease of Laminaria japonica," International Journal of Systematic Bacteriology, vol. 48, no. 3, pp. 769–774, 1998.

18. Yamamoto, H. Takemoto, K. Kuno et al., "Cycloprodigiosin hydrochloride, a new H^+/Cl^- symporter, induces apoptosis in human and rat hepatocellular cancer cell lines in vitro and inhibits the growth of hepatocellular carcinoma xenografts in nude mice," Hepatology, vol. 30, no. 4, pp. 894–902, 1999.

19. S. M. Lewis and W. A. Corpe, "Prodigiosin-producing bacteria from marine sources," Applied Microbiology, vol. 12, no. 1, pp. 13–17, 1964.

20. G. Gauthier, M. Gauthier, and R. Christen, "Phylogenetic analysis of the genera Alteromonas, Shewanella, and Moritella using genes coding for small-subunit rRNA sequences and division of the genusAlteromonas into two genera, Alteromonas (emended) and Pseudoalteromonas gen. nov., and proposal of twelve new species combinations," International

Journal of Systematic Bacteriology, vol. 45, no. 4, pp. 755–761, 1995.

21. K. Kawauchi, K. Shibutani, H. Yagisawa et al., "A possible immunosuppressant, cycloprodigiosin hydrochloride, obtained from Pseudoalteromonas denitrificans," Biochemical and Biophysical Research Communications, vol. 237, no. 3, pp. 543–547, 1997.

22. H. S. Kim, M. Hayashi, Y. Shibata et al., "Cycloprodigiosin hydrochloride obtained fromPseudoalteromonas denitrificans is a potent antimalarial agent," Biological and Pharmaceutical Bulletin, vol. 22, no. 5, pp. 532–534, 1999.

23. J. E. Lazaro, J. Nitcheu, R. Z. Predicala et al., "Heptyl prodigiosin, a bacterial metabolite, is antimalarial in vivo and non-mutagenic in vitro," Journal of Natural Toxins, vol. 11, no. 4, pp. 367–377, 2002.

24. H. K. Lee, J. Chun, E. Y. Moon et al., "Hahella chejuensis gen. nov., sp. nov., an extracellular- polysaccharide-producing marine bacterium," International Journal of Systematic and Evolutionary Microbiology, vol. 51, no. 2, pp. 661–666, 2001.

25. W. Y. Shieh, Y. W. Chen, S. M. Chaw, and H. H. Chiu, "Vibrio ruber sp. nov., a red, facultatively anaerobic, marine bacterium isolated from sea water," International Journal of Systematic and Evolutionary Microbiology, vol. 53, no. 2, pp. 479–484, 2003.

26. H. Yi, Y. H. Chang, H. W. Oh, K. S. Bae, and J. Chun, "Zooshikella ganghwensis gen. nov., sp. nov., isolated from tidal flat sediments," International Journal of Systematic and Evolutionary Microbiology, vol. 53, no. 4, pp. 1013–1018, 2003.

27. Kim, J. S. Lee, Y. K. Park et al., "Biosynthesis of antibiotic prodiginines in the marine bacteriumHahella chejuensis KCTC 2396," Journal of Applied Microbiology, vol. 102, no. 4, pp. 937–944, 2007.

28. T. Nakashima, M. Kurachi, Y. Kato, K. Yamaguchi, and T. Oda, "Characterization of bacterium isolated from the sediment at coastal area of Omura bay in Japan and several biological activities of pigment produced by this isolate," Microbiology and Immunology, vol. 49, no. 5, pp. 407–415, 2005.

29. D. A. Austin and M. O. Moss, "Numerical taxonomy of red-pigmented bacteria isolated from a lowland river, with the description of a new taxon, Rugamonas rubra gen. nov., sp. nov.," Journal of General Microbiology, vol. 132, no. 7, pp. 1899–1909, 1986.

30. M. O. Moss, "A note on a prodigiosin-producing pseudomonad isolated from a lowland river," Journal of Applied Bacteriology, vol. 55, no. 2, pp. 373–375, 1983.

31. L. I. Sly and M. H. Hargreaves, "Two unusual budding bacteria isolated from a swimming pool," Journal of Applied Bacteriology, vol. 56, no. 3, pp. 479–486, 1984.

32. S. Yada, M. Ohba, K. Enomoto, et al., "Analysis of bacterial species in the Muroto deep seawater," Deep Ocean Water Resources, vol. 4, no. 2, pp. 47–56, 2003 (Japanese).

33. Oren and F. Rodríguez-Valera, "The contribution of halophilic Bacteria to the red coloration of saltern crystallizer ponds," Federation of European Materials Societies Microbiology Ecology, vol. 36, no. 2-3, pp. 123–130, 2001.

34. N. Misawa, Y. Satomi, K. Kondo et al., "Structure and functional analysis of a marine bacterial carotenoid biosynthesis gene cluster and astaxanthin biosynthetic pathway proposed at the gene level," Journal of Bacteriology, vol. 177, no. 22, pp. 6575–6584, 1995.

35. J. H. Lee, Y. S. Kim, T. J. Choi, W. J. Lee, and Y. T. Kim, "Paracoccus haeundaensis sp. nov., a Gram-negative, halophilic, astaxanthin-producing bacterium," International Journal of Systematic and Evolutionary Microbiology, vol. 54, no. 5, pp. 1699–1702, 2004.

36. D. Rettori and N. Durán, "Production, extraction and purification of violacein: an antibiotic pigment produced by Chromobacterium violaceum," World Journal of Microbiology and Biotechnology, vol. 14, no. 5, pp. 685–688, 1998.

37. N. Durán and C. F. M. Menck, "Chromobacterium violaceum: a review of pharmacological and industiral perspectives," Critical Reviews in Microbiology, vol. 27, no. 3, pp. 201–222, 2001. ·

38. C. Sánchez, A. F. Braña, C. Méndez, and J. A. Salas, "Reevaluation of the violacein biosynthetic pathway and its relationship to indolocarbazole biosynthesis," ChemBioChem, vol. 7, no. 8, pp. 1231–1240, 2006.

39. S. Asamizu, Y. Kato, Y. Igarashi, and H. Onaka, "VioE, a prodeoxyviolacein synthase involved in violacein biosynthesis, is responsible for intramolecular indole rearrangement," Tetrahedron Letters, vol. 48, no. 16, pp. 2923–2926, 2007.

40. C. J. Balibar and C. T. Walsh, "In vitro biosynthesis of violacein from L-tryptophan by the enzymes VioA-E from Chromobacterium violaceum," Biochemistry, vol. 45, no. 51, pp. 15444–15457, 2006.

41. N. Durán, G. Z. Justo, C. V. Ferreira, P. S. Melo, L. Cordi, and D. Martins, "Violacein: properties and biological activities," Biotechnology and Applied Biochemistry, vol. 48, no. 3-4, pp. 127–133, 2007.

42. C. R. Andrighetti-Fröhner, R. V. Antonio, T. B. Creczynski-Pasa, C. R. M. Barardi, and C. M. O. Simões, "Cytotoxicity and potential antiviral evaluation of violacein produced by Chromobacterium violaceum,"Memórias do Instituto Oswaldo Cruz, vol. 98, no. 6, pp. 843–848, 2003.

43. C. Matz, P. Deines, J. Boenigk et al., "Impact of violacein-producing bacteria on survival and feeding of bacterivorous nanoflagellates," Applied and Environmental Microbiology, vol. 70, no. 3, pp. 1593–1599, 2004.

44. R. M. Brucker, R. N. Harris, C. R. Schwantes et al., "Amphibian Chemical defense: antifungal metabolites of the microsymbiont Janthinobacterium lividum on the salamander Plethodon cinereus," Journal of Chemical Ecology, vol. 34, no. 11, pp. 1422–1429, 2008.

45. M. H. Becker, R. M. Brucker, C. R. Schwantes, R. N. Harris, and K. P. C. Minbiole, "The bacterially produced metabolite violacein is associated with survival of amphibians infected with a lethal fungus,"Applied and Environmental Microbiology, vol. 75, no. 21, pp. 6635–6638, 2009.

46. R. N. Harris, R. M. Brucker, J. B. Walke et al., "Skin microbes on frogs prevent morbidity and mortality caused by a lethal skin fungus," International Society for Microbial Ecology Journal, vol. 3, no. 7, pp. 818–824, 2009.

47. R. D. Hamilton and K. E. Austin, "Physiological and cultural characteristics of Chromobacterium marinum sp. n.," Antonie van Leeuwenhoek, vol. 33, no. 3, pp. 257–264, 1967.

48. M. J. Gauthier, "Morphological, physiological, and biochemical characteristics of some violet-pigmented bacteria isolated from seawater," Canadian Journal of Microbiology, vol. 22, no. 2, pp. 138–149, 1976.

49. M. J. Gauthier, J. M. Shewan, D. M. Gibson, and J. V. Lee, "Taxonomic position and seasonal variations in marine neritic environment of some gram-negative antibiotic-producing bacteria," Journal of General Microbiology, vol. 87, no. 2, pp. 211–218, 1975.

50. S. Yada, Y. Wang, Y. Zou et al., "Isolation and characterization of two groups of novel marine bacteria producing violacein," Marine Biotechnology, vol. 10, no. 2, pp. 128–132, 2008.

51. S. Hakvåg, E. Fjærvik, G. Klinkenberg et al., "Violacein-producing Collimonas sp. from the sea surface microlayer of costal waters in Trøndelag, Norway," Marine Drugs, vol. 7, no. 4, pp. 576–588, 2009.

52. S. A. MacCarthy, T. Sakata, D. Kakimoto, and R. M. Johnson, "Production and isolation of purple pigment by Alteromonas luteoviolacea," Bulletin

of the Japanese Society of Scientific Fisheries, vol. 51, no. 3, pp. 479–484, 185.

53. N. J. Novick and M. E. Tyler, "Isolation and characterization of Pseudoalteromonas luteoviolacea strains with sheathed flagella," International Journal of Systematic Bacteriology, vol. 35, no. 1, pp. 111–113, 1985.

54. J. M. Turner and A. J. Messenger, "Occurrence, biochemistry and physiology of phenazine pigment production," Advances in Microbial Physiology, vol. 27, no. C, pp. 211–275, 1986.

55. L. S. Pierson and E. A. Pierson, "Metabolism and function of phenazines in bacteria: impact on the behavior of bacteria in the environment and biotechnological process," Applied Microbiology and Biotechnology, vol. 86, no. 6, pp. 1659–1670, 2010.

56. J. Gibson, A. Sood, and D. A. Hogan, "Pseudomonas aeruginosa-Candida albicans interactions: localization and fungal toxicity of a phenazine derivative," Applied and Environmental Microbiology, vol. 75, no. 2, pp. 504–513, 2009.

57. J. B. Laursen and J. Nielsen, "Phenazine natural products: biosynthesis, synthetic analogues, and biological activity," Chemical Reviews, vol. 104, no. 3, pp. 1663–1685, 2004.

58. S. Saha, R. Thavasi, and S. Jayalakshmi, "Phenazine pigments from Pseudomonas aeruginosa and their application as antibacterial agent and food colourants," Research Journal of Microbiology, vol. 3, no. 3, pp. 122–128, 2008.

59. Nansathit, S. Apipattarakul, C. Phaosiri, P. Pongdontri, S. Chanthai, and C. Ruangviriyachai, "Synthesis, isolation of phenazine derivatives and their antimicrobial activities," Walailak Journal of Science and Techology, vol. 6, no. 1, pp. 79–91, 2009.

60. H. Ran, D. J. Hassett, and G. W. Lau, "Human targets of Pseudomonas aeruginosa pyocyanin," Proceedings of the National Academy of Sciences of the United States of America, vol. 100, no. 2, pp. 14315–14320, 2003.

61. W. Lau, H. Ran, F. Kong, D. J. Hassett, and D. Mavrodi, "Pseudomonas aeruginosa pyocyanin is critical for lung infection in mice," Infection and Immunity, vol. 72, no. 7, pp. 4275–4278, 2004.

62. M. W. Tan, S. Mahajan-Miklos, and F. M. Ausubel, "Killing of Caenorhabditis elegans by Pseudomonas aeruginosa used to model mammalian bacterial pathogenesis," Proceedings of the National Academy of Sciences of the United States of America, vol. 96, no. 2, pp. 715–720, 1999.

63. R. P. Maskey, I. Kock, E. Helmke, and H. Laatsch, "Isolation and structure determination of phenazostatin D, a new phenazine from a marine actinomycete isolate Pseudonocardia sp. B6273,"Zeitschrift für Naturforschung, vol. 58b, no. 7, pp. 692–694, 2003.

64. D. Li, F. Wang, X. Xiao, X. Zeng, Q. Q. Gu, and W. Zhu, "A new cytotoxic phenazine derivative from a deep sea bacterium Bacillus sp.," Archives of Pharmacal Research, vol. 30, no. 5, pp. 552–555, 2007.

65. K. Pusecker, H. Laatsch, E. Helmke, and H. Weyland, "Dihydrophencomycin methyl ester, a new phenazine derivative from a marine streptomycete," Journal of Antibiotics, vol. 50, no. 6, pp. 479–483, 1997.

66. R. Wilson, T. Pitt, G. Taylor, et al., "Pyocyanin and 1-hydroxyphenazine produced by Pseudomonas aeruginosa inhibit the beating of human respiratory cilia in vitro," Journal of Clinical Investigation, vol. 79, no. 1, pp. 221–229, 1987.

67. M. Akagawa-Matsushita, T. Itoh, Y. Katayama, H. Kuraishi, and K. Yamasato, "Isoprenoid quinone composition of some marine Alteromonas, Marinomonas, Deleya, Pseudomonas and Shewanella species,"Journal of General Microbiology, vol. 138, no. 11, pp. 2275–2281, 1992.

68. R. P. Maskey, E. Helmke, and H. Laatsch, "Himalomycin A and B: isolation and structure elucidation of new fridamycin type antibiotics from a marine Streptomyces isolate," Journal of Antibiotics, vol. 56, no. 11, pp. 942–949, 2003.

69. P. Z. Margalith, Pigment Microbiology, Chapman & Hall, London, UK, 1992.

70. J. Koyama, "Anti-infective quinone derivatives of recent patents," Recent Patents on Anti-Infective Drug Discovery, vol. 1, no. 1, pp. 113–125, 2006.

71. Li, R. P. Maskey, S. Qin et al., "Chinikomycin A and B: isolation, structure elucidation and biological activity of novel antibiotics from a marine Streptomyces sp. isolate M045," Journal of Natural Products, vol. 68, no. 3, pp. 349–353, 2005.

72. C. F. Norton and G. E. Jones, "A marine isolate of Pseudomonas nigrifacience. II. Characterization of its blue pigment," Archiv für Mikrobiologie, vol. 64, no. 4, pp. 369–376, 1969.

73. Kobayashi, Y. Nogi, and K. Horikoshi, "New violet 3,3'-bipyridyl pigment purified from deep-sea microorganism Shewanella violacea DSS12," Extremophiles, vol. 11, no. 2, pp. 245–250, 2007.

74. C. Holmström, S. Egan, A. Franks, S. McCloy, and S. Kjelleberg, "Antifouling activities expressed by marine surface associated Pseudoalteromonas species," Federation of European Materials Societies Microbiology Ecology, vol. 41, no. 1, pp. 47–58, 2002.

75. S. Egan, S. James, C. Holmström, and S. Kjelleberg, "Correlation between pigmentation and antifouling compounds produced by Pseudoalteromonas tunicate," Environmental Microbiology, vol. 4, no. 8, pp. 433–442, 2002.

76. Franks, P. Haywood, C. Holmström, S. Egan, S. Kjelleberg, and N. Kumar, "Isolation and structure elucidation of a novel yellow pigment from the marine bacterium Pseudoalteromonas tunicata,"Molecules, vol. 10, no. 10, pp. 1286–1291, 2005.

77. Carté and D. J. Faulkner, "Defensive metabolites from three nembrothid nudibranchs," Journal of Organic Chemistry, vol. 48, no. 14, pp. 2314–2318, 1983.

78. N. Lindquist and W. Fenical, "New tambjamines class alkaloids from the marine ascidian Atapozoa sp. and its nudibranch predators. Origin of the tambjamines in Atapozoa," Experientia, vol. 47, no. 5, pp. 504–506, 1991.

79. J. Blackman and C. P. Li, "New tambjamine alkaloids from the marine bryozoan Bugula dentate,"Australian Journal of Chemistry, vol. 47, no. 8, pp. 1625–1629, 1994.

80. D. M. Pinkerton, M. G. Banwell, M. J. Garson et al., "Antimicrobial and cytotoxic activities of synthetically derived tambjamines C and E-J, BE-18591, and a related alkaloid from the marine bacterium Pseudoalteromonas tunicata," Chemistry & Biodiversity, vol. 7, no. 5, pp. 1311–1324, 2010.

81. C. Granato, J. H. H. L. de Oliveira, M. H. R. Seleghim et al., "Produtos naturais da ascidia Botrylloides giganteum, das esponjas Verongula gigantea, Ircinia felix, Cliona delitrix e do nudibrânquio Tambja eliora, da costa do Brasil," Quimica Nova, vol. 28, no. 2, pp. 192–198, 2005.

82. D. M. Pinkerton, M. G. Banwell, and A. C. Willis, "Total syntheses of tambjamines C, E, F, G, H, I and J, BE-18591, and a related alkaloid from the marine bacterium Pseudoalteromonas tunicata," Organic Letters, vol. 9, no. 24, pp. 5127–5130, 2007.

83. S. I. Kotob, S. L. Coon, E. J. Quintero, and R. M. Weiner, "Homogentisic acid is the primary precursor of melanin synthesis in Vibrio cholerae, a Hyphomonas strain, and Shewanella colwelliana," Applied and Environmental Microbiology, vol. 61, no. 4, pp. 1620–1622, 1995.

84. C. Ruzafa, A. Sanchez-Amat, and F. Solano, "Characterization of the melanogenic system in Vibrio cholerae, ATCC 14035," Pigment Cell Research, vol. 8, no. 3, pp. 147–152, 1995. ·

85. E. P. Ivanova, E. A. Kiprianova, V. V. Mikhailov et al., "Characterization and identification of marineAlteromonas nigrifaciens strains and emendation of the description," International Journal of Systematic Bacteriology, vol. 46, no. 1, pp. 223–228, 1996.

86. W. C. Fuqua and R. M. Weiner, "The melA gene is essential for melanin biosynthesis in the marine bacterium Shewanella colwelliana," Journal of General Microbiology, vol. 139, no. 5, pp. 1105–1114, 1993.

87. Sanchez-Amat, C. Ruzafa, and F. Solano, "Comparative tyrosine degradation in Vibrio cholerae strains. The strain ATCC 14035 as a prokaryotic melanogenic model of homogentisate-releasing cell,"Comparative Biochemistry and Physiology—Part B, vol. 119, no. 3, pp. 557–562, 1998.

88. Y. Kahng, B. S. Chung, D. H. Lee, J. S. Jung, J. H. Park, and C. O. Jeon, "Cellulophaga tyrosinoxydanssp. nov., a tyrosinase-producing bacterium isolated from seawater," International Journal of Systematic and Evolutionary Microbiology, vol. 59, no. 4, pp. 654–657, 2009.

89. W. Fenical and P. R. Jensen, "Marine microorganisms: a new biomedical resource," in Marine Biotechnology, D. H. Attaway and O. R. Zaborsky, Eds., pp. 419–457, Plenum Press, New York, NY, USA, 1993.

90. H. Claus and H. Decker, "Bacterial tyrosinases," Systematic and Applied Microbiology, vol. 29, no. 1, pp. 3–14, 2006.

91. F. Solano, E. García, E. P. de Egea, and A. Sanchez-Amat, "Isolation and characterization of strain MMB-1 (CECT 4803), a novel melanogenic marine bacterium," Applied and Environmental Microbiology, vol. 63, no. 9, pp. 3499–3506, 1997.

92. P. J. Proteau, W. H. Gerwick, F. Garcia-Pichel, and R. Castenholz, "The structure of scytonemin, an ultraviolet sunscreen pigment from the sheaths of cyanobacteria," Experientia, vol. 49, no. 9, pp. 825–829, 1993.

93. C. S. Stevenson, E. A. Capper, A. K. Roshak et al., "Scytonemin—a marine natural product inhibitor of kinases key in hyperproliferative inflammatory diseases," Inflammation Research, vol. 51, no. 2, pp. 112–114, 2002.

94. H.-P. Grossart, M. Thorwest, I. Plitzko, T. Brinkhoff, M. Simon, and A. Zeeck, "Production of a blue pigment (glaukothalin) by marine Rheinheimera spp.," International Journal of Microbiology, vol. 2009, Article ID 701735, 7 pages, 2009.

95. Wagner-Döbler, W. Beil, S. Lang, M. Meiners, and H. Laatsch, "Integrated approach to explore the potential of marine microorganisms for the production of bioactive metabolites," Advances in Biochemical Engineering/Biotechnology, vol. 74, pp. 207–238, 2002.

96. N. R. Williamson, H. T. Simonsen, R. A. A. Ahmed et al., "Biosynthesis of the red antibiotic, prodigiosin, in Serratia: identification of a novel 2-methyl-3-n-amylpyrrole (MAP) assembly pathway, definition of the terminal condensing enzyme, and implications for undecylprodigiosin biosynthesis in Streptomyces,"Molecular Microbiology, vol. 56, no. 4, pp. 971–989, 2005.

97. N. R. Williamson, P. C. Fineran, F. J. Leeper, and G. P. C. Salmond, "The biosynthesis and regulation of bacterial prodiginines," Nature Reviews Microbiology, vol. 4, no. 12, pp. 887–899, 2006.

98. K. P. Harris, N. R. Williamson, H. Slater et al., "The Serratia gene cluster encoding biosynthesis of the red antibiotic, prodigiosin, shows species- and strain-dependent genome context variation,"Microbiology, vol. 150, no. 11, pp. 3547–3560, 2004.

99. H. Jeong, J. H. Yim, C. Lee et al., "Genomic blueprint of Hahella chejuensis, a marine microbe producing an algicidal agent," Nucleic Acids Research, vol. 33, no. 22, pp. 7066–7073, 2005.

100. M. Cerdeño, M. J. Bibb, and G. L. Challis, "Analysis of the prodiginine biosynthesis gene cluster ofStreptomyces coelicolor A3(2): new mechanisms for chain initiation and termination in modular multienzymes," Chemistry and Biology, vol. 8, no. 8, pp. 817–829, 2001.

101. M.-J. Ding and R. P. Williams, "Biosynthesis of prodigiosin by white strains of Serratia marcescensisolated from patients," Journal of Clinical Microbiology, vol. 17, no. 3, pp. 476–480, 1983. ·

102. Shinoda, T. Hasegawa, H. Sato et al., "Biosynthesis of violacein: a genuine intermediate, protoviolaceinic acid, produced by VioABDE, and insight into VioC function," Chemical Communications, no. 40, pp. 4140–4142, 2007.

103. S. Hirano, S. Asamizu, H. Onaka, Y. Shiro, and S. Nagano, "Crystal structure of VioE, a key player in the construction of the molecular skeleton of violacein," Journal of Biological Chemistry, vol. 283, no. 10, pp. 6459–6466, 2008.

104. S. Ryan, C. J. Balibar, K. E. Turo, C. T. Walsh, and C. L. Drennan, "The violacein biosynthetic enzyme VioE shares a fold with lipoprotein transporter proteins," Journal of Biological Chemistry, vol. 283, no. 10, pp. 6467–6475, 2008.

105. G. S. Byng, D. C. Eustice, and R. A. Jensen, "Biosynthesis of phenazine pigments in mutant and wild-type cultures of Pseudomonas aeruginosa," Journal of Bacteriology, vol. 138, no. 3, pp. 846–852, 1979.

106. D. V. Mavrodi, R. F. Bonsall, S. M. Delaney, M. J. Soule, G. Phillips, and L. S. Thomashow, "Functional analysis of genes for biosynthesis of pyocyanin and phenazine-1-carboxamide from Pseudomonas aeruginosa PAO1," Journal of Bacteriology, vol. 183, no. 21, pp. 6454–6465, 2001.

107. C. Burke, T. Thomas, S. Egan, and S. Kjelleberg, "The use of functional genomics for the identification of a gene cluster encoding for the biosynthesis of an antifungal tambjamine in the marine bacteriumPseudoalteromonas tunicate," Environmental Microbiology, vol. 9, no. 3, pp. 814–818, 2007.

108. D. López-Serrano, F. Solano, and A. Sanchez-Amat, "Identification of an operon involved in tyrosinase activity and melanin synthesis in Marinomonas mediterranea," Gene, vol. 342, no. 1, pp. 179–187, 2004.

109. S. K. Kelley, V. E. Coyne, D. D. Sledjeski, W. C. Fuqua, and R. M. Weiner, "Identification of a tyrosinase from a periphytic marine bacterium," Federation of European Materials Societies Microbiology Letters, vol. 67, no. 3, pp. 275–279, 1990.

110. T. Soule, V. Stout, W. D. Swingley, J. C. Meeks, and F. Garcia-Pichel, "Molecular genetics and genomic analysis of scytonemin biosynthesis in Nostoc punctiforme ATCC 29133," Journal of Bacteriology, vol. 189, no. 12, pp. 4465–4472, 2007.

111. E. P. Balskus and C. T. Walsh, "An enzymatic cyclopentyl[b]indole formation involved in scytonemin biosynthesis," Journal of the American Chemical Society, vol. 131, no. 41, pp. 14648–14649, 2009.

112. E. P. Balskus and C. T. Walsh, "Investigating the initial steps in the biosynthesis of cyanobacterial sunscreen scytonemin," Journal of the American Chemical Society, vol. 130, no. 46, pp. 15260–15261, 2008.

113. C. Holmström, S. James, S. Egan, and S. Kjelleberg, "Inhibition of common fouling organisms by pigmented marine bacterial isolates with special reference to the role of pigmented bacteria," Biofouling, vol. 10, no. 1–3, pp. 251–259, 1996.

114. L. Lemos, A. E. Toranzo, and J. L. Barja, "Antibiotic activity of epiphytic bacteria isolated from intertidal seaweeds," Microbial Ecology, vol. 11, no. 2, pp. 149–163, 1985.

115. Bhatnagar and S. K. Kim, "Immense essence of excellence: marine microbial bioactive compounds,"Marine Drugs, vol. 8, no. 10, pp. 2673–2701, 2010.

116. Griffiths, W. R. Sistrom, G. Cohen-Bazire, and R. Y. Stanier, "Function of carotenoids in photosynthesis," Nature, vol. 176, no. 4495, pp. 1211–1214, 1955.

117. S. Rottem and O. Markowitz, "Carotenoids act as reinforcers of the Acholeplasma laidlawii lipid bilayer,"Journal of Bacteriology, vol. 140, no. 3, pp. 944–948, 1979.

118. C. Holmström, D. Rittschof, and S. Kjelleberg, "Inhibition of settlement by larvae of Balanus amphitriteand Ciona intestinalis by a surface-colonizing marine bacterium," Applied and Environmental Microbiology, vol. 58, no. 7, pp. 2111–2115, 1992.

119. S. Egan, S. James, C. Holmström, and S. Kjelleberg, "Inhibition of algal spore germination by the marine bacterium Pseudoalteromonas tunicata," Federation of European Microbiological Societies Microbiology Ecology, vol. 35, no. 1, pp. 67–73, 2001.

120. S. Egan, T. Thomas, C. Holmström, and S. Kjelleberg, "Phylogenetic relationship and antifouling activity of bacterial epiphytes from the marine alga Ulva lactuca," Environmental Microbiology, vol. 2, no. 3, pp. 343–347, 2000.

121. R. J. Andersen, M. S. Wolfe, and D. J. Faulkner, "Autotoxic antibiotic production by a marineChromobacterium," Marine Biology, vol. 27, no. 4, pp. 281–285, 1974.

122. M. J. Gauthier and G. N. Flatau, "Antibacterial activity of marine violet-pigmented Alteromonas with special reference to the production of brominated compounds," Canadian Journal of Microbiology, vol. 22, no. 11, pp. 1612–1619, 1976.

123. C. Holmström and S. Kjelleberg, "The effect of external biological factors on settlement of marine invertebrates and new antifouling technology," Biofouling, vol. 8, no. 2, pp. 147–160, 1994.

124. C. Holmström, P. Steinberg, V. Christov, G. Christie, and S. Kjelleberg, "Bacteria immobilised in gels: improved methodologies for antifouling and biocontrol applications," Biofouling, vol. 15, no. 1–3, pp. 109–117, 2000.

125. Imai, Y. Ishida, K. Sakaguchi, and Y. Hata, "Algicidal marine bacteria isolated from northern Hiroshima Bay, Japan," Fisheries Science, vol. 61, no. 4, pp. 628–636, 1995.

126. C. Lovejoy, J. P. Bowman, and G. M. Hallegraeff, "Algicidal effects of a novel marine Pseudoalteromonasisolate (class Proteobacteria, gamma subdivision) on harmful algal bloom species of the generaChattonella, Gymnodinium, and Heterosigma," Applied and Environmental

Microbiology, vol. 64, no. 8, pp. 2806–2813, 1998.

127. S. Maki, D. Rittschof, and R. Mitchell, "Inhibition of larval barnacle attachment to bacterial films: an investigation of physical properties," Microbial Ecology, vol. 23, no. 1, pp. 97–106, 1992. ·

128. S. A. Mary, S. V. Mary, D. Rittschof, and R. Nagabhushanam, "Bacterial-barnacle interaction: potential of using juncellins and antibiotics to alter structure of bacterial communities," Journal of Chemical Ecology, vol. 19, no. 10, pp. 2155–2167, 1993.

129. B. Bruhn, L. Gram, and R. Belas, "Production of antibacterial compounds and biofilm formation byRoseobacter species are influenced by culture conditions," Applied and Environmental Microbiology, vol. 73, no. 2, pp. 442–450, 2007.

130. S. Angell, B. J. Bench, H. Williams, and C. M. H. Watanabe, "Pyocyanin isolated from a marine microbial population: synergistic production between two distinct bacterial species and mode of action," Chemistry & Biology, vol. 13, no. 12, pp. 1349–1359, 2006.

131. H. B. Bode, "No need to be pure: mix the cultures!," Chemistry and Biology, vol. 13, no. 12, pp. 1245–1246, 2006.

132. Bromberg, G. Z. Justo, M. Haun, N. Durán, and C. V. Ferreira, "Violacein cytotoxicity on human blood lymphocytes and effect on phosphatases," Journal of Enzyme Inhibition and Medicinal Chemistry, vol. 20, no. 5, pp. 449–454, 2005.

133. Fürstner, K. Reinecke, H. Prinz, and H. Waldmann, "The core structures of roseophilin and the prodigiosin alkaloids define a new class of protein tyrosine phosphatase inhibitors," ChemBioChem, vol. 5, no. 11, pp. 1575–1579, 2004.

134. T. Sato, H. Konno, Y. Tanaka et al., "Prodigiosins as a new group of H^+/Cl^- symporters that uncouple proton translocators," Journal of Biological Chemistry, vol. 273, no. 34, pp. 21455–21462, 1998.

135. S. Ohkuma, T. Sato, M. Okamoto et al., "Prodigiosins uncouple lysosomal vacuolar-type ATPase through promotion of H^+/Cl^- symport," Biochemical Journal, vol. 334, no. 3, pp. 731–741, 1998.

136. L. Seganish and J. T. Davis, "Prodigiosin is a chloride carrier that can function as an anion exchanger,"Chemical Communications, no. 46, pp. 5781–5783, 2005.

137. Kawauchi, K. Tobiume, K. Iwashita et al., "Cycloprodigiosin hydrochloride activates the Ras-PI3K-Akt pathway and suppresses protein synthesis inhibition-induced apoptosis in PC12 cells," Bioscience,

Biotechnology and Biochemistry, vol. 72, no. 6, pp. 1564–1570, 2008.

138. D. S. Bhakuni and D. S. Rawat, Bioactive Marine Natural Products, Springer, New York, NY, USA, 2005.

139. V. Giri, N. Anandkumar, G. Muthukumaran, and G. Pennathur, "A novel medium for the enhanced cell growth and production of prodigiosin from Serratia marcescens isolated from soil," BMC Microbiology, vol. 4, no. 11, pp. 1–10, 2004.

140. Y.-H. Wei and W. C. Chen, "Enhanced production of prodigiosin-like pigment from Serratia marcescensSMΔR by medium improvement and oil-supplementation strategies," Journal of Bioscience and Bioengineering, vol. 99, no. 6, pp. 616–622, 2005.

141. Y.-H. Wei, W.-J. Yu, and W.-C. Chen, "Enhanced undecylprodigiosin production from Serratia marcescens SS-1 by medium formulation and amino-acid supplementation," Journal of Bioscience and Bioengineering, vol. 100, no. 4, pp. 466–471, 2005.

142. X. Jiang, H. S. Wang, C. Zhang, K. Lou, and X. H. Xing, "Reconstruction of the violacein biosynthetic pathway from Duganella sp. B2 in different heterologous hosts," Applied Microbiology and Biotechnology, vol. 86, no. 4, pp. 1077–1088, 2010.

143. Debbab, A. H. Aly, W. H. Lin, and P. Proksch, "Bioactive compounds from marine bacteria and fungi,"Microbial Biotechnology, vol. 3, no. 5, pp. 544–563, 2010.

144. J. W. Blunt, B.-R. Copp, W. P. Hu, M. H. G. Munro, P. T. Northcote, and M. R. Prinsep, "Marine natural products," Natural Product Reports, vol. 26, no. 2, pp. 170–244, 2009.

145. K. Jha and X. Zi-Rong, "Biomedical compounds from marine organisms," Marine Drugs, vol. 2, no. 3, pp. 123–146, 2004.

Chapter 5

IMPACTS OF TEMPERATURE ON THE STABILITY OF TROPICAL PLANT PIGMENTS AS SENSITIZERS FOR DYE SENSITIZED SOLAR CELLS

Aiman Yusoff,[1] N. T. R. N. Kumara,[2] Andery Lim,[1] Piyasiri Ekanayake,[2] andKushan U. Tennakoon[1]

[1]Faculty of Science & Institute for Biodiversity & Environmental Research, Universiti Brunei Darussalam, Tungku Link, Gadong, BE1410, Brunei Darussalam

[2]Applied Physics Program, Faculty of Science, Universiti Brunei Darussalam, Jalan Tungku Link, Gadong BE1410, Brunei Darussalam

ABSTRACT

Natural dyes have become a viable alternative to expensive organic sensitizers because of their low cost of production, abundance in supply, and eco-friendliness. We evaluated 35 native plants containing anthocyanin pigments as potential sensitizers for DSSCs. Melastoma malabathricum (fruit pulp), Hibiscus rosa-sinensis(flower), and Codiaeum variegatum (leaves) showed the highest absorption peaks. Hence, these were used to determine anthocyanin content and stability based on the impacts of storage temperature. Melastoma malabathricum fruit pulp exhibited the highest anthocyanin content (8.43 mg/L) followed by H. rosa-sinensisand C. variegatum. Significantly greater stability of extracted anthocyanin pigment was shown when all three were stored at 4°C. The highest half-life periods for anthocyanin in M. malabathricum, H. rosa-sinensis, and C. variegatum were 541, 571, and 353 days at 4°C. These were rapidly decreased to 111, 220, and 254 days when stored at 25°C. The photovoltaic efficiency of M. malabathricum was1.16%, while the values for H. rosa-sinensisand C. variegatum were 0.16% and 1.08%, respectively. Hence, M. malabathricum fruit pulp extracts can be further evaluated as an alternative natural sensitizer for DSSCs.

INTRODUCTION

Dye sensitized solar cell (DSSC) is a new derivative of a solar cell, developed by Grätzel [1]. It is based on semiconductor electrode-adsorbed dye, a counter electrode, and an electrolyte containing iodide and triiodide ions [2]. This device is capable of generating energy by converting the light absorbed into electrical energy.

Numerous metal complexes and organic dyes have been used and utilized as sensitizers [3]. Previously, it has been reported that the highest efficiency from a metal as sensitizer has been achieved from a compound containing Ruthenium, with a total of 11-12% efficiency [4]. Recent findings have found that perovskite sensitized solar cells have achieved a power conversion efficiency of approximately 15% [5]. Although such results provide better efficiency and high durability, the advantages are often offset by their high cost of production, complicated synthetic routes, environmental impact, and the tendency to undergo degradation in presence of water [6].

In contrast, the natural organic dyes are widely available and involve simple preparation, nontoxic, and complete biodegradation [7]. The use of nontoxic natural pigments as sensitizer would definitely enhance the environmental and economic benefits of this alternative form of solar energy conversion [8]. Due to these reasons, natural dyes are becoming attractive inexpensive candidates for renewable energy resources. The natural dye sensitizer may still produce very low efficiency, but with continuous advanced studies and research, improvisation of the efficiency of DSSCs has become a reality and hopeful.

Anthocyanins are the most abundant, naturally occurring flavonoid pigments which often give a bright red, blue, or violet color to plant petals, fruits, and stems [9]. Sometimes, they are present in a range of tissues including roots, tubers, and stems [4]. Since anthocyanin shows the red to blue color of the visible spectrum, it is considered as one of the best sensitizers for wide bandgap semiconductors [3].

The performance of the cell mainly depends on the dye used as sensitizer [10]. Optimizing the structure of a natural dye is necessary to improve DSSC efficiency [4]. Although anthocyanin pigments are abundant in plants, isolated anthocyanin pigments are highly instable and degradable [11]. Their stability is affected by several factors including pH, storage temperature, and sunlight exposure levels [12]. Hence, it is important to evaluate the optimum conditions required to maintain the anthocyanin stability over a long period of time.

Storage temperature plays a critical role for anthocyanin stability [13]. Investigating the effects of storage temperature on anthocyanin degradation will be highly beneficial because one of the vital steps in the procedure of manufacturing DSSCs involves storage of the extracted pigments.

In this study, a range of plants grown in Brunei Darussalam were tested for anthocyanin pigments. Special emphasis was paid to study the stability of promising pigments stored under different storage temperature regimes. Potential dye extracts were further tested as natural sensitizers in DSSCs.

MATERIALS AND METHODS

Plant Materials

Brightly red/purple colored plant parts (flowers, fruits, tubers, and leaves) were harvested to determine the presence of anthocyanin (Table 1).

Table 1: List of plants studied to determine the anthocyanin content

Number	Family	Species	Plant part analyzed for pigments
1	Anacardiaceae	Mangiferaindica L.	Leaves
2	Myrtaceae	Syzygium campanulatum	Leaves
3	Lamiaceae	Coleus blumei	Leaves
4	Amaranthaceae	Alternantheradentata var 1	Leaves
5	Amaranthaceae	Alternantheradentata var 2	Leaves
6	Euphorbiaceae	Acalyphawilkesiana	Leaves
7	Euphorbiaceae	Codiaeumvariegatum	Leaves
8	Agavaceae	Cordylineterminalis	Leaves
9	Heliconiaceae	Heliconiarostrata	Flowers
10	Malvaceae	Hibiscus rosa-sinensis	Flowers
11	Convolvulaceae	Ipomoea sp.	Flowers
12	Nyctaginaceae	Bougainvillea spp.	Flowers
13	Leguminosae	Caesalpinia pulcherrima	Flowers
14	Bignoniaceae	Jacaranda obtusifolia	Flowers
15	Papilionaceae	Andiramernis	Flowers
16	Lythraceae	Lagerstroemia sp.	Flowers
17	Verbenaceae	Durantaerecta/repens	Flowers
18	Melastomataceae	Melastomamalabathricum	Fruit pulp
19	Dilleniaceae	Dilleniasuffruticosa	Fruits
20	Palmaceae	Licuala orbicularis	Fruits
21	Solanaceae	Solanumtuberosum	Tubers
22	Amaranthaceae	Spinacia oleracea	Stem
23	Dioscoreaceae	Dioscorea villosa	Tubers
24	Costaceae	Costuswoodsonii	Flowers
25	Heliconiaceae	Heliconiarostrata	Flowers
26	Verbenaceae	Durantaerecta	Flowers
27	Clusiaceae	Garciniamangostana	Fruits
28	Fabaceae	Delonixregia	Flowers
29	Nepenthaceae	Nepenthes rafflesiana	Modified leaves
30	Nepenthaceae	Nepenthes ampullaria	Modified leaves
31	Amaranthaceae	Gomphrenaglobosa	Flowers
32	Myrtaceae	Rhodomyrtustomentosa	Flowers
33	Musaceae	Musa paradisiacal	Flowers
34	Leguminosae	Mimosa pudica	Flowers
35	Bignoniaceae	Tabebuiapentaphylla	Flowers

Anthocyanin Extraction

The anthocyanin extractions of the above plant parts were made following the procedure described by Rodriguez-Soana and Wrolstad [14]. 5 g of each freshly collected plant samples was used to extract the anthocyanin pigments. The pigments were initially extracted using 150 mL of 70% ethanol (w/v%) and stored overnight at 4°C. On the following day, the extraction was mixed thoroughly by using a magnetic stirrer for two hours under air-conditioned room temperature (25°C). The extraction was filtered using Whatman's ashless 110 mm filter paper to remove any solid residues. Subsequently, the extracts were centrifuged at 4500 rpm using a Denley BS400 (UK) centrifuge machine for five minutes to separate all residues. Lastly the supernatant of the ethanolic extracts was gently mixed with equal volumes of petroleum ether to separate polar and nonpolar pigments. The final ethanolic extract was assumed to carry only the polar anthocyanin pigments. This component was carefully poured to a 10 mL glass bottle, tightly stoppered and wrapped in aluminum foil to avoid exposure to light and treatments for different temperature regimes.

Plant Screening for Anthocyanin Pigments

Screening of separated anthocyanin pigments was done by measuring their absorbance spectra using UV-vis spectrophotometer (Shimadzu UV-1800, Japan). Before the commencement of absorbance measurements, each of the samples was treated with 45 μL of concentrated HCl [15]. This acidification process converts anthocyanin derivatives to anthocyanidin that gives absorption spectra in the region of 490–550 nm [11, 15, 16]. Plant extracts that showed higher absorption spectra were selected for further investigations to evaluate the impacts of varying temperature regimes. All measurements were done in three replicates per sample.

Determination of Anthocyanin Content

To finalize the sample selection for DSSCs, those extracts that showed the highest UV-vis absorbance reading were chosen, and their anthocyanin contents were determined following the pH differential method described by Giusti and Wrolstad [11]. The results were expressed as micrograms per gram fresh weight.

Anthocyanin content was calculated according to the following equation:

$$\text{Anthocyanin pigment content} = \frac{A \times MW \times DF \times 10^3}{\varepsilon \times L},$$

(1)

where $A = (A_{520nm} - A_{700nm})$ pH 1.0 $- (A_{520nm} - A_{700nm})$ pH 4.5, MW (Molecular Weight) = 449.2 g/mol, for cyanidin-3-glucoside, DF = Dilution factor, $\varepsilon = 26900$ $Lmol^{-1}cm^{-1}$, 10^3 is the factor for converting g to mg, and L is the assumed path length in cm.

Aliquots of plant extracts were brought to pH 1 and 4.5 and allowed to equilibrate for one hour. The absorbance of each equilibrated solution was then measured at 520 nm (λ_{max}) and 700 nm for haze correction. Spectroscopic absorbance readings were repeated against 70% ethanol as the reference. All measurements were done in three replicates per sample.

The MW used in this formula corresponds to the predominant anthocyanin in the sample. In some cases, predominant anthocyanin in a material may be known and could be different from cyanidin-3-glucoside. However, throughout the years, there has been a lack of uniformity in the values of absorptivity of purified anthocyanin, mainly due to difficulties of obtaining pure crystalline anthocyanin in adequate quantities [11,17]. Since there is a huge variety of anthocyanin spread in nature, it has been suggested that if the major anthocyanin is unknown, it can be expressed as cyanidin-3-glucoside because that is the most abundant anthocyanin in nature [11, 12, 17–20].

Impacts of Storage Temperature on Anthocyanin Stability

The anthocyanin extracts of M. malabathricum, H. rosa-sinensis, and C. variegatum were stored in a tightly stoppered glass bottle fully covered with aluminum foil to avoid exposure to light. Extracts were stored at three different storage temperatures, namely, 4°C, −20°C, and 25°C, to evaluate the stability during storage. In order to determine the anthocyanin contents, the spectroscopic absorbance of the extracts were initially determined for three consecutive days followed by weekly measurements over a period of four months from September 2012 to January 2013.

Degradation Rate of Anthocyanin during Storage

The first-order reaction constant rate (k) and half-life ($t_{1/2}$) were calculated using the following equation [21]:

$$\ln\left(\frac{C_t}{C_o}\right) = -k \times t,$$

$$t_{1/2} = \ln(0.5) \times k^{-1}, \tag{2}$$

where C_0 is the initial monomeric anthocyanin content and C_t is the monomeric anthocyanin content after minute storage at a given temperature.

Photovoltaic Test of DSSC

The preparations of TiO_2 anode are described elsewhere [22]. The anodes were dipped in the dye extract for overnight at room temperature (25°C) and dried out [15]. The cell was assembled using Dyesol's Test Cell Assembly Machine with the Surlyn (50 μm, Dyesol). The electrolyte solution containing tetra-butylammonium iodide (TBAI; 0.5 M)/I_2 (0.05 M), acetonitrile, and ethylene carbonate (6 : 4, v/v) [16] was introduced through a predrilled hole in platinum counter electrode. The cell was kept under irradiation of about 3-4 h for light soaking.

Finally I-V characteristic of the DSSC was measured under 1 sun level (DYESOL Solar Simulator LP-156B). The effective irradiated area of solar cell was 0.25 cm². The performance of DSSC sensitized with anthocyanin pigments extracted from M. malabathricum, H. rosa-sinensis, and C. variegatum was evaluated by short circuit current (J_{sc}), open circuit voltage (V_{oc}), fill factor (ff), and energy conversion efficiency (η).

The absorbance spectra of the dye adsorbed on TiO_2 electrodes were also measured. Before the commencement of absorbance measurements, each of the TiO_2 electrodes were dipped in the dye extract overnight at room temperature (25°C) and air dried.

RESULTS AND DISCUSSION

Plant Selection for DSSCs

As shown in Figure 1, the maximum absorbance of anthocyanin varied significantly in different species.Jacaranda obtusifolia, Licuala orbicularis, Spinacia oleracea, and Durantaerecta flower extracts showed no absorbance at 520 nm; hence it can be concluded that they do not possess anthocyanin. Among the rest, 17 other plant extracts showed maximum absorbance of 0.1 or lower and therefore they were not selected to further investigations. On the other hand, the remaining sample extracts showed absorbance maxima greater than 0.1. However, only three species, each representing fruit, flower, and leaves (Melastoma malabathricum, Hibiscus rosa-sinensis, and Codiaeum variegatum), which showed that highest absorbance maxima were selected for further investigations.

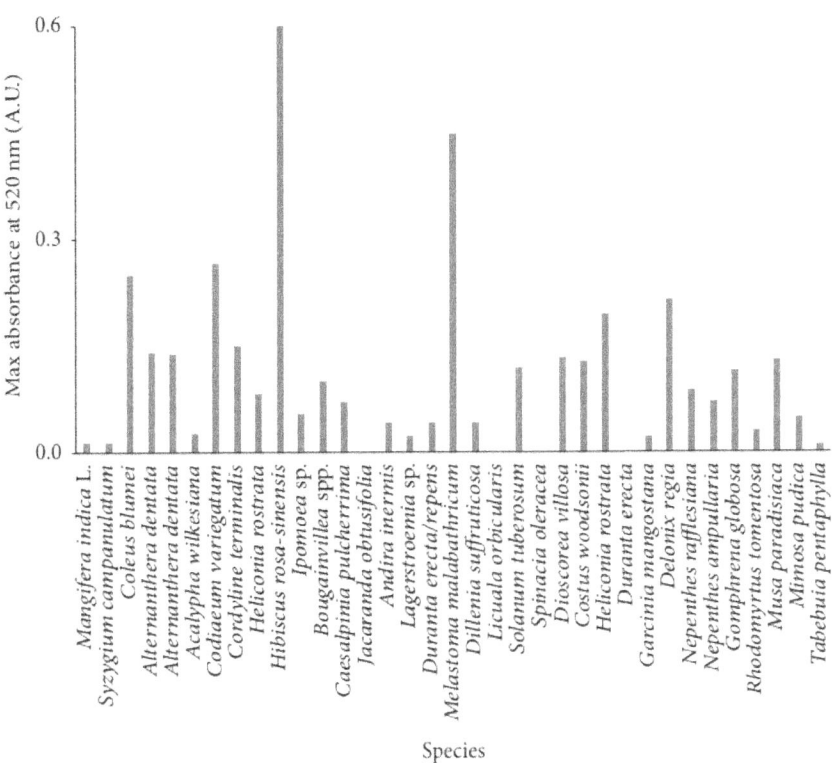

Figure 1: The absorbance spectra of anthocyanin pigments extracted from study species (n=35) observed at 520 nm during the initial screening for the presence of anthocyanin pigments.

Determination of Anthocyanin Content of Selected Plant Extracts for the Evaluation of DSSCs

Table 2 showed that among the samples investigated after preliminary screening, the highest anthocyanin concentration was found to be in the fruit pulp of M. malabathricum (8.43 mg L^{-1}), followed by H. rosa-sinensis(4.63 mg L^{-1}) then C. variegatum (2.22 mg L^{-1}).

Table 2: Anthocyanin content of promising species that showed higher absorbance reading at 520 nm during the preliminary screening process

Study species	Plant part used for pigment extraction	Anthocyanin content (mg/L fresh weight)[*]
Hibiscus rosa-sinensis	Flower	4.63
Melastomamalabathricum	Fruit pulp	8.43
Codiaeumvariegatum	Leaf	2.22

[*]$n = 3$.

The Absorbance Spectrum

All three extracts showed prominent peaks at 490–550 nm after the extracts were acidified with HCl (Figure2(a)). This result indicated and proved once again that more anthocyanidin presents in the extracts [11, 15, 16].

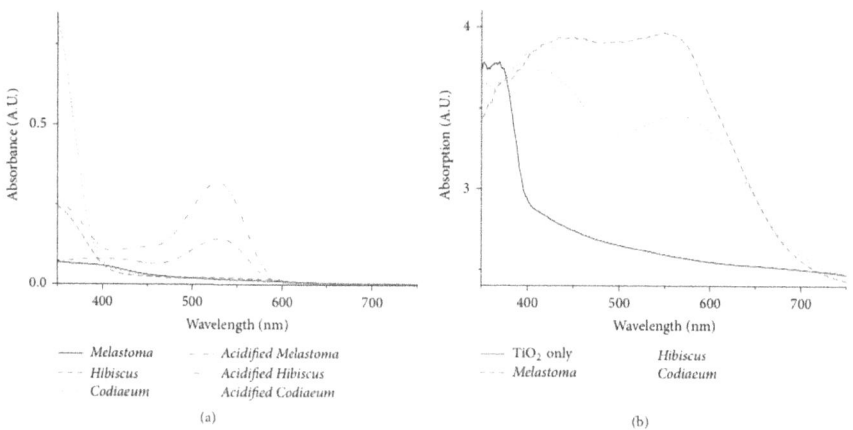

Figure 2: (a) The absorption spectra of the extracts of Melastoma malabathricum, Hibiscus rosa-sinensis, and Codiaeum variegatum in original and acidified extract and (b) absorption spectra of Melastoma malabathricum, Hibiscus rosa-sinensis, and Codiaeum variegatum dye onto TiO$_2$ film.

On the other hand, Figure 2(b) showed that M. malabathricum extract exhibited the best absorbance after being adsorbed into the TiO$_2$ electrode. This extract also gave the best efficiencies in DSSCs, while C. variegatum inTiO$_2$ gave the second best absorbance, followed by H. rosa-sinensis. The absorbance results of the dye adsorbed TiO$_2$ electrodes were consistent with I-V characteristics data.

The Effect of Storage Temperature on Anthocyanin Stability

The storage temperature had a strong influence on the degradation of anthocyanins extracted from all three extracts (see Figure 3 and Table 3).

Table 3: Kinetic parameters of anthocyanin degradation in M. malabathricum fruit pulp, H. rosa-sinensis flowers, and C. variegatum leaves at three different storage temperatures

Species	Original pH	Temp./°C	$k/10^{-3}$ (day⁻¹)	$t_{1/2}$ (day)
Melastoma malabathricum	pH 5.23	25	6.261	110.71
		4	1.282	540.77
		−20	1.286	539.13
Hibiscus rosa-sinensis	pH 5.73	25	3.154	219.74
		4	1.34	571.19
		−20	2.061	336.37
Codiaeum variegatum	pH 5.93	25	2.726	254.25
		4	1.964	352.86
		−20	1.708	405.72

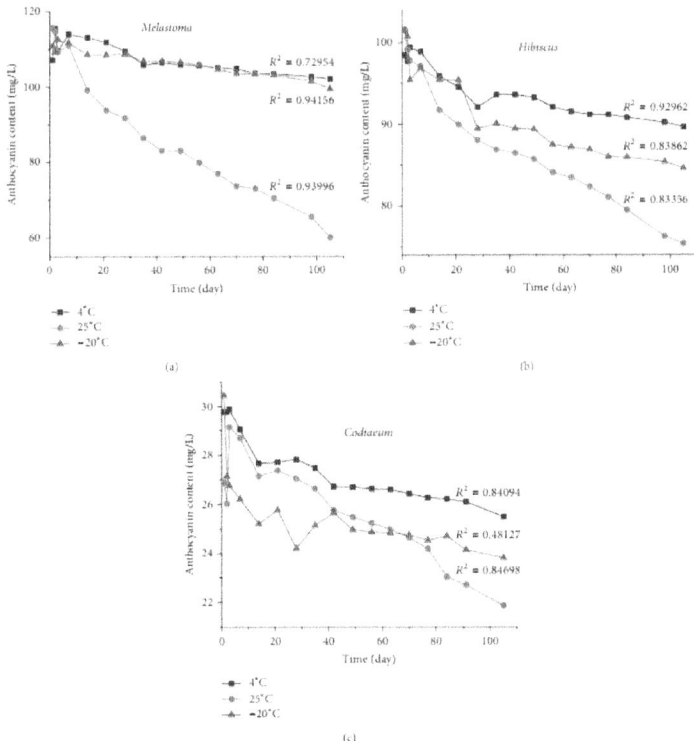

Figure 3: Degradation of anthocyanin pigments extracted from M. malabathricum fruit pulp (a), H. rosa-sinensis flowers, (b) and C. variegatum leaves (c) at three different storage temperatures (−20°C, 4°C and 25°C) over a three-month period.

The most distinctive pattern that was found in all three species was that anthocyanin pigments decreased progressively when stored at 25°C over a three-month period. However the stability of all three pigments was relatively high when the temperature was maintained at 4°C.

The degradation rates are represented by the half-life values; the higher the number, the more stable the anthocyanin extract. Result showed significantly greater stability of anthocyanin in all three species when they were stored at 4°C, and storage at 25°C resulted in much faster degradation. The highest half-life periods for anthocyanin in M. malabathricum, H. rosa-sinensis, and C. variegatum were 540.77, 571.19, and 352.86 days at 4°C, respectively, and it decreased rapidly to 110.71, 219.74, and 254.25 days at 25°C over a period of three months.

Similar results were reported by Janna et al. [23], who also studied the stability of Melastoma malabathricumand found that the suitable storage condition for anthocyanin pigment is acidic solution in dark and low temperature (4°C). The result of this investigation was also consistent with other similar studies where they found that anthocyanin pigments degrade faster as the temperature increases to 25°C and the stability is maintained at low temperatures (i.e., 4°C) [12, 21, 23].

A previous study on the anthocyanin degradation in black carrot showed that the $t_{1/2}$ value in shalgam drinks maintained at 4 and 25°C were 34 and 11 weeks, respectively [24]. A similar study also found that the $t_{1/2}$ value of monomeric anthocyanin of black carrot showed a distinct difference of 71.8 and 18.7 weeks, respectively, when maintained at 4 and 20°C, respectively [21]. Our investigation showed that frozen anthocyanin extracts maintained at −20°C also ensure a good stability over a period of three months; however, the best storage temperature was still 4°C.

The Efficiency of Natural Dye

The current-voltage characteristics of the DSSCs sensitized with the anthocyanin pigment extracted from M. malabathricum fruit pulp, H. rosa-sinensis flowers, and C. variegatum leaves are shown in Figure 4. The conversion efficiencies (η) of DSSCs were 1.16, 0.16, and 1.08%, respectively (Table 4). The highest effciency was obtained from DSSC sensitized with M. malabathricum fruit pulp extract with the open curcuit voltage ($V_{oc} = 0.383$ V), short curcuit current density ($I_{sc} = 6.17$ mA/cm^2), and fill factor (ff = 0.44)

Table 4: The photoelectric parameters of DSSCs sensitized with natural dye extracted from the fruit pulp of M. malabathricum, H. rosa-sinensis flowers, and C. variegatum

Sensitizer	I_{sc} (mA cm^{-2})	V_{oc} (V)	ff	η (%)
Melastoma malabathricum	6.17	0.383	0.44	1.16
Hibiscus rosa-sinensis	3.31	0.145	0.30	0.16
Codiaeumvariegatum	4.03	0.435	0.55	1.08

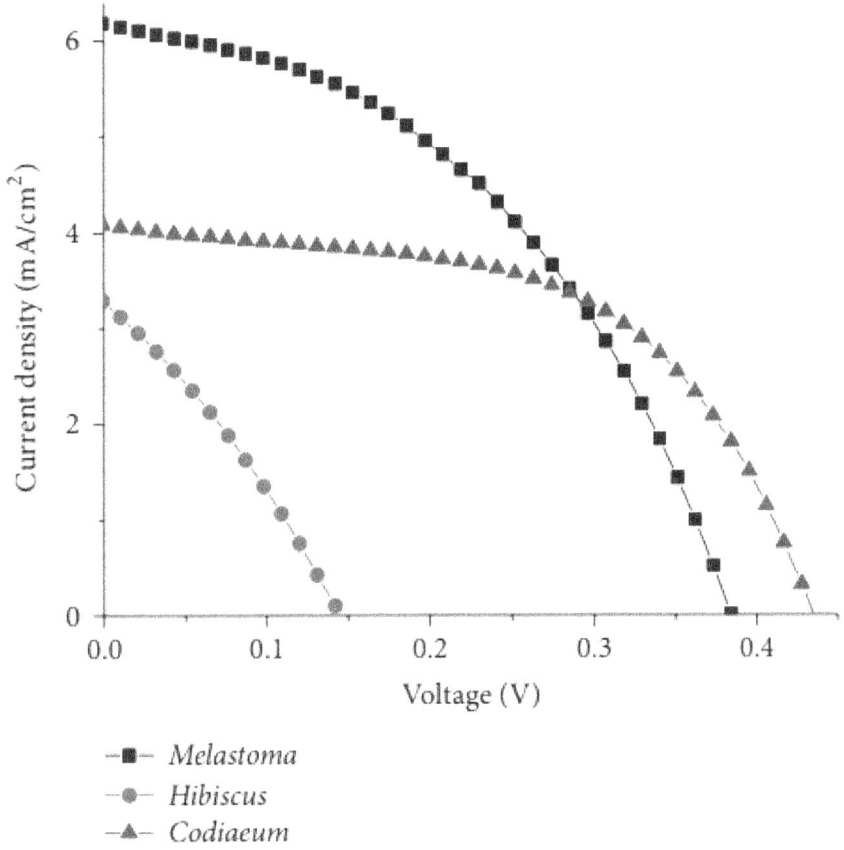

—■— *Melastoma*
—●— *Hibiscus*
—▲— *Codiaeum*

Figure 4: Current-voltage characteristics of the DSSCs sensitized with anthocyanins extracted from Melastoma malabathricum, Hibiscus rosa-sinensis, and Codiaeum variegatum.

Natural pigments extracted from fruits and vegetables such as chlorophyll and anthocyanins have been extensively investigated as sensitizers for DSSCs. By far, the best performance reported was obtained from beet roots with an efficiency of 2.71% [25, 26].

Other studies include Punicagranatum, Hibiscus sabdariffa, pomegranate juice, wild Silicon prickly pear (Opuntia vulgaris), Rhoeospathacea, Mangosteen pericarp, red turnip, Ficus reusa, and Hibiscus surattensis with conversion efficiencies of 1.86, 1.6, 1.5, 2.06, 1.49, 1.17, 1.70, 1.18, and 1.14%, respectively [6, 7, 27–31].

Our study has shown that extract from M. malabathricum yielded the highest efficiency, 1.16%. The result is encouraging and the methods employed to maintain its stability is extremely promising. High efficiency obtained in the fruit pulps of M. malabathricum can be attributed to the carbonyl and hydroxyl groups of anthocyanin molecules present [3, 6, 7, 25]. This ability favours photoelectric conversion as it allows effective binding with the surface of TiO_2 porous film. Further improvements in refinement of extraction and application methods will no doubt increase the efficiency of this dye in DSSCs.

CONCLUSION

Out of the 35 different species that were tested for the presence of anthocyanin pigments, Melastoma malabathricum, Hibiscus rosa-sinensis, and Codiaeum variegatum were selected as potential candidates in DSSCs. Among the three species, M. malabathricum extract exhibited the highest anthocyanin content. Based on the studies of anthocyanin stability on storage temperature, 4°C was the best to ensure pigment stability during storage. Among the three different species investigated, dye obtained from M. malabathricum fruit pulp also gave the highest efficiency. The photovoltaic performance of this dye was encouraging (1.16%). With further refinement of extraction and application methods, the efficiency of this dye can be further improved. Furthermore, due to the simple and cost-effective preparation techniques involved in the dye extraction of this species, it makes a promising alternative sensitizer for DSSCs.

ACKNOWLEDGMENTS

Universiti Brunei Darussalam (UBD) Research Grant UBD/PNC2/2/RG/1(176) and Brunei Research Council Science and Technology Research Grant (S & T 17) are acknowledged for financial support.

REFERENCES

1. M. Grätzel, "Dye-sensitized solar cells," Journal of Photochemistry and Photobiology C, vol. 4, no. 2, pp. 145–153, 2003.

2. H. Zhou, L. Wu, Y. Gao, and T. Ma, "Dye-sensitized solar cells using 20 natural dyes as sensitizers," Journal of Photochemistry and Photobiology

A, vol. 219, no. 2-3, pp. 188–194, 2011.·

3.   S. Hao, J. Wu, Y. Huang, and J. Lin, "Natural dyes as photosensitizers for dye-sensitized solar cell," Solar Energy, vol. 80, no. 2, pp. 209–214, 2006.

4.   M. R. Narayan, "Review: dye sensitized solar cells based on natural photosensitizers," Renewable and Sustainable Energy Reviews, vol. 16, no. 1, pp. 208–215, 2012.·

5.   J. Burschka, N. Pellet, S. J. Moon et al., "Sequential deposition as a route to high-performance perovskite-sensitized solar cells," Nature, vol. 499, no. 7458, pp. 316–319, 2013.

6.   R. Hernandez-Martinez, M. Estevez, S. Vargas, F. Quintanilla, and R. Rodriguez, "Natural pigment-based dye-sensitized solar cells," Journal of Applied Research and Technology, vol. 10, no. 1, pp. 38–47, 2012.

7.   H. Zhou, L. Wu, Y. Gao, and T. Ma, "Dye-sensitized solar cells using 20 natural dyes as sensitizers,"Journal of Photochemistry and Photobiology A, vol. 219, no. 2-3, pp. 188–194, 2011.·

8.   D. Zhang, S. M. Lanier, J. A. Downing, J. L. Avent, J. Lum, and J. L. McHale, "Betalain pigments for dye-sensitized solar cells," Journal of Photochemistry and Photobiology A, vol. 195, no. 1, pp. 72–80, 2008. ·

9.   E. Młodzińska, "Survey of plant pigments: molecular and environmental determinants of plant colors,"Acta Biologica Cracoviensia Series Botanica, vol. 51, no. 1, pp. 7–16, 2009.

10.   K. Wongcharee, V. Meeyoo, and S. Chavadej, "Dye-sensitized solar cell using natural dyes extracted from rosella and blue pea flowers," Solar Energy Materials and Solar Cells, vol. 91, no. 7, pp. 566–571, 2007. ·

11.   M. M. Giusti and R. E. Wrolstad, "Characterization and measurement of anthocyanin by UV-visible spectroscopy," in Current Protocols in Food Analytical Chemistry, pp. F1.2.1–F1.2.13, John Wiley & Sons, New York, NY, USA, 2001.

12.   Castañeda-Ovando, M. D. L. Pacheco-Hernández, M. E. Páez-Hernández, J. A. Rodríguez, and C. A. Galán-Vidal, "Chemical studies of anthocyanins: a review," Food Chemistry, vol. 113, no. 4, pp. 859–871, 2009.

13.   Patras, N. P. Brunton, B. K. Tiwari, and F. Butler, "Stability and degradation kinetics of bioactive compounds and colour in strawberry jam during storage," Food and Bioprocess Technology, vol. 4, no. 7, pp. 1245–1252, 2011.

14. L. E. Rodriguez-Saona and R. E. Wrolstad, "Extraction, isolation and purifications of anthoyanins," inCurrent Protocols in Food Analytical Chemistry, pp. F1.1.1–F1.1.11, John Wiley & Sons, New York, NY, USA, 2001.

15. N. T. R. N. Kumara, P. Ekanayake, A. Lim, M. Iskandar, and L. C. Ming, "Study of the enhancement of cell performance of dye sensitized solar cells sensitized with Nephelium lappaceum (F: Sapindaceae),"Journal of Solar Energy Engineering, vol. 135, no. 3, Article ID 031014, 5 pages, 2013.

16. N. T. R. N. Kumara, P. Ekanayake, A. Lim et al., "Layered co-sensitization for enhancement of conversion efficiency of natural dye sensitized solar cells," Journal of Alloys and Compounds, vol. 581, pp. 186–191, 2013.

17. J. Lee, K. W. Barnes, T. Eisele et al., "Determination of total monomeric anthocyanin pigment content of fruit juices, beverages, natural colorants, and wines by the pH differential method: collaborative study,"Journal of AOAC International, vol. 88, no. 5, pp. 1269–1278, 2005.

18. P. M. Dey and J. B. Harborne, Plant Phenolics Methods in Plant Biochemistry, Academic Press, London, UK, 2nd edition, 1993.

19. F. J. Francis, "Food colorants: anthocyanins," Critical Reviews in Food Science and Nutrition, vol. 28, no. 4, pp. 273–314, 1989.

20. J.-M. Kong, L.-S. Chia, N.-K. Goh, T.-F. Chia, and R. Brouillard, "Analysis and biological activities of anthocyanins," Phytochemistry, vol. 64, no. 5, pp. 923–933, 2003.

21. Kirca, M. Özkan, and B. Cemeroğlu, "Effects of temperature, solid content and pH on the stability of black carrot anthocyanins," Food Chemistry, vol. 101, no. 1, pp. 212–218, 2006.

22. P. Ekanayake, M. R. R. Kooh, N. T. R. N. Kumara et al., "Combined experimental and DFT–TDDFT study of photo-active constituents of Canarium odontophyllum for DSSC application," Chemical Physics Letters, vol. 585, pp. 121–127, 2013.

23. O. A. Janna, A. Khairul, M. Maziah, and Y. Mohd, "Flower pigment analysis of Melastoma malabathricum,"African Journal of Biotechnology, vol. 5, no. 2, pp. 170–174, 2006.

24. N. Turker, S. Aksay, and H. I. Ekiz, "Effect of storage temperature on the stability of anthocyanins of a fermented black carrot (Daucus carota var. L.) beverage: shalgam," Journal of Agricultural and Food Chemistry, vol. 52, no. 12, pp. 3807–3813, 2004.

25. M. Shahid, Shahid-ul-Islam, and F. Mohammad, "Recent advancements

in natural dye applications: a review," Journal of Cleaner Production, vol. 53, pp. 310–331, 2013.

26. Sandquist and J. L. McHale, "Improved efficiency of betanin-based dye-sensitized solar cells," Journal of Photochemistry and Photobiology A, vol. 221, no. 1, pp. 90–97, 2011.

27. G. Calogero, J.-H. Yum, A. Sinopoli, G. Di Marco, M. Grätzel, and M. K. Nazeeruddin, "Anthocyanins and betalains as light-harvesting pigments for dye-sensitized solar cells," Solar Energy, vol. 86, no. 5, pp. 1563–1575, 2012.

28. H. Hug, M. Bader, P. Mair, and T. Glatzel, "Biophotovoltaics: natural pigments in dye-sensitized solar cells," Applied Energy, vol. 115, pp. 216–225, 2014.

29. G. Calogero, G. Di Marco, S. Cazzanti et al., "Efficient dye-sensitized solar cells using red turnip and purple wild sicilian prickly pear fruits," International Journal of Molecular Sciences, vol. 11, no. 1, pp. 254–267, 2010.

30. M. H. Bazargan, "Performance of nano structured dye-sensitized solar cell utilizing natural sensitizer operated with platinum and carbon coated counter electrodes," Digest Journal of Nanomaterials & Biostructures, vol. 4, no. 4, pp. 723–727, 2009.

31. W. H. Lai, Y. H. Su, L. G. Teoh, and M. H. Hon, "Commercial and natural dyes as photosensitizers for a water-based dye-sensitized solar cell loaded with gold nanoparticles," Journal of Photochemistry and Photobiology A, vol. 195, no. 2-3, pp. 307–313, 2008.

Chapter 6

ULTRASTRUCTURE AND PIGMENTS OF IVY (HEDERA HELIX L.) VARIETIES WITH GREEN AND VARIEGATED LEAVES

Giorgio Manenti and Giuliano Tedesco

Istituto di Scienze Botaniche, Universita di Milano, Milano, Italy

INTRODUCTION

The chloroplast mutants have been largely studied from the ultrastructural, biochemical and genetic standpoint (WALLES 1971), but the information available on the variegated woody plants is very limited (VASIL'Ev 1962; SuN 1966; UEDA et al. 1966; VALANNE and VALANNE 1972; MANENTI 1975). We have studied the histology and the ultrastructure of the leaves of two variegated varieties of ivy (Hedera helix L.), in comparison with the leaves of a green variety. At the same time we have studied also the chlorophyll and carotenoid content.

MATERIALS AND METHODS

Young and well developed leaves of the following ivy varieties were examined: 1) a variety with green leaves (Fig. 1c); 2) a variety(« Aureomarginata »,Fig. 1b) with leaves showing a green centre, a white-yellow margin and intermediate pale green areas; a variety (« Goldheart », Fig. 1a) with leaves showing a white-yellow centre and a green margin.

Light and Electron Microscopy.

Small fragments of leaves were fixed in 3% glutaraldehyde in 0.1 M phosphate buffer, pH 6.9, at 4°C and postfixed in 1% buffered Os04 at 4°C. After dehydration with increasing alcohol concentrations, the samples were stained with uranyl acetate and embedded in epon-araldite mixture. 1 (J.m-thick sections, obtained with an LKB Pyramitone, were mounted on slides with a mixture of alcohol and glycerol and observed in the phase contrast under a Leitz Ortholux light-microscope.

Figure. 1: Leaves of the three ivy varieties, x 1 1a) « Goldheart »; 1b) « Aureomarginata »; 1c) variety with green leaves.

Ultrathin sections, obtained with an LKB Ultrotome II, were stained with lead citrate (REYNOLDS 1963) and observed with both an AEI EM6B electron microscope at 60 Kv and an Hitachi HU IIB electron microscope at 75 Kv.

Extraction and Estimation of Pigments.

Pigments were extracted with 85% acetone. Chlorophylls were estimated in the acetone extract by ARNON's equation (1949). Approximate carotenoid content was estimated by VON WETTSTEIN's equation (1957). Carotenes and xanthophylls were determined according to GooDWIN's method (1955). The ratio xanthophylls carotenes was calculated by graphic integration of the areas determined by their relative absorption spectra comprised between 360 and 500 nm.

RESULTS

Light Microscopy.

Green variety: the mesophyll is made of two layers of palisade cells and 5 to 7 layers of spongy cells (Fig. 2). « Aureomarginata » variety: the leaf blade is always thinner than in

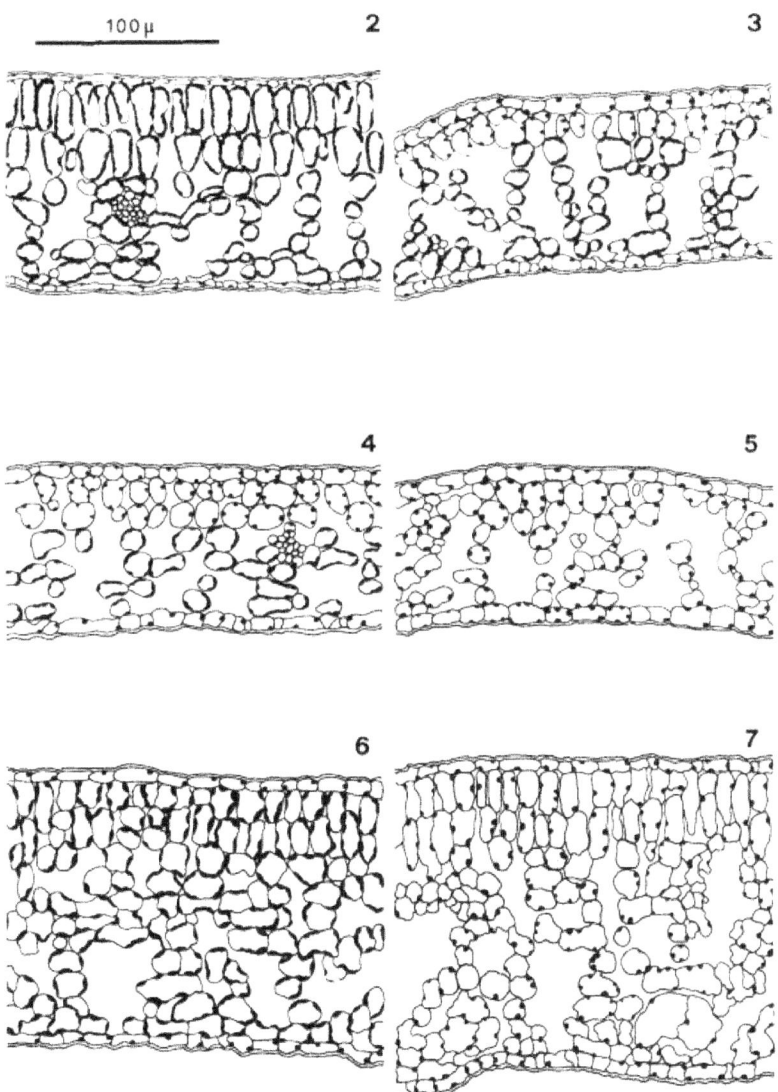

Figures. 2: Cross sections of leaves, x 310.

2) Variety with green leaves. 3) « Aureomarginata », green area. 4) « Aureomarginata », light green area. 5) « Aureomarginata », white-yellow area. 6) « Goldheart », green area. 7) « Goldheart », white-yellow area.

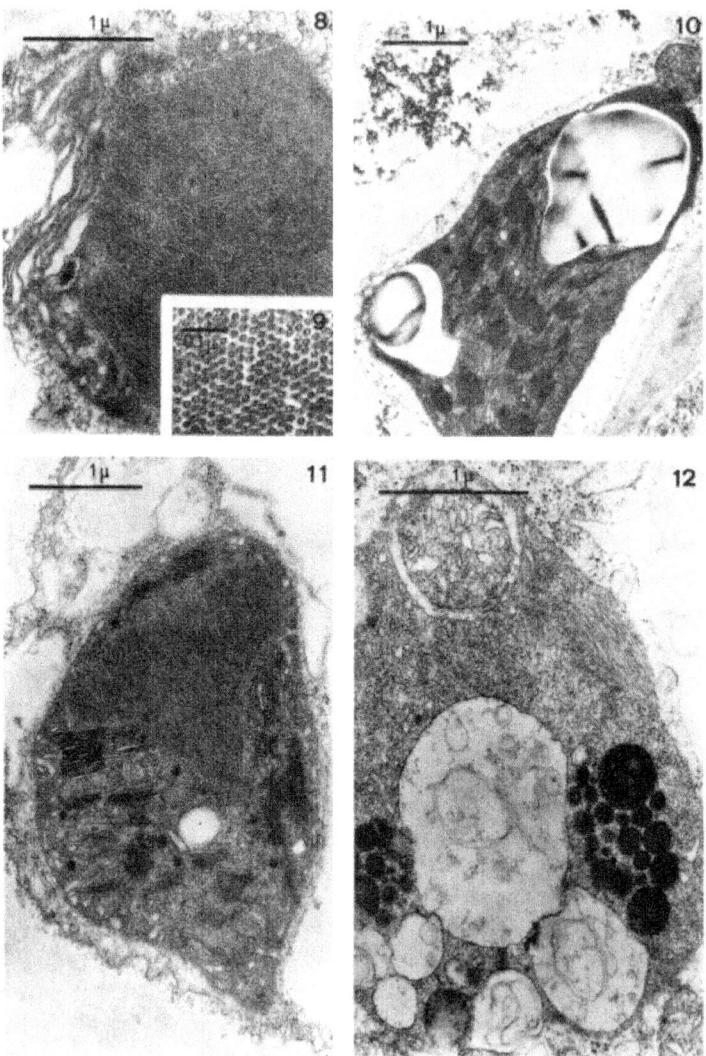

Figures. 3: Ultrastructure of the variety with green leaves.

8) Chloroplast in the epidermh, x 23,000. 9) Paracrystalline body in a chloroplast of the epidermis, x 82,100. 10) Chloroplast in the palisade mesophyll, x 14,000. Figs. 11-17. - Ultrastructure of « Aureomarginata ». 11) Chloroplast in the epidermis, x 20,100. 12) Plastid in a cell of the upper subepidermal layer of green areas, x 26,000. 13) Chloroplast in a cell of an inner layer of the mesophyll

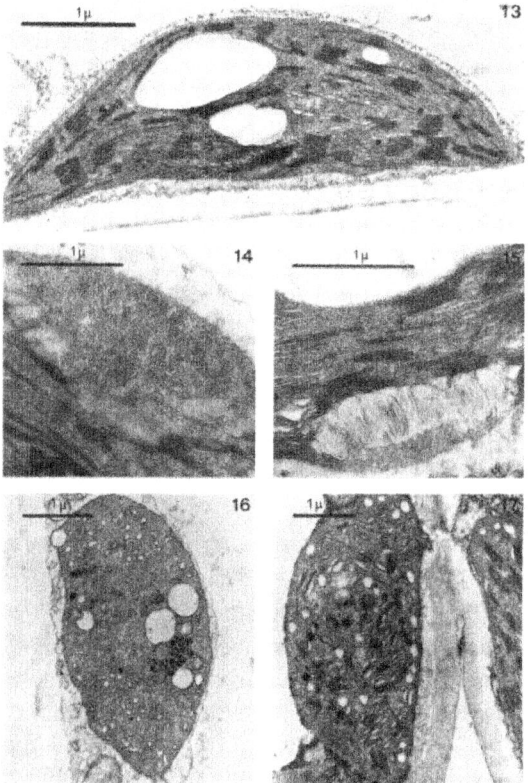

of green areas, x 27,300. 14 and 15) Portions of chloroplasts of the in‚ner layers of the mesophyll of green areas; notice the paracrystalline body and the altered chloroplast membrane. Fig. 14: x 24,000; Fig. 15: x 28,200. 16) Plastid in a cell of the upper second layer of the mesophyll of light green areas, x 16,800. 17) Chloroplast in a cell of the mesophyll of the peripheral areas of young leaves, x 14,600. green ivy. In the green areas (Fig. 3) the palisade cells are isodiametric and form two discontinous layers; the outer layer contains roundish plastids, similar to those of the epidermis, the inner layer contains lenticular plastids; the spongy mesophyll is made of isodiametric cells, containing lenticular plastids. In the pale green areas (Fig. 4), the palisade mesophyll is formed by two layers of isodiametric cells, containing roundish plastids; the spongy mesophyll cells contain lenticular plastids. In the white-yellow areas (Fig. 5), palisade mesophyll is not clearly identifiable and all mesophyll cells show roundish plastids. « Goldheart » variety: leaf blade (Figs. 6 and 7) is always much thicker than in the other two varieties, due to the increased number of spongy mesophyll layers. Palisade mesophyll is made of a double layer of elongated cells. In the green, marginal areas (Fig. 6) all the mesophyll cells possess lenticular plastids, while in the epidermis they are roundish. In the central white-yellow areas all the cells possess roundish plastids (Fig. 7).

Electron microscopy.

Green variety: the plastids of epidermal cells (Fig. 8) have a scarcely developed lamellar system and often show swollen thylakoids. Most of the volume is occupied by a paracrystalline body made of microtubules of 160-170 A diameter (Fig. 9). The mesophyll plastids show the typical chloroplast structures (Fig. 10). « Aureomarginata » variety: the plastids of epidermal cells have the same aspect as described in green ivy (Fig. 11). In the green areas the upper subepidermal layer shows markedly altered plastids (Fig. 12), which are completely devoid of thylakoids and contain many vescicles and osmiophylic globules. All the other mesophyll layers, including the lower subepidermal layer, have normal chloroplasts (Fig. 13). In pale green areas the doublelayered palisade mesophyll shows deeply altered plastids (Fig. 16), similar to those of the upper subepidermal layer of the green areas. The other mesophyll layers show normal chloroplasts. In the white-yellow areas all the mesophyll cells contain deeply altered plastids, as already described in Figs. 12 and 16. The plastids in all the layers of leaf blade contain a paracrystalline body similar to the one described in epidermal plastids of the green ivy (Figs. 11, 12, 14, 15 and 16). In young leaves, the leaf blade margin is characterized by a greenyellow color. In this case all the mesophyll cells possess normal chloroplasts. With ageing, the margin of the leaf blade loses color and the chloroplasts undergo a progressive degeneration of lamellar system (Fig. 17), which leads to the complete disappearance of photosynthetic lamellae. « Goldheart » variety: epidermal plastids show a normal lamellar system (Fig, 18). In the green, marginal areas the mesophyll chloroplasts appear normal (Fig. 19), while in white-yellow areas they are deeply altered (Fig. 21). All the plastids contain the paracrystalline body (Figs. 18, 20, 21, 22 and 23). In young leaves, in the central yellow-green areas the mesophyll plastids show single or paired thylakoids, without grana (Figs. 23 and 24). With ageing these areas become white-yellow and concomitantly the thylakoids degenerate.

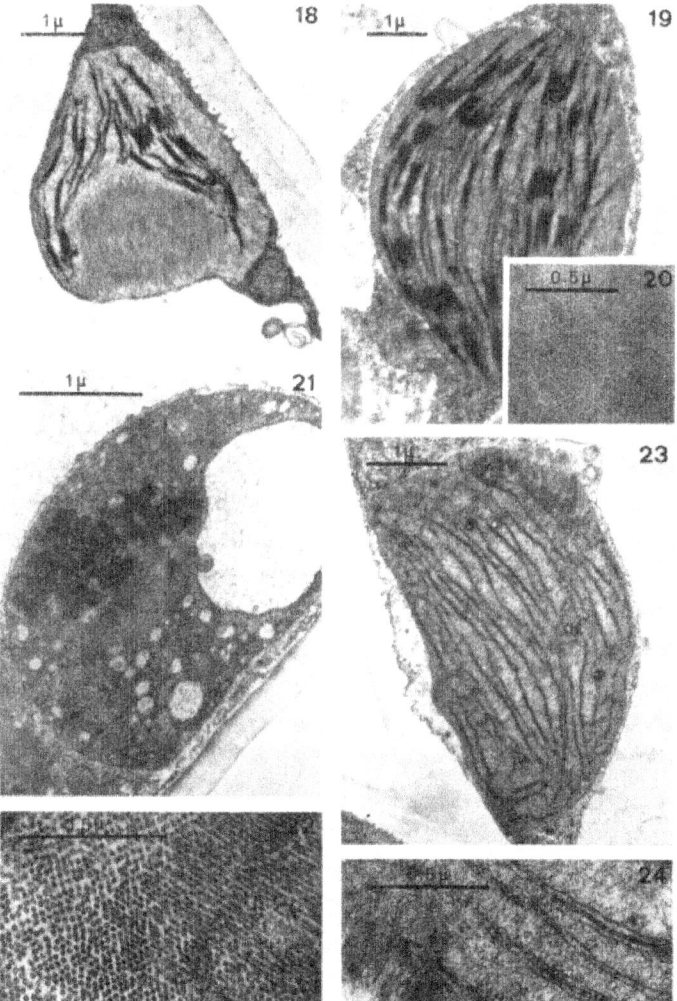

Pigments.

The absorption spectra of acetone extracts from green areas of « Aureomarginata » and « Goldheart » are similar to those obtained from the variety with green leaves (Fig. 25). Nevertheless, « Aureomarginata » has a pigment content clearly lower than the other two varieties. The pigment content of the white-yellow areas of variegated varieties is low as aspected; however, the curve is similar to that observed in the green areas. The white-yellow portion of « Goldheart » has a pigment content lower than the white-yellow portion of « Aureomarginata ». The absorption spectra of carotenes (Fig. 26) show the same shape in all the three varieties, no matter which leaf area is examined. On the contrary, the curve of xanthophylls (Fig. 27) obtained from green areas of « Aureomarginata » shows absorption maxima at wavelenghts different from those of the other two varieties. Absorption maxima of xanthophylls from white-yellow areas are the same as in corresponding green portions, but they are markedly different each other. Chlorophyll content in the green areas of « Aureomarginata » (Table I) is about half that of green portions of the pther two varieties. The whiteyellow portion of « Goldheart » has an amount of chlorophylls about 1/100 that of the green area of the same variety, while in « Aureomarginata » the white-yellow portion has a content of more than 1/10 that of the green portion. As regarding carotenoid content, the ratios between white-yellow and green areas are 1/5 in « Aureomarginata » and about 1/15 in « Goldheart ».

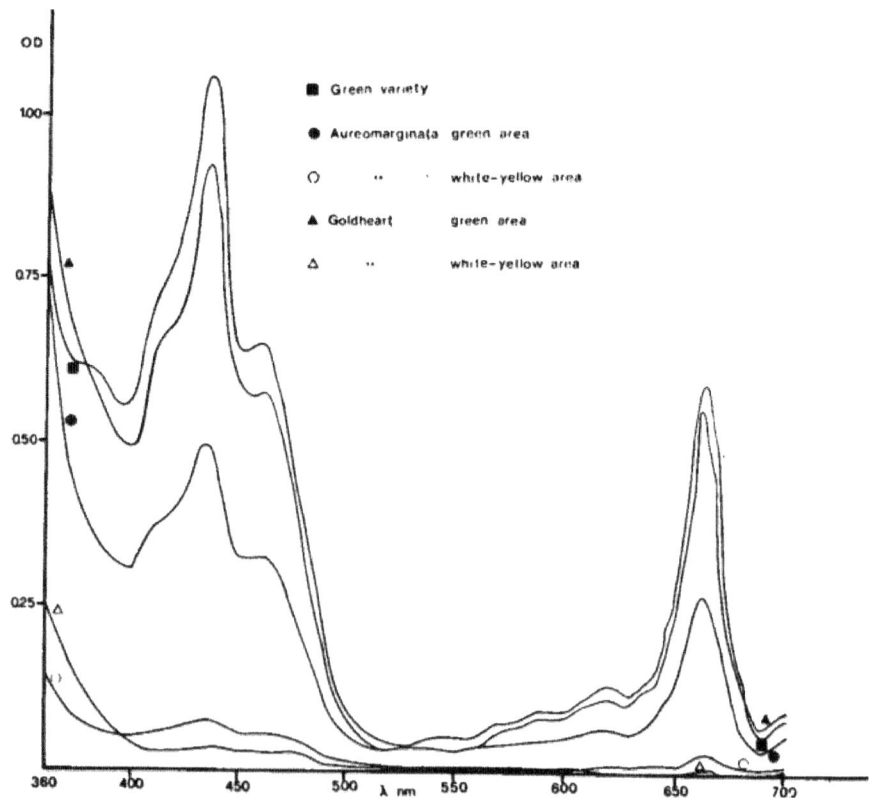

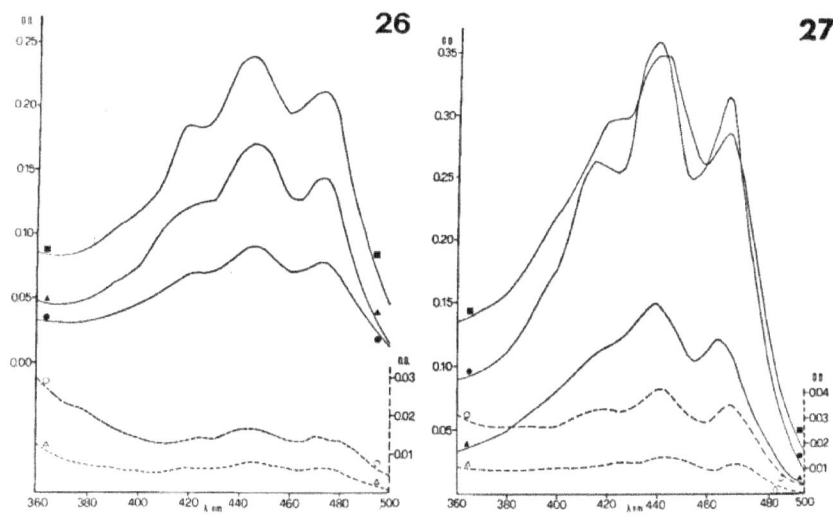

Table I: Chlorophyll and carotenoid content, chlorophyllsfcarotenoids and xantho-phyllsfcarotenes ratios of the three ivy varieties

	Total chlorophylls mg/g f.w.	Carotenoids mg/g f.w.	Chlorophylls carotenoids	Xanthophylls carotenes
Variety with green leaves	2.70	0.72	3.72	1.45
« Aureomarginata »				
green area	1.30	0.31	4.14	3.56
white-yellow area	0.15	0.06	2.27	2.84
« Goldheart »				
green area	2.41	0.61	3.93	0.79
white-yellow area	0.02	0.04	0.52	1.59

The three vanetles have almost similar chlorophylls/carotenoids ratios in the green portions. In white-yellow areas the ratio is lower than in green areas of the same variety: about 1/2 in « Aureomarginata » and about 1/8 in « Goldheart » rispectively. Concerning the xanthophylls/carotenes ratio, there are considerable differences between the two varieties with variegated leaves. In « Aureomarginata » the ratios are similar in green and in white-yellow areas; on the contrary, « Goldheart » has, in white-yellow portions, a ratio about double that of the green portions.

DISCUSSION

The first problem is which kind of chimaera the two ivy vanetles are. We can formulate two hypothesis about the origin of the variegation for « Aureomarginata » variety, according to the hystologic characteristics of leaf blade. First hypothesis: << Aureomarginata » is a mesochimaera of GWG type, where the « white » subepidermal layer is replaced by normal inner layers. Second hypothesis: « Aureomarginata » is a periclinal chimaera of << green over white » type where the cell layers with normal chloroplasts originate by periclinal division of the epidermis and replace some «white» inner layers. The validity of this second hypothesis depends on the possibility that in the apical meristem the epidermis divides periclinally only in the area where the lower layers of leaf will have their origin. Regarding this matter, we remember that in the apical meristem cells divide along directions determined by cellular dimensions and by external pressure on cells (DERMEN 1969). The first hypothesis seems less probable, because the phenomenon of perforation of an external layer by an inner one is generally believed to be very rare (TILNEY-BASSET 1963). So, in the leaves of « Aureomarginata », the mesophyll layers with normal plastids should be a result of periclinal divisions of the lower epidermis. The hystological interpretation of leaves in « Goldheart » seems easier; in fact the green, marginal area of the mesophyll most probably originate from the epidermis by periclinal divisions, as it happens, generally, in the periclinal « green over white » chimaeras (TILNEY-BAs SET 1963). Therefore, both the chimaeras seem to be of periclinal « green over white » type, although the leaf phenotype is considerably different in the two varieties. Nevertheless, the analogies between the leaves of the two varieties are limited to the type of chimaera, because the two varieties show big differences in their structure, ultrastructure and pigment composition. First of all, the mesophyll of « Aureomarginata » shows a considerable hypoplasy, while the mesophyll of « Goldheart » is normally developed (Figs. 3, 4, 5, 6 and 7). As regards chloroplast ultrastructure, the mesophyll of very young leaves of the « Aureomarginata » variety shows almost normal plastids in all the layers, i.e. with stroma lamellae and grana. Successively, a degeneration process interests, in the green areas, the first superior subepidermal layer, and, in the pale green areas, also the second one. This degeneration process starts in partially developed leaves, causes swelling of the thylakoids, total disappearance of the lamellar system, and appearance of many vescicles and osmiophylic globules in the plastids. Similar degenerations have been observed by GEROLA and DAs su (1960) in tobacco plants grown in nutritional deficiency. In the green areas, « Aureomarginata » has a pigment content clearly lower than the variety with green leaves, because the cells of the upper subepidermal layer contain altered plastids. The white-yellow portions of this variety show a pigment content markedly lower than the green portions; in spite of this, the curve shapes are the same both in the green and in the whiteyellow areas, both in the case of absorption spectra of acetonic extracts and of carotenes and xanthophylls. Moreover, the ratios chloropylls/carotenoids and xanthophylls/carotenes are the same in the green areas as in the white-yellow ones. The similarity of pigments between green and white-yellow areas, and the normal ultrastructure of all the chloroplasts in young leaves, related to histological observations, suggest that chloroplast degeneration in « Aureomarginata » is not caused by a block in the synthesis of a specific component of the plastid, but is caused by the deficiency of some factor having general importance in cell differentiation and metabolism and, therefore, necessary also for plastid development. In white-yellow areas of young leaves the chloroplast development is possible because this component can be supplied, at least in part, by the green portion of leaf blade; when the leaf reaches maturity, the amount produced by the green portion is perhaps no longer sufficient and, therefore, chloroplast degeneration occurs. Researches are in course to control this hypothesis. In « Goldheart » plants the palisade tissue is well developed both in the green and in the white-yellow areas. Mesophyll of white-yellow central portions of young leaves presents chloroplasts without grana, similar to those observed by BACHMANN et al. (1967) in a mutant of maize deficient in carotenoids. During leaf development, the lamellae degenerate and, at the end, the plastids are similar to those of white-yellow areas of « Aureomarginata ». In white-yellow portions of « Goldheart », the chlorophylls/carotenoids and xanthophylls/carotenes ratios are completely different from those of green areas. In particular, the xanthophyll content of white-yellow areas is higher than carotenes, the contrary occurring in green areas. Pigments distribution and histological observations suggest that in this case the alterations concern particularly the photosynthetic apparatus. In all the cells of the . two varieties with variegated leaves a paracrystalline body has been observed, both in normal and in altered plastids. On the contrary, in the variety with green leaves this paracrystalline structure has been detected only in plastids of epidermal cells. This structure is composed by microtubules having a diameter of 160-170 A, regularly oriented and without limiting membrane, but, apparently, related to '.:he photosynthetic membranes (Figs. 14 and 15). Although different kinds of paracrystalline inclusions have been described in the plastids (SHUMWAY, WEIER and STOCKING 1967; LARSSON, CoLLIN and ALBERTSSON 1973; LEE and THOMPSON 1973; DE GREEF and VERBELEN 1973, 1974; BRANDAO and SALEMA 1974; EsAu 1975), the ultrastructure of the paracrystalline body that we have observed is similar, in our opinion, only to the microtubules seen in chloroplasts of Sedum sp. (BRANDAO and SA LEMA 197 4) and to the paracrystalline bodies observed by LEE and THOMPSON (197 3) in plastids of Kalanchoe. All the authors cited above consider these paracrystalline bodies composed mainly by the Fraction I protein of the plastid stroma (i.e. ribulose 1 ,5-diphosphate carboxilase) and caused by an osmotic unbalance of the cell. This interpretation has been confirmed, in particular, by DE GREEF and VERBELEN (1974); they observed crystalline inclusions in bean plastids following treatment with 2,4-DNP. Since 2,4-DNP is an uncoupler of oxydative phosphorilation and the osmotic pressure of the cell is controlled by ATP-mediated selective permeability of membranes, these authors suggest that the inhibition of the energetic metabolism causes an alteration of membrane permeability; later on paracrystalline structures can originate. This interpretation could apply also in the case of our inclusions. In fact, in the variety with green leaves these bodies are present only in epidermic cells, while in variegated varieties they occur both in normal and altered plastids of all the mesophyll cells. The varieties with variegated leaves are characterized by the presence of many cells unable to photosynthetize, therefore the content of available ATP is probably low; consequently, membrane permeability could be altered, osmotic pressure modified and crystal formation induced. In many cases we been observed alterations of the plastid membranes (Figs. 15 and 23). Although we have not carried out physiological studies on the ongm of these paracrystalline bodies, we think that the one we have already discussed could be useful working hypotheses for further researches.

ACKNOWLEDGEMENTS

The authors are grateful to Prof. F. M. GEROLA for his helpful advice and suggestions throughout this work and to Mrs. M. FoNTANA, R. BosoNI and A. GRIPPO for technical assistence.

REFERENCES

1.   Arnon D. I., 1949.—Copper enzymes in isolated chloroplasts. Plant Physiol., 24: 1–12.

2.   Bachmann M. D., Robertson D. S., Bowen C. C. and Anderson J. C., 1967.—Chloroplast development in pigment deficient mutants of maize. J. Ultrastruct. Res., 21: 41–60.

3.   Brandao I. and Salema R., 1974.—Microtubules in chloroplasts of a higher plant (Sedum sp.). J. Submicr. Cytol., 6: 381–390.

4.   De Greef J. A. and Verbelen J. P., 1973.—Physiological stress and crystallites in leaf plastids of Phaseolus vulgaris L. Ann. Bot., 37: 593–596.

5. De Greef J. A. and Verbelen J. P., 1974.—Plastid stroma crystals induced by 2,4-DNP treatment of Phaseolus vulgaris L. Proceedings of the third International Congress on photosynthesis. ed. by M. Avron. Elsevier Scientific Publishing Company (Amsterdam), 2131–2137.

6. Dermen H., 1969.—Directional cell division in shoot apices. Cytologia, 34: 541–558.

7. Esau K., 1975.—Crystalline inclusion in thilakoids of spinach chloroplasts. J. Ultrastruct. Res., 53: 235–243.

8. Gerola F. M. e Dassù G., 1960.—Variazioni delle infrastrutture dei cloroplasti di tabacco (Nicotiana tabacum) durante l'imbianchimento sperimentalmente indotto delle foglie e il loro successivo rinverdimento. Caryologia, 13: 398–410. [Taylor & Francis Online]

9. Goodwin T. W., 1955.—Carotenoids. In: Modem Methods of plant analysis, 3. Springer Verlag, Berlin.

10. Larsson C., Collin C. and Albertsson P., 1973.—The fine structure of chloroplast stroma crystals. J. Ultrastruct. Res, 45: 50–58.

11. Lee R. E. and Thompson A., 1973.—The stromacentre of plastids of Kalanchöe pinnata Persoon. J. Ultrastruct. Res., 42: 451–456.

12. Manenti G., 1975.—The structure of variegated leaves of Acer negundo L.: a light and electron microscope study. Israel J. Bot., 24: 61–70.

13. Reynolds E. S., 1963.—The use of lead citrate at high pH as an electron opaque stain in electron microscopy. J. Cell Res, 17: 208–212.

14. Shumway L.K., Weier T. E. and Stocking C.R., 1967.—Crystalline structures in Vicia faba chloroplasts. Planta, 76: 182–189.

15. Sun C.N., 1966.—Ultrastructure of plastids in normal and variegated leaves in Abutilon striatum. Cytologie, 31: 452–456.

16. Tilney-Basset R. A.E., 1963.—The structure of periclinal chimaeras. Heredity, 18: 265–285.

17. Ueda R, Toyama S. and Saikawa M, 1966.—Electron microscope studies of the development of plastids in variegated leaves of the Japanese Spindle-Tree. Proceedings of Sixth International Congress for Electr. Micr. Ed. by Ryozi Uyeda-Maruzen Co. LTD (Tokyo), 379–380.

18. Valanne N. and Valanne T, 1972.—Structure of plastids of a variegated Betula pubescens mutans. Canad. J. Bot, 50: 1835–1839.

19. Vasil'ev A.E., 1962.—The structure of leaves of sectorial chimaeras shoots (Populus spp.). Bot. Zh, 47: 1661 1666.

20. Walles B, 1971.—Plastid inheritance and mutations. In: Structure and function of chloroplasts. ed. by M. Gibbs-Springer-Verlag, Berlin, Heidelberg, New York.

21. Von Wettstein D, 1957.—Chlorophyll-letale und der submikroskopische Formwechsel der Piastiden. Exp. Cell Res, 12: 427–506.

Chapter 7

UV-B RADIATION IMPACTS SHOOT TISSUE PIGMENT COMPOSITION IN ALLIUM FISTULOSUM L. CULTIGENS

Kristin R. Abney,[1] Dean A. Kopsell,[1] Carl E. Sams,[1] Svetlana Zivanovic,[2] andDavid E. Kopsell[3]

[1]Plant Sciences Department, The University of Tennessee, 2431 Joe Johnson Drive, Knoxville, TN 37996, USA

[2]Department of Food Science and Technology, The University of Tennessee, 2605 River Drive, Knoxville, TN 37996, USA

[3]Department of Agriculture, Illinois State University, Normal, IL 61790, USA

ABSTRACT

Plants from the Allium genus are valued worldwide for culinary flavor and medicinal attributes. In this study, 16 cultigens of bunching onion (Allium fistulosum L.) were grown in a glasshouse under filtered UV radiation (control) or supplemental UV-B radiation [$7.0\,\mu mol \cdot m^{-2} \cdot s^{-2}$ ($2.68\,W \cdot m^{-2}$)] to determine impacts on growth, physiological parameters, and nutritional quality. Supplemental UV-B radiation influenced shoot tissue carotenoid concentrations in some, but not all, of the bunching onions. Xanthophyll carotenoid pigments lutein and β-carotene and chlorophylls a and b in shoot tissues differed between UV-B radiation treatments and among cultigens. Cultigen "Pesoenyj" responded to supplemental UV-B radiation with increases in the ratio of zeaxanthin + antheraxanthin to zeaxanthin + antheraxanthin + violaxanthin, which may indicate a flux in the xanthophyll carotenoids towards deepoxydation, commonly found under high irradiance stress. Increases in carotenoid concentrations would be expected to increase crop nutritional values.

INTRODUCTION

Fruits and vegetables have varying levels of phytonutrients, in addition to vitamins and minerals. Two important classes of phytonutrients are carotenoid and chlorophyll pigments. The primary carotenoids found in leaf tissue of most plant species include zeaxanthin, antheraxanthin, violaxanthin, lutein, β-carotene, and neoxanthin [1]. Chlorophylls are the dominant pigments in plants and serve primary roles in photosynthesis. These compounds are very effective antioxidants and help prevent certain types of cancers and aging eye diseases like macular eye degeneration [2, 3]. However, the chemical structures of these pigments also give them the ability to donate electrons and effectively become prooxidants under certain conditions [4]. Alliumspecies contain chlorophyll and carotenoid pigments in shoot tissues [5] and also contain different levels of sulfur-containing compounds which prevent certain cancers [6]. While all higher plants contain chlorophylls and carotenoids, genetic variations for pigment accumulations exist both within and among plant species. Within any given crop species, there can be multiple landraces, accessions and cultivars, or, collectively, cultigens. These variations are important to advancements in plant development programs for increased nutrition, disease prevention, or other factors. However, cultigens will react differently under almost any given stress.

The absorption of light by chlorophyll and antenna pigments and the transfer of excitation energy to the reaction centers of PSII and PSI are the initial steps in photosynthesis. The photosynthetic apparatus has the ability to react to many different environmental stimuli, especially changes in light intensity. Under conditions of high light stress, photosynthetic systems are saturated, and excess energy needs to be diverted to avoid potential damage [7]. Carotenoids are unsaturated long chain polycarbons which protect the photosynthetic apparatus from high light excitation by quenching free radicals, functioning in nonphotochemical quenching, and dissipating excess thermal energy [7, 8]. The xanthophyll cycle, or violaxanthin cycle, is the mechanism by which plants regulate light energy available for photosynthesis. In intense light situations, violaxanthin is rapidly and reversibly converted to zeaxanthin, via antheraxanthin. Zeaxanthin is a direct quencher of chlorophyll excited states and can prevent photooxidative stress and lipid peroxidation [7].

Higher amounts of UV-A (380–320 nm) and UV-B radiation (280–320 nm) may influence the accumulation of plant compounds used to combat light stress. Carotenoid metabolites not only protect plants from excess UV radiation but also can protect humans from UV radiation when translocated to subdermal skin tissues [9]. What remains uncertain is the impact of increased

UV radiation on growth and development and nutritional values of cultivated crops [10]. Previous studies have demonstrated impacts of UV radiation on plant performance, cellular structures, and pigment accumulations. In a study by Yuan et al. [11], 20 cultivars of wheat (Triticum aestivum L.) were grown under UV-B radiation stress to determine possible detrimental influences. Most wheat cultivars responded negatively to UV-B radiation; however, several cultivars showed increases in plant height and biomass. Structural changes like ruptured chloroplast envelopes have been noted in UV-sensitive rice cultivars (Oryza sativa L.) when exposed to UV stress [12]. Increases in UV radiation can delay flowering and harvest times among different cultigens of bush beans (Phaseolus vulgarisL.) [13]. Bush beans grown under UV radiation showed decreases in fruit size and yield when compared to cultivars not grown under UV radiation stress. Tomatoes (Solanum lycopersicum cv. DRW 5981) grown using UV-B blocking filters showed increases in lycopene and β-carotene, while fruits of the same variety showed decreases in lycopene, phytoene, and phytofluene when grown without UV-B blocking filters [14]. The tomato cultivar "HP1" accumulated more than twice the amount of lycopene in fruit tissues when grown under no UV-B radiation. Results from such studies demonstrate impacts on nutritional quality from excess UV radiation.

Allium species are valued worldwide for culinary flavor and medicinal attributes. Plants in this genus have been important to multiple cultures for centuries. Alliums have high levels of nutritionally important secondary plant metabolites which convey numerous health benefits. For example, bulb onions (Allium cepa L.) contain high levels of flavonoids [15], S-alk(en)yl-L-cysteine sulfoxides [16], and a variety of volatile antioxidant compounds [17]. However, no studies to date have measured the impact of UV radiation on the production of nutritionally important pigments in Alliums. Allium fistulosum is consumed, in part, for its shoot tissues as well as pseudostems. Carotenoid and chlorophyll compounds are present in the shoot tissues ofA. fistulosum [5, 18]. Therefore, the objectives of this project were to examine both environmental and genetic responses to elevated UVB radiation among a large subset of A. fistulosum cultigens. Responses were noted for plant height, shoot tissue biomass, photochemical efficiency (F_v/F_m), and concentrations of carotenoid and chlorophyll pigments in the shoot and pseudostem tissues.

METHODS AND MATERIALS

Plant Culture

On December 16, 2008, seeds of 16 different A. fistulosum cultigens were sown in 15 cm pots holding soilless media in a glasshouse in Knoxville,

TN USA (35°96'N Lat.), which blocked UV-wavelengths (280–380 nm). The photosynthetically active radiation (PAR) in the glasshouse averaged 540 μmol·m^{-2}·s^{-2} (Apogee Nanologger model ANL, Apogee Instruments, Inc., Roseville, CA USA). The cultigens included eight accessions [PI 274254-05GI, PI 462345-05GI (Jionji Negi), PI 546343-90U01 (GA-C 76), PI 546228-06GI (Improved Beltsville Bunching), PI 280562-04GI (Pesoenyj), PI 436539-06GI (Zhang Qui Da Cong), PI 462357-06GI (Shounan), and G 30393-06GI] from the USDA-ARS National Plant Germplasm Repository (Geneva, NY USA); four cultivars ("Long White Bunching," "Feast, Performer," and "Parade") from Seedway, LLC (Hall, NY, USA); and four cultivars ("White Spear," "Evergreen Hardy White," "Deep Purple," and "Ishikura Improved F1") from Johnny's Selected Seeds (Winslow, ME, USA). The seedlings were watered daily for the duration of the experiment. On January 10, 2009, the seedlings were thinned to two plants per pot and fertilized with a nutrient solution containing (mg·L^{-1}): N (105), P (91.5), K (117.3), Ca (80.2), Mg (24.6), S (32.0), Fe (0.5), B (0.25), Mo (0.005), Cu (0.01), Mn (0.25), and Zn (0.025) [19]. Each pot was fertilized once a week for the duration of the experiment with 100 mL of nutrient solution. The experimental design was a split plot arranged as a randomized complete block. UV-B treatments were the main plots, and A. fistulosum cultigens were the subplots. Six individual plants per cultigen composed a replication, with four replications randomly assigned to each UV-B treatment.

Supplemental UV-B radiation (313 nm) was provided by banks of commercially available UV-B 313 lamps (Q-Panel Lab Products, Cleveland, OH, USA), and treatment began on January 27, 2009 delivering 7.0 μmol·m^{-2}·s^{-2} (2.68 W·m^{-2}) of UV-B (Spectroradiometer Model SPEC-UV/PAR, Apogee Instruments, Inc., Roseville, CA, USA) to the treated plants. To control pests in the greenhouse, the beneficial insect species of Hypoaspis miles and Neoseiulus cucumeris were used to control thrips, while Orius insidiosus was used to help control aphids. These insects were first released on January 23, 2009, and were released every two weeks thereafter.

On March 3, 2009, all of the bunching onion cultigens were harvested. Six plants were harvested from each replication. Fresh weights and plant heights were taken and averaged for each replication. One measure of F_v/F_m was taken from each of the harvest plants at the midpoint of plant height using a modulated fluorometer (OS1-F1 Modulated Fluorometer, Opti-Sciences, Hudson, NH, USA). The F_v/F_m value is an indication of photoinhibition and overall plant health. All plants were harvested, and pseudostem and leaf tissue were separated. The samples were immediately placed in a −20∘ C freezer before being moved to a −80∘ C freezer within 8 h.

Pigment Extraction and Determination

Tissue pigments were extracted according to Kopsell et al. [20] and analyzed The Scientific World Journal 3 according to Kopsell et al. [5]. The samples were freeze-dried and ground with a mortar and pestle with liquid nitrogen. A 0.10 g subsample was rehydrated with 0.8 mL of ultrapure H_2O. The samples were incubated for 20 min, before 0.8 mL of ethyl-β-8 -apo-carotenotate (Sigma Chemical Co., St. Louis, MO, USA) was added as an internal standard to establish extraction efficiency. For pigment extraction, 2.5 mL of tetrahydrofluran was added to the sample. Using a Potter-Elvehjem tissue grinding tube (Kontes, Vineland, NJ, USA), the samples was homogenized in an ice bath to dissipate heat generated from maceration. The tubes were then centrifuged in a clinical centrifuge (Centrific Model 225, Fisher Scientific, Pittsburg, PA, USA) for 3 min at 500_{gn}. The supernatant was removed, and the pellet was rehydrated with 2.0 mL tetrahydrofluran. This procedure was repeated twice more until the supernatant was colorless. The combined supernatants were reduced to 0.5 mL under a stream of nitrogen gas and brought to a final volume of 5 mL with methanol. The samples were then filtered through a 0.2 μm Econofilter PTFE 25/20 polytetrafluoroethylene filter (Agilent Technologies, Wilmington, DE, USA) using a 5 mL syringe. A 1.5 mL aliquot was put into an amber vial and capped prior to high performance liquid chromatography (HPLC) analysis.

An Agilent 1200 series HPLC unit with a photodiode array detector (Agilent Technologies, Palo Alto, CA, USA) was used for pigment separation (Figure 1). The column used was a 250 × 4.6 mm i.d., 5 μm analytical scale polymeric RP-C_{30}, with a 10 × 4.0 mm i.d. guard cartridge and holder (ProntoSIL, MAC-MOD Analytical Inc., Chadds Ford, PA, USA), which allowed for effective separation of chemically similar compounds. The column was maintained at 30°C using a thermostated column compartment. All separations were achieved isocratically using a binary mobile phase of 11% methyl tert-butyl ether (MTBE), 88.99% MeOH, and 0.01% triethylamine (TEA) (v/v/v). The flow rate was 1.0 mL/min, with a run time of 53 min. There was a 2 min equilibration prior to the next injection. Eluted compounds from a 10 μL injection loop were detected at 453 nm (carotenoids, internal standard, chlorophyll b) and 652 nm (chlorophyll a). Data were collected, recorded, and integrated using ChemStation Software (Agilent Technologies). Peak assignments for each pigment were performed by comparing retention times and line spectra obtained from the photodiode array detection using external standards. Standards included antheraxanthin, neoxanthin, lutein, violaxanthin, zeaxanthin, and β-carotene, chlorophyll a and chlorophyll b (ChromaDex Inc., Irvine, CA, USA). The concentrations of the external standards were determined spectrophotometrically using a procedure by Davies and Köst [21]. Pigment data is presented on a fresh mass (FM) basis.

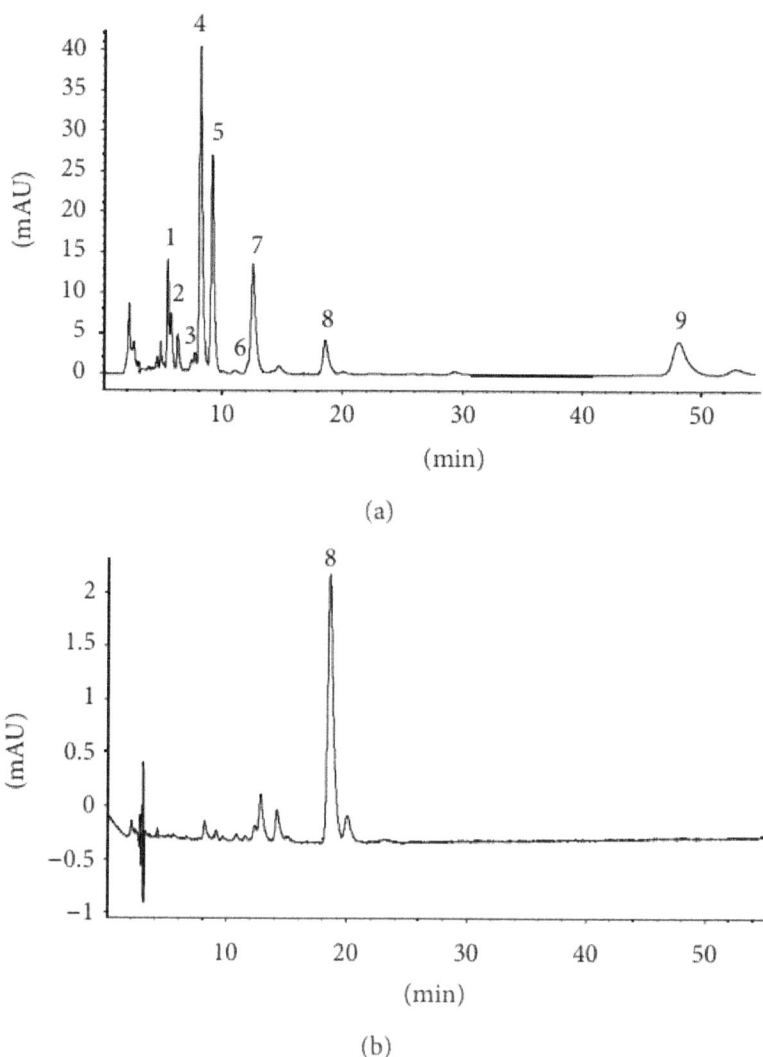

Figure 1: HPLC chromatogram of Allium fistulosum L. leaf (a) and pseudostem (b) tissues at 453 nm. Retention times (min) for the pigments were (1) violaxanthin, 5.52 min; (2) neoxanthin, 5.81 min; (3) antheraxanthin, 7.59 min; (4) chlorophyll b, 8.51 min; (5) lutein, 9.33 min; (6) zeaxanthin, 11.31 min; (7) chlorophyll a, 13.90 min; (8) ethyl-β-8-apo-carotenoate (internal standard), 19.32 min; and (9) β-carotene, 48.46 min. HPLC conditions are described in the text.

Statistical Analyses

Statistical analyses were completed using the GLM procedure of SAS (v. 9.1, SAS Institute, Cary, NC, USA). Cultigen means within each treatment were separated by least significant difference (LSD) at $\alpha = 0.05$. Differences between cultigens means between treatments were detected by using Student's t-test ($P = 0.05$) using JMP (v 7.0.1, SAS Institute).

RESULTS AND DISCUSSION

Shoot Tissue Biomass

Significant differences were found among cultigens ($F = 6.67$, $P < 0.001$) for shoot tissue height, but no differences were found between the UV-B treatments or the interaction between the cultigen and UV-B radiation treatment (Table 1). Only one cultigen (GA-C 76) differed significantly between UV-B radiation treatments for shoot tissue height. "Long White Bunching" demonstrated the greatest growth in shoot tissue height under both UV-B radiation treatments, while "G 30393- 06GI" had the shortest final shoot tissue height. There were differences in shoot tissue FM between UV-B radiation treatments ($F = 238.10$, $P < 0.001$) and among cultigens ($F = 11.09$, $P < 0.001$), but no difference in the treatment by cultigen interaction (Table 1). "Deep Purple," "Feast," "GAC 76," "Ishikura Improved F1," "Improved Beltsville Bunching," "Jionji," "Long White Bunching," "Parade," "Performer," "Pesoenyj," "Shounan," "White Spear," "274254-05GI," and "G 30393-06GI" all showed decreases in shoot tissue FM with exposure to the UV-B radiation treatment. Significant decreases in shoot tissue biomass from the UV-B treatment would indicate a radiational stress had occurred in the bunching onion cultigens in the current study. The cultigens with the greatest shoot tissue FM accumulations were "LongWhite Bunching" and "Improved Beltsville Bunching" (Table 1).

Table 1: Mean values[a] for shoot tissue height (cm) and fresh biomass (g) for Allium fistulosum L. cultigens grown under supplemental UV-B (313 nm) light [7.0 mol·m^{-2}·s^{-2} (2.68 W·m^{-2}); UV-B] or UV-filtered (control) light in a glasshouse in Knoxville, TN, USA (35°96′N Lat.)

Cultigen	Shoot tissue height (cm)			Shoot tissue fresh biomass (g)		
	UV-B	Control	Pr > |t|[b]	UV-B	Control	Pr > |t|
Deep Purple	44.71 ± 5.49	45.19 ± 4.52	ns	64.54 ± 17.02	97.80 ± 17.44	P = 0.034
Evergreen Hardy White	39.11 ± 1.96	41.05 ± 5.38	ns	38.65 ± 7.80	70.78 ± 8.08	ns
Feast	38.58 ± 3.24	39.96 ± 2.40	ns	48.79 ± 15.99	90.46 ± 13.85	P = 0.008
GA-C 76	38.82 ± 1.91	44.13 ± 3.70	P = 0.043	41.90 ± 2.66	73.00 ± 15.53	P = 0.008
Ishikura improved F1	39.63 ± 1.12	41.17 ± 3.13	ns	54.93 ± 9.22	99.80 ± 9.47	P = 0.001
Improved Beltsville Bunching	44.26 ± 1.69	48.06 ± 2.19	ns	76.86 ± 9.82	110.42 ± 19.39	P = 0.021
Jionji Negi	36.20 ± 1.82	38.63 ± 1.54	ns	38.05 ± 4.73	62.04 ± 8.23	P = 0.002
Long White Bunching	49.68 ± 2.09	50.38 ± 4.62	ns	74.07 ± 8.73	116.01 ± 20.24	P = 0.009
Parade	42.13 ± 2.60	36.03 ± 9.40	ns	53.89 ± 10.73	90.81 ± 16.87	P = 0.010
Performer	39.37 ± 2.36	39.42 ± 3.10	ns	54.74 ± 11.08	86.25 ± 16.66	P = 0.020
Pesoenyj	39.44 ± 4.23	44.54 ± 4.03	ns	29.22 ± 9.30	58.76 ± 6.37	P = 0.002
Shounan	36.72 ± 3.02	37.14 ± 2.54	ns	39.17 ± 3.56	63.32 ± 4.02	P = 0.001
White Spear	40.06 ± 2.79	42.42 ± 1.72	ns	53.16 ± 14.41	87.42 ± 10.58	P = 0.009
Zhang Qui Da Cong	36.62 ± 2.88	37.87 ± 3.80	ns	53.57 ± 19.51	84.72 ± 16.67	ns
274254-05GI	42.48 ± 1.86	43.79 ± 4.40	ns	47.63 ± 4.34	84.65 ± 6.98	P = 0.001
G 30393-06GI	36.62 ± 3.40	35.67 ± 1.31	ns	50.96 ± 8.63	81.69 ± 12.01	P = 0.006
LSD$_{0.05}$[c]	5.77	5.80		10.72	19.33	

[a]Composition of n = 6 plant samples from 4 replications ± standard deviation. [b]Significance based on paired Student's t-test among treatments; ns: not significant. [c]LSD for differences between cultivar means α = 0.05.

Shoot Tissue Carotenoid Pigment Concentrations

No carotenoid pigments were measured in the pseudostem tissues of any of the bunching onion cultigens (Figure 1; data not shown). Kopsell et al. [5] also reported no carotenoid pigmentation present in bunching onion pseudostem tissues. Shoot tissue zeaxanthin differed significantly among the bunching onion cultigens ($F = 4.07$; $P < 0.001$) (Table 2). However, there were no significant changes in shoot tissue zeaxanthin in response to UV-B treatment, or the interaction of the UV-B treatments and cultigens. Only the cultigens of "G 30393-06GI" and "Feast" showed an increase in shoot tissue zeaxanthin under the supplemental UV-B radiation, as compared to control. The ranges of zeaxanthin concentrations in the bunching onions under supplemental UV-B were from 0.08 mg/100 g FM for "Deep Purple" and "White Spear" to 0.16 mg/100 g FM for "Improved Beltsville Bunching". Cultigen "Pesoenyj" had the highest concentration of zeaxanthin among plants grown without supplemental UV radiation at 0.19 mg/100 g FM, while "Feast" and "Evergreen Hardy White" had the lowest zeaxanthin concentrations at 0.07 mg/100 g FM. Increases in zeaxanthin could be an indication that the plants experienced radiational stress from the UV-B treatment. Plant responses through increased zeaxanthin concentrations would be expected to help dissipate excess energy from the photosystems [7].

Table 2: Mean values[a] for shoot tissue zeaxanthin, violaxanthin, and antheraxanthin (mg/100 g fresh mass) for Allium fistulosum L. cultigens grown under supplemental UV-B (313 nm) light [7.0 mol·m^{-2}·s^{-2} (2.68 W·m^{-2}); UV-B] or UV-filtered (control) light in a glasshouse in Knoxville, TN, USA (35°96′N Lat.)

Cultigen	Zeaxanthin			Violaxanthin			Antheraxanthin		
	UV-B	Control	Pr > |t|[b]	UV-B	Control	Pr > |t|	UV-B	Control	Pr > |t|
				mg/100 g fresh mass					
Deep Purple	0.08 ± 0.02	0.10 ± 0.03	ns	1.25 ± 0.36	0.88 ± 0.42	ns	1.07 ± 0.31	0.78 ± 0.23	ns
Evergreen Hardy White	0.10 ± 0.03	0.07 ± 0.02	ns	1.73 ± 0.61	1.20 ± 0.25	ns	1.03 ± 0.18	0.78 ± 0.14	ns
Feast	0.11 ± 0.01	0.07 ± 0.01	P = 0.010	1.35 ± 0.82	0.66 ± 0.58	ns	1.15 ± 0.27	0.92 ± 0.19	ns
GA-C 76	0.12 ± 0.02	0.08 ± 0.02	ns	2.04 ± 0.19	1.34 ± 0.20	P = 0.002	1.92 ± 0.65	1.27 ± 0.47	ns
Ishikura Improved F1	0.10 ± 0.05	0.09 ± 0.02	ns	1.75 ± 0.94	1.10 ± 0.38	ns	0.81 ± 0.50	0.59 ± 0.13	ns
Improved Beltsville Bunching	0.16 ± 0.02	0.15 ± 0.04	ns	1.62 ± 0.54	1.25 ± 0.14	ns	0.99 ± 0.38	0.72 ± 0.21	ns
Jionji Negi	0.12 ± 0.06	0.13 ± 0.03	ns	1.54 ± 0.25	1.51 ± 0.35	ns	1.20 ± 0.25	0.78 ± 0.50	ns
Long White Bunching	0.12 ± 0.03	0.11 ± 0.03	ns	0.89 ± 0.18	0.81 ± 0.42	ns	0.82 ± 0.13	0.89 ± 0.14	ns
Parade	0.10 ± 0.02	0.11 ± 0.02	ns	1.23 ± 0.50	1.02 ± 0.61	ns	0.86 ± 0.16	0.79 ± 0.24	ns
Performer	0.09 ± 0.02	0.10 ± 0.02	ns	0.59 ± 0.52	1.45 ± 0.60	ns	0.87 ± 0.06	0.85 ± 0.32	ns
Pesoenyj	0.12 ± 0.03	0.19 ± 0.06	ns	1.93 ± 0.33	2.35 ± 0.82	ns	1.38 ± 0.30	1.35 ± 0.52	ns
Shounan	0.13 ± 0.05	0.09 ± 0.01	ns	1.87 ± 0.81	1.04 ± 0.36	ns	1.18 ± 0.40	0.63 ± 0.30	ns
White Spear	0.08 ± 0.02	0.08 ± 0.03	ns	1.00 ± 0.60	0.77 ± 0.47	ns	0.74 ± 0.13	0.60 ± 0.09	ns
Zhang Qui Da Cong	0.11 ± 0.04	0.12 ± 0.02	ns	1.29 ± 0.48	1.30 ± 0.41	ns	0.81 ± 0.30	0.74 ± 0.06	ns
274254-05GI	0.13 ± 0.04	0.15 ± 0.03	ns	1.64 ± 0.42	1.27 ± 0.52	ns	0.79 ± 0.14	0.71 ± 0.33	ns
G 30393-06GI	0.13 ± 0.03	0.08 ± 0.02	P = 0.016	0.60 ± 0.51	0.50 ± 0.43	ns	1.07 ± 0.35	0.67 ± 0.14	ns
LSD$_{0.05}$[c]	ns	0.04		0.74	0.70		0.44	0.44	

[a] Composition of n = 6 plant samples from 4 replications ± standard deviation. [b] Significance based on paired Students t-test among treatments; ns: not significant. [c] LSD for differences between cultivar means α = 0.05.

Shoot tissue violaxanthin responded significantly to both UV-B radiation treatment ($F = 6.76$; $P = 0.0109$) and cultigen ($F = 4.42$, $P < 0.001$), but not to the interaction between treatment and cultigen (Table 2). Many of the bunching onion cultigens showed higher concentrations of violaxanthin in response to UV-B radiational supplementation. However, only one cultigen had significant increases in violaxanthin concentrations (GA-C 76) in response to UV-B radiation treatment. Increases in violaxanthin in bunching onions grown under UV-B radiation may suggest that these cultigens may not be as susceptible to UV-B radiational damage as the other cultigens. Violaxanthin concentrations under supplemental UV-B radiation ranged from 2.04 mg/100 g FM for "GA-C 76" to 0.59 mg/100 g FM for "Performer." Cultigen "Pesoenyj" had the highest concentrations of violaxanthin (2.35 mg/100 g FM) for bunching onions grown without supplemental UV-B radiation, while "G 30393-06GI" had the lowest violaxanthin concentrations (0.53 mg/100 g FM).

Antheraxanthin, the intermediate compound in xanthophyll cycle, responded significantly to changes in UV-B radiation treatments ($F = 16.61$; $P < 0.0001$) and cultigens ($F = 4.68$; $P < 0.001$). The majority of cultigens had higher antheraxanthin concentrations in response to the UVB radiation treatment; however, no cultigens had significantly higher levels as compared to the control treatment (Table 2). The ranges for antheraxanthin concentrations in bunching onions grown under UV-B radiation treatment were from 1.38

mg/100 g FM for "Pesoenyj" to 0.79 mg/100 g FM for "274254-05GI." In the plants grown without UV-B radiation, "Pesoenyj" had the highest antheraxanthin concentrations (1.35 mg/100 g FM), while "Ishikura Improved F1" had the lowest concentrations (0.59 mg/100 g FM). While changes in this compound cannot directly tell which way the xanthophyll cycle is fluxing, increases or decreases may help predict potential energy flow.

Neoxanthin concentrations responded significantly to UV-B radiation treatment (F = 12.13; P = 0.0008), cultigen (F = 3.20; P = 0.0003), and the interaction of UV radiation treatment and cultigen (F = 2.27; P = 0.0092). There were significant increases in neoxanthin from the UV-B treatment for the cultigens "Feast," "GA-C 76," and "G 30393- 06GI" when compared to the control treatment (Table 3). "Feast" showed the highest concentrations of neoxanthin under UV-B radiation treatment (1.86 mg/100 g FM), while "Deep Purple" had the lowest concentration of neoxanthin (0.73 mg/100 g FM). "Pesoenyj" showed the highest neoxanthin concentration (1.96 mg/100 g FM) compared to the other cultigens grown under the control treatment. "Hardy Evergreen White" had the lowest of all of the cultigens not grown under supplemental UV-B radiation at 0.40 mg/100 g FM.

Table 3: Mean values[a] for shoot tissue neoxanthin, lutein, and β-carotene (mg/100 g fresh mass) for Allium fistulosum L. cultigens grown under supplemental UV-B (313 nm) light [7.0 mol·m^{-2}·s^{-2} (2.68 W·m^{-2}); UV-B] or UV-filtered (control) light in a glasshouse in Knoxville, TN, USA (35°96′N Lat.)

Cultigen	Neoxanthin			Lutein			β-carotene		
	UV-B	Control	Pr > |t|[b]	UV-B	Control	Pr > |t|	UV-B	Control	Pr > |t|
				mg/100 g fresh mass					
Deep Purple	0.73 ± 0.51	1.04 ± 0.59	ns	5.04 ± 1.48	5.38 ± 0.73	ns	1.07 ± 0.81	1.09 ± 0.30	ns
Evergreen Hardy White	0.74 ± 0.47	0.40 ± 0.18	ns	7.10 ± 2.86	5.10 ± 1.49	ns	0.88 ± 0.71	0.64 ± 0.23	ns
Feast	2.09 ± 0.48	0.79 ± 0.57	P = 0.013	7.66 ± 0.90	4.11 ± 0.54	P = 0.001	1.85 ± 0.85	1.04 ± 0.74	ns
GA-C 76	1.53 ± 0.32	0.66 ± 0.18	P = 0.003	7.66 ± 0.38	5.57 ± 0.67	P = 0.002	1.48 ± 0.27	1.17 ± 0.09	ns
Ishikura Improved F1	0.63 ± 0.43	0.63 ± 0.30	ns	6.35 ± 3.18	4.80 ± 1.20	ns	2.20 ± 2.59	0.78 ± 0.41	ns
Improved Beltsville Bunching	0.82 ± 0.89	0.60 ± 0.16	ns	6.95 ± 1.34	5.62 ± 0.65	ns	1.64 ± 0.95	1.39 ± 0.30	ns
Jionji Negi	0.92 ± 0.25	0.91 ± 0.36	ns	7.35 ± 1.65	6.21 ± 1.49	ns	1.74 ± 1.20	2.29 ± 1.32	ns
Long White Bunching	1.76 ± 0.26	1.47 ± 0.35	ns	6.00 ± 0.58	5.05 ± 1.17	ns	1.69 ± 0.29	1.94 ± 0.56	ns
Parade	1.46 ± 0.81	0.87 ± 0.36	ns	6.36 ± 1.17	6.04 ± 1.47	ns	1.26 ± 0.14	2.54 ± 1.23	ns
Performer	1.14 ± 0.90	0.75 ± 0.47	ns	6.33 ± 0.96	5.60 ± 2.36	ns	1.28 ± 0.48	2.38 ± 1.65	ns
Pesoenyj	1.03 ± 0.19	1.96 ± 0.78	ns	8.01 ± 1.21	9.23 ± 2.59	ns	1.87 ± 0.20	3.45 ± 2.39	ns
Shounan	1.32 ± 0.71	0.54 ± 0.32	ns	7.66 ± 2.83	5.03 ± 1.26	ns	2.80 ± 1.46	1.10 ± 0.54	ns
White Spear	1.21 ± 0.62	0.93 ± 0.63	ns	6.08 ± 0.94	4.70 ± 1.37	ns	1.49 ± 0.51	1.08 ± 0.60	ns
Zhang Qui Da Cong	0.75 ± 0.42	0.65 ± 0.29	ns	5.65 ± 1.44	5.31 ± 1.47	ns	1.05 ± 0.24	1.20 ± 0.57	ns
274254-05GI	0.84 ± 0.37	0.72 ± 0.45	ns	6.18 ± 0.79	5.33 ± 1.82	ns	1.81 ± 0.86	1.86 ± 1.57	ns
G 30393-06GI	1.86 ± 0.44	0.85 ± 0.62	P = 0.040	6.02 ± 1.18	4.42 ± 0.69	ns	1.86 ± 0.77	1.28 ± 0.67	ns
LSD$_{0.05}$[c]	0.78	0.67		ns	2.14		ns	1.54	

[a]Composition of n = 6 plant samples from 4 replications ± standard deviation. [b]Significance based on paired Student's t-test among treatments; ns: not significant. [c]LSD for differences between cultivar means α = 0.05.

The bunching onions showed significant changes in lutein in response to UV-B treatment (F = 17.89; P < 0.0001) and cultigen (F = 2.34; P = 0.0070). The majority of cultigens had higher lutein concentrations in response to the UV-B radiation treatment; however, only "Feast" and "GA-C 76" had significantly higher lutein (Table 3). "Pesoenyj" had the highest concentrations of lutein both with and without supplemental UV-B radiation at 8.01 and 9.23 mg/100 g FM, respectively. "Deep Purple" had the lowest concentration of lutein among bunching onions grown with supplemental UV-B radiation at 5.04 mg/100 g FM, and "Feast" had the lowest amount of lutein for bunching onions grown without supplemental UV-B radiation at 4.11 mg/100 g FM. Lutein acts as an accessory pigment and is the predominant carotenoid in photosystem (PS) II [7]. Research shows UV radiation will impact PSII functioning to a greater extent than PSI [22]. Therefore, increases in lutein concentrations for the cultigens in the current study may indicate increased radiational stress within PSII from the supplemental UV-B treatment.

Concentrations of β-carotene showed no changes in response to UV treatment or cultigen (Table 3). "Pesoenyj" had the highest concentrations of β-carotene in bunching onions grown without UV-B radiation, and "Ishikura Improved F1" had the lowest concentrations. The range of shoot tissue β-carotene levels for cultigens grown under supplemental UV-B radiation were 2.80 mg/100 g FM for "Shounan" to 0.88 mg/100 g FM for "Evergreen Hardy White." For the cultigens that were not grown under supplemental UV-B radiation, the ranges for β-carotene concentration were 3.45 mg/100 g FM (Pesoenyj) and 0.64 mg/100 g FM (Evergreen Hardy White). Reported mean value for β-carotene in shoot tissues of A. fistulosum is 0.60 mg/100 g FM, while the mean values for lutein and zeaxanthin are 1.14 mg/100 g FM [23]. Umehara et al. [24] reported β-carotene values in the leaves of A. fistulosum L. cultigen "Kujyoasagikei" to be 4.63 mg/100 g FM. β-carotene is an accessory pigment and is the predominant carotenoid in PSI. β-carotene is present in PSII, but mostly in regions around the reaction center [7]. Since there were no impacts on β-carotene concentrations in the current study, it is possible that PSI is not under as much stress from the UV-B treatments imposed in this study [22].

The xanthophyll cycle pigments (zeaxanthin, antheraxanthin, and violaxanthin) are important for the dissipation of excess absorbed light, performed almost exclusively by ZEA. Photosynthetic rates are reduced under many environmental stressors, which increase the need for dissipation of excess absorbed light energy [7]. The ratio of zeaxanthin + antheraxanthin to zeaxanthin + antheraxanthin + violaxanthin (ZA/ZAV) responded significantly to cultigen ($F = 3.01$; $P = 0.0006$), but not to UV-B radiation treatment or the interaction between treatment and cultigen. Significant increases in response to supplemental UV-B were found for "Pesoenyj." "G 30393-06GI" had the highest ZA/ZAV ratio of cultigens grown under supplemental UV-B radiation, and "Ishikura Improved F1" had the lowest ZA/ZAV ratio at 0.34.

For the cultigens not grown under UV-B radiation, "Feast" had the highest ZA/ZAV ratio at 0.65, while "Jionji Negi" had the lowest ZA/ZAV ratio at 0.35 (Table 4). Changes in the ZA/ZAV ratio can identify fluxes within the xanthophyll energy dissipation cycle. An increase in ZA/ZAV ratio shows a decrease in violaxanthin, which could mean these compounds are undergoing deepoxydation because of high light energy [7]. A study by Niyogi et al. [25] helped demonstrate the importance of this photoprotective mechanistic cycle. In this study, mutant Arabidopsis thaliana was unable to undergo deepoxydation and converts violaxanthin to zeaxanthin, which resulted in an increased sensitivity to higher light levels. While the Niyogi et al. [25] study did not specifically look at how UV-B radiation affected xanthophyll cycle functioning, energy from UV wavelengths is higher than energy from PAR wavelengths and could be expected to change the flux between the xanthophyll pigments.

Table 4: Mean values[a] for the ratio of zeaxanthin + antheraxanthin to zeaxanthin + antheraxanthin + violaxanthin (Z + A/A + Z + V) and the ratio of chlorophyll a to chlorophyll b (chlorophyll a/chlorophyll b) in shoot tissues for Allium fistulosum L. cultigens grown under supplemental UV-B (313 nm) light [$7.0 \, mol \cdot m^{-2} \cdot s^{-2}$ ($2.68 \, W \cdot m^{-2}$); UV-B] or UV-filtered (control) light in a glasshouse in Knoxville, TN, USA (35°96'N Lat.)

Cultigen	Z + A/A + Z + V			Chlorophyll a/chlorophyll b		
	UV-B	Control	Pr > |t|[b]	UV-B	Control	Pr > |t|
Deep Purple	0.48 ± 0.01	0.51 ± 0.19	ns	1.25 ± 0.90	1.50 ± 0.23	ns
Evergreen Hardy White	0.41 ± 0.06	0.41 ± 0.01	ns	1.15 ± 0.56	0.91 ± 0.16	ns
Feast	0.51 ± 0.16	0.65 ± 0.21	ns	2.14 ± 0.43	1.07 ± 0.63	P = 0.031
GA-C 76	0.49 ± 0.07	0.49 ± 0.10	ns	1.58 ± 0.18	0.93 ± 0.51	ns
Ishikura Improved F1	0.34 ± 0.03	0.40 ± 0.08	ns	1.69 ± 1.06	1.45 ± 0.70	ns
Improved Beltsville Bunching	0.41 ± 0.03	0.41 ± 0.07	ns	1.04 ± 0.92	1.29 ± 0.50	ns
Jionji Negi	0.46 ± 0.07	0.35 ± 0.15	ns	1.85 ± 0.70	2.21 ± 0.50	ns
Long White Bunching	0.52 ± 0.04	0.58 ± 0.17	ns	1.86 ± 0.06	2.25 ± 0.68	ns
Parade	0.45 ± 0.10	0.50 ± 0.24	ns	1.77 ± 0.33	2.16 ± 0.48	ns
Performer	0.66 ± 0.17	0.40 ± 0.02	ns	1.60 ± 0.11	2.07 ± 1.19	ns
Pesoenyj	0.44 ± 0.03	0.39 ± 0.01	P = 0.022	1.40 ± 0.42	1.95 ± 0.80	ns
Shounan	0.42 ± 0.03	0.40 ± 0.03	ns	1.78 ± 0.52	1.08 ± 0.53	ns
White Spear	0.50 ± 0.21	0.51 ± 0.22	ns	1.46 ± 0.42	1.08 ± 0.62	ns
Zhang Qui Da Cong	0.42 ± 0.05	0.41 ± 0.08	ns	1.11 ± 0.50	1.32 ± 0.46	ns
274254-05GI	0.36 ± 0.06	0.40 ± 0.00	ns	1.80 ± 0.71	2.14 ± 0.81	ns
G 30393-06GI	0.69 ± 0.20	0.55 ± 0.12	ns	1.84 ± 0.31	1.32 ± 0.64	ns
LSD$_{0.05}$[c]	0.15	0.20		ns	0.96	

[a] Composition of n = 6 plant samples from 4 replications ± standard deviation. [b] Significance based on paired Student's t-test among treatments; ns: not significant. [c] LSD for differences between cultivar means α = 0.05.

Kopsell et al. [5] grew many of the same bunching onion cultigens under field conditions in Knoxville, TN, USA, and Geneva, NY, USA, and reported similar levels of shoot tissue β-carotene and neoxanthin as found in the current study; however, values for violaxanthin, antheraxanthin, lutein, chlorophyll , and chlorophyll were much higher in the current study than previously reported. Differences in shoot tissue pigments for cultigens among the two studies may be attributed to differences in growing conditions (field versus glasshouse) and the time of year the cultigens were evaluated (summer versus winter).

Epidemiological data supports the positive association between increased dietary intake of plant foods high in carotenoids and greater carotenoid tissue concentrations with lower risks of certain chronic diseases. Many of these disease suppressing abilities can be attributed to the antioxidant properties of carotenoids. One of the most important physiological functions of carotenoids in human nutrition is as vitamin A precursors. Provitamin A carotenoid compounds (β-carotene, α-carotene, and cryptoxanthins) support the maintenance of healthy epithelial cell differentiation, normal reproductive performance, and visual functions [26]. Both provitamin A carotenoids and nonprovitamin A carotenoids (lutein, zeaxanthin, and lycopene) function as free radical scavengers, enhance the immune response, suppress cancer development, and protect eye tissues [27]. Humans cannot synthesize carotenoids and therefore must rely on dietary sources to provide sufficient levels. Studies indicate that high intakes of a variety of vegetables, providing a mixture of carotenoids, were more strongly associated with reduced cancer and eye disease risk than intake of individual carotenoid supplements [28]. There is clear evidence that cultural practices that maintain or enhance tissue carotenoid levels would be beneficial to humans when regularly consumed in the diet.

Shoot Tissue Chlorophyll Pigment Concentrations

No chlorophyll pigments were measured in the pseudostem tissues of any of the bunching onion cultigens (Figure 1; data not shown). Chlorophyll a responded significantly to UV radiation treatments ($F = 4.35$; $P = 0.0398$), but not to cultigens or the interaction between treatment and cultigen. "Feast" had the highest concentration of chlorophyll a at 59.56 mg/100 g FM for cultigens grown under supplemental UV-B radiation, while "Deep Purple" had the lowest at 27.75 mg/100 g FM. For the cultigens grown without supplemental UV-B radiation, "Pesoenyj" had the highest concentration of chlorophyll a at 63.27 mg/100 g FM, while "Evergreen HardyWhite" had the lowest at 16.52 mg/100 g FM (Table 5). Values for chlorophyll a for cultigens are in close agreement with Dissanayake et al. [29] who reported values of ~75.00 mg chlorophyll a/100 g FM for the A. fistulosum cultigen "Kujyo-hoso."

Table 5: Mean values[a] for shoot tissue chlorophyll a, chlorophyll b, and total chlorophyll (chlorophyll a + chlorophyll b) (mg/100 g fresh mass) for Allium fistulosum L. cultigens grown under supplemental UV-B (313 nm) light [7.0 mol·m^{-2}·s^{-2} (2.68 W·m^{-2}); UV-B] or UV-filtered (control) light in a glasshouse in Knoxville, TN, USA (35°96′N Lat.)

Cultigen	Chlorophyll a			Chlorophyll b			Total chlorophyll		
	UV-B	Control	Pr > |t|[b]	UV-B	Control	Pr > |t|	UV-B	Control	Pr > |t|
				mg/100 g fresh mass					
Deep Purple	27.8 ± 22.1	25.3 ± 5.6	ns	20.6 ± 4.4	16.9 ± 2.4	ns	48.3 ± 25.6	42.1 ± 7.5	ns
Evergreen Hardy White	31.4 ± 21.1	16.5 ± 3.2	ns	25.5 ± 5.1	18.2 ± 0.7	ns	56.9 ± 26.1	34.7 ± 3.5	ns
Feast	59.6 ± 20.3	19.0 ± 10.5	P = 0.012	27.3 ± 4.0	17.7 ± 1.9	P = 0.005	86.8 ± 24.3	36.7 ± 11.0	P = 0.009
GA-C 76	47.0 ± 16.4	17.2 ± 10.6	ns	29.2 ± 7.1	18.1 ± 1.3	P = 0.021	76.2 ± 23.3	35.4 ± 11.5	P = 0.020
Ishikura Improved F1	49.9 ± 53.6	23.1 ± 5.9	ns	25.6 ± 11.6	17.6 ± 4.7	ns	74.5 ± 65.1	40.7 ± 2.6	ns
Improved Beltsville Bunching	23.9 ± 28.4	20.6 ± 8.9	ns	18.5 ± 7.6	15.8 ± 1.0	ns	42.4 ± 36.0	36.1 ± 9.8	ns
Jionji Negi	48.3 ± 23.4	47.2 ± 18.9	ns	25.5 ± 3.1	20.9 ± 4.8	ns	73.8 ± 26.0	68.0 ± 23.4	ns
Long White Bunching	36.9 ± 6.6	37.2 ± 7.4	ns	19.7 ± 2.8	17.1 ± 2.7	ns	56.5 ± 9.2	54.3 ± 6.3	ns
Parade	38.6 ± 13.2	44.4 ± 18.8	ns	21.2 ± 4.1	19.9 ± 4.4	ns	59.8 ± 17.2	64.3 ± 23.0	ns
Performer	36.8 ± 3.1	49.6 ± 31.9	ns	23.0 ± 2.1	20.9 ± 7.1	ns	59.8 ± 4.8	70.5 ± 38.9	ns
Pesoenyj	36.8 ± 10.2	63.3 ± 36.9	ns	26.3 ± 2.4	29.7 ± 8.7	ns	26.3 ± 2.4	93.0 ± 45.6	ns
Shounan	44.3 ± 24.9	20.6 ± 19.1	ns	24.3 ± 8.3	16.6 ± 7.6	ns	68.6 ± 32.6	37.2 ± 26.7	ns
White Spear	30.4 ± 9.9	18.2 ± 9.6	ns	20.8 ± 1.9	17.8 ± 1.0	P = 0.033	51.2 ± 10.7	35.4 ± 8.9	ns
Zhang Qui Da Cong	25.3 ± 18.3	27.3 ± 13.5	ns	20.4 ± 7.4	19.3 ± 4.2	ns	45.6 ± 25.6	47.1 ± 17.6	ns
274254-05GI	36.6 ± 23.1	40.0 ± 18.1	ns	18.9 ± 5.3	18.7 ± 4.9	ns	55.5 ± 28.3	58.6 ± 21.5	ns
G 30393-06GI	39.8 ± 11.3	22.6 ± 13.8	ns	21.3 ± 4.1	17.0 ± 3.7	ns	61.1 ± 15.1	39.6 ± 16.0	ns
LSD$_{0.05}$[c]	ns	26.0		ns	6.8		ns	31.6	

[a] Composition of n = 6 plant samples from 4 replications ± standard deviation. [b]Significance based on paired Student's t-test among treatments; ns: not significant. [c] LSD for differences between cultivar means α = 0.05.

The bunching onions showed significant differences in chlorophyll *b* caused by UV-B treatment (*F* = 19.04; *P* < 0.0001) and cultigen (*F* = 2.08; *P* = 0.0179), but there were no influences from their interaction. Values for chlorophyll *b* for cultigens are in close agreement with Dissanayake et al. [29] who reported values of ~17.00 mg chlorophyll b/100 g FM for the A. fistulosum cultigen "Kujyo-hoso." Significant increases in chlorophyll *b* in response to UV-B radiation were found for cultigens "Feast," "GA-C 76," and "Shounan" (Table 5). The concentrations of chlorophyll *b* for cultigens grown under supplemental UV-B radiation ranged from 29.24 mg/100 g FM for "GA-C 76" to 18.49 mg/100 g FM for "Improved Beltsville Bunching." For cultigens grown without supplemental UV-B radiation, chlorophyll *b* concentrations ranged from 29.74 mg/100 g FM for "Pesoenyj" to 15.78 mg/100 g FM for "Improved Beltsville Bunching.

Concentrations of total chlorophyll (chlorophyll a + b) in bunching onions were found to differ between UV-B treatments (*F* = 6.82; *P* = 0.0105), but not among cultigens. "Feast" and "GA-C 76" were the only bunching onion cultigens to show differences between UV-B treatments (Table 5). Total chlorophyll concentrations ranged from 88.82 mg/100 g FM for "Feast" to 45.62 mg/100 g FM for "Zhang Qui Da Cong" for bunching onions grown under supplemental UVB radiation. For the plants grown without UV-B radiation, ranges from total chlorophyll varied from 93.01 mg/100 g FM for "Pesoenyj" to 34.74 mg/100 g FM for "Evergreen Hardy White."

The ratio of chlorophyll a to chlorophyll b in the bunching onions showed significant changes based on cultigen ($F = 2.26$; $P = 0.0094$), but not for UV-B radiation treatments. In general, cultigens were evenly divided in their responses to UV-B radiation, with half the cultigens displaying higher chlorophyll a/chlorophyll b under the supplemental UV-B radiation treatment (Table 4). However, only the cultigen "Feast" had a significantly higher chlorophyll a/chlorophyll b ratio under UV-B radiation. "Long White Bunching" had the highest chlorophyll a/chlorophyll b ratio in the bunching onions grown without supplemental UV-B at 2.25, and "GAC 76" had the lowest ratio at 0.91. Under UV-B radiation treatment, "Feast" has the highest chlorophyll a/chlorophyll b ratio at 2.14, while "Improved Beltsville Bunching" has the lowest ratio at 1.04.

Shoot Tissue Photochemical Efficiency (F_v/F_m)

Photochemical efficiency (F_v/F_m) showed significant differences between UV treatments ($F = 13.89$, $P = 0.0003$) and cultigen ($F = 2.11$, $P = 0.0152$), but no difference due to treatment and cultigen interaction (data not shown). Values for F_v/F_m for all of the cultigens evaluated in the study averaged 0.82. One previous study by Tsormpatsidis et al. [30] showed that while "Lollo Rosso" lettuce (Lactuca sativa L.) had decreased vegetative growth under UV light treatments, there was no difference in photochemical efficiency. By contrast, when the agronomic crop wheat was exposed to UV radiation, decreases in F_v/F_m occurred under the UV light treatment [31]. None of the cultigens in this study showed differences in F_v/F_m; however, most of the cultigens differed in shoot tissue fresh biomass when exposed to UV-B radiation.

CONCLUSION

Data from multiple studies demonstrates cultigens within a given plant species can react differently under variable stress conditions. Most often, harsh stress conditions negatively impact plant biomass. In the current study, decreases in bunching onion shoot tissue biomass confirmed that a radiational stress from the UV-B treatment had occurred. The bunching onion cultigens demonstrated genetic variability in response to UV-B radiation (Tables 1–5). Changes in plant pigments associated with light harvesting and photoprotection can be expected when bunching onion cultigens experience greater levels of UV-B radiation in the growing environment. In the current study, the cultigens with the greatest stimulation in carotenoid pigments from UV-B exposure were "Feast" and the accession G 30393-06GI. Data presented here may be valuable to improve abiotic stress tolerance to increasing UV-B radiation for specialty crop breeding programs.

ACKNOWLEDGMENTS

Mention of trade names or commercial products in this publication is solely for the purpose of providing specific information and does not imply recommendation or endorsement by the University of Tennessee Institute of Agriculture. Authors would like to acknowledge funding and support for this work by the University of Tennessee Agricultural Experiment Station.

REFERENCES

1. G. Sandmann, "Genetic manipulation of carotenoid biosynthesis: strategies, problems and achievements," Trends in Plant Science, vol. 6, no. 1, pp. 14–17, 2001.

2. J. T. Landrum and R. A. Bone, "Lutein, zeaxanthin, and the macular pigment," Archives of Biochemistry and Biophysics, vol. 385, no. 1, pp. 28–40, 2001.

3. M. G. Ferruzzi and J. Blakeslee, "Digestion, absorption, and cancer preventative activity of dietary chlorophyll derivatives," Nutrition Research, vol. 27, no. 1, pp. 1–12, 2007.

4. Y. Endo, R. Usuki, and T. Kaneda, "Antioxidant effects of chlorophyll and pheophytin on the autoxidation of oils in the dark. II. The mechanism of antioxidative action of chlorophyll," Journal of the American Oil Chemists› Society, vol. 62, no. 9, pp. 1387–1390, 1985.

5. D. A. Kopsell, C. E. Sams, D. E. Deyton, K. R. Abney, D. E. Kopsell, and L. Robertson, "Characterization of nutritionally important carotenoids in bunching onion," HortScience, vol. 45, no. 3, pp. 463–465, 2010.

6. K. A. Steinmetz and J. D. Potter, "Vegetables, fruit, and cancer. II. Mechanisms," Cancer Causes and Control, vol. 2, no. 6, pp. 427–442, 1991.

7. B. Demmig-Adams, A. M. Gilmore, and W. W. Adams III, "In vivo functions of carotenoids in higher plants," The FASEB Journal, vol. 10, no. 4, pp. 403–412, 1996.

8. R. Croce, S. Weiss, and R. Bassi, "Carotenoid-binding sites of the major light-harvesting complex II of higher plants," The Journal of Biological Chemistry, vol. 274, no. 42, pp. 29613–29623, 1999.

9. J. A. Mares-Perlman, A. E. Millen, T. L. Ficek, and S. E. Hankinson, "The body of evidence to support a protective role for lutein and zeaxanthin in

delaying chronic disease. Overview," Journal of Nutrition, vol. 132, no. 3, pp. 517S–524S, 2002.

10. G. J. F. MacDonald, Ed., Biological Impacts of Increased Intensities of Solar Ultraviolet Radiationedition, National Academy of Sciences, Washington, DC, USA, 1st edition, 1973.

11. L. Yuan, Z. Yanqun, C. Haiyan, C. Jianjun, Y. Jilong, and H. Zhide, "Intraspecific responses in crop growth and yield of 20 wheat cultivars to enhanced ultraviolet-B radiation under field conditions,"Field Crops Research, vol. 67, no. 1, pp. 25–33, 2000.

12. M. Caasi-Lit, M. I. Whitecross, M. Nayudu, and G. J. Tanner, "UV-B irradiation induces differential leaf damage, ultrastructural changes and accumulation of specific phenolic compounds in rice cultivars,"Australian Journal of Plant Physiology, vol. 24, no. 3, pp. 261–274, 1997.

13. M. Saile-Mark and M. Tevini, "Effects of solar UV-B radiation on growth, flowering and yield of Central and Southern European bush bean cultivars (Phaseolus vulgaris L.)," Plant Ecology, vol. 128, no. 1-2, pp. 114–125, 1997.

14. D. Giuntini, G. Graziani, B. Lercari, V. Fogliano, G. F. Soldatini, and A. Ranieri, "Changes in carotenoid and ascorbic acid contents in fruits of different tomato genotypes related to the depletion of UV-B radiation," Journal of Agricultural and Food Chemistry, vol. 53, no. 8, pp. 3174–3181, 2005.

15. M. Marotti and R. Piccaglia, "Characterization of flavonoids in different cultivars of onion (Allium cepaL.)," Journal of Food Science, vol. 67, no. 3, pp. 1229–1232, 2002.

16. D. A. Kopsell and W. M. Randle, "Selenium affects the S-alk(en)yl cysteine sulfoxides among short-day onion cultivars," Journal of the American Society for Horticultural Science, vol. 124, no. 3, pp. 307–311, 1999.

17. M. Takahashi and T. Shibamoto, "Chemical compositions and antioxidant/ anti-inflammatory activities of steam distillate from freeze-dried onion (Allium cepa L.) sprout," Journal of Agricultural and Food Chemistry, vol. 56, no. 22, pp. 10462–10467, 2008.

18. Denny and J. Buttriss, Synthesis Report no. 4, Plant Foods and Health: Focus on Plant Bioactive, European Food Information Resource (EuroFIR) Consortium. EuroFIR Project/British Nutrition Foundation, Institute of Food Research,, Norwich, UK, 2007.

19. D. R. Hoagland and D. I. Arnon, "The water culture method for growing plants without soil," California Agricultural Experiment Station Circular, vol. 347, pp. 4–32, 1950.

20. D. A. Kopsell, D. E. Kopsell, M. G. Lefsrud, J. Curran-Celentano, and L. E. Dukach, "Variation in lutein, β-carotene, and chlorophyll concentrations among Brassica oleracea cultigens and seasons,"HortScience, vol. 39, no. 2, pp. 361–364, 2004.

21. B. H. Davies and H. P. Köst, "Chromatographic methods for the separation on carotenoids," in Plant Pigments. Fat Soluble Pigments, H. P. Köst, G. Zweig, and J. Sherma, Eds., vol. 1 of Handbook of Chromatography, pp. 1–85, CRC Press, Boca Raton, Fla, USA, 1988.

22. V. G. Kakani, K. R. Reddy, D. Zhao, and K. Sailaja, "Field crop responses to ultraviolet-B radiation: a review," Agricultural and Forest Meteorology, vol. 120, no. 1–4, pp. 191–218, 2003.·

23. U.S. Department of Agriculture, USDA National Nutrient Database for Standard Release, SR25,http://ndb.nal.usda.gov.

24. M. Umehara, T. Sueyoshi, K. Shimomura et al., "Interspecific hybrids between Allium fistulosum and Allium schoenoprasumreveal carotene-rich phenotype," Euphytica, vol. 148, no. 3, pp. 295–301, 2006.

25. K. K. Niyogi, A. R. Grossman, and O. Björkman, "Arabidopsis mutants define a central role for the xanthophyll cycle in the regulation of photosynthetic energy conversion," Plant Cell, vol. 10, no. 7, pp. 1121–1134, 1998.

26. G. F. Combs, "Vitamin A," in The Vitamins: Fundamental Aspects in Nutrition and Health, pp. 107–153, Academic Press, San Diego, Calif, USA, 2nd edition, 1998.

27. K. J. Yeum and R. M. Russell, "Carotenoid bioavailability and bioconversion," Annual Review of Nutrition, vol. 22, pp. 483–504, 2002.

28. E. J. Johnson, B. R. Hammond, K. J. Yeum et al., "Relation among serum and tissue concentrations of lutein and zeaxanthin and macular pigment density," American Journal of Clinical Nutrition, vol. 71, no. 6, pp. 1555–1562, 2000.

29. P. K. Dissanayake, N. Yamauchi, and M. Shigyo, "Chlorophyll degradation and resulting catabolite formation in stored Japanese bunching onion (Allium fistulosum L.)," Journal of the Science of Food and Agriculture, vol. 88, no. 11, pp. 1981–1986, 2008.

30. E. Tsormpatsidis, R. G. C. Henbest, F. J. Davis, N. H. Battey, P. Hadley, and A. Wagstaffe, "UV irradiance as a major influence on growth, development

and secondary products of commercial importance in Lollo Rosso lettuce "Revolution" grown under polyethylene films," Environmental and Experimental Botany, vol. 63, no. 1–3, pp. 232–239, 2008. ·

31. X. C. Lizana, S. Hess, and D. F. Calderini, "Crop phenology modifies wheat responses to increased UV-B radiation," Agricultural and Forest Meteorology, vol. 149, no. 11, pp. 1964–1974, 2009.

Chapter 8

NATURAL PIGMENTS FROM PLANTS USED AS SENSITIZERS FOR TIO2BASED DYE-SENSITIZED SOLAR CELLS

Reena Kushwaha, Pankaj Srivastava, and Lal Bahadur

Department of Chemistry, Faculty of Science, Banaras Hindu University, Varanasi 221005, India

ABSTRACT

Four natural pigments, extracted from the leaves of teak (Tectona grandis), tamarind (Tamarindus indica), eucalyptus (Eucalyptus globulus), and the flower of crimson bottle brush (Callistemon citrinus), were used as sensitizers for TiO_2 based dye-sensitized solar cells (DSSCs). The dyes have shown absorption in broad range of the visible region (400–700 nm) of the solar spectrum and appreciable adsorption onto the semiconductor (TiO_2) surface. The DSSCs made using the extracted dyes have shown that the open circuit voltages (j_{sc}) varied from 0.430 to 0.610 V and the short circuit photocurrent densities (V_{oc}) ranged from 0.11 to 0.29 mA cm^{-2}. The incident photon-to-current conversion efficiencies (IPCE) varied from 12–37%. Among the four dyes studied, the extract obtained from teak has shown the best photosensitization effects in terms of the cell output.

INTRODUCTION

Harvesting energy from sunlight using photovoltaic technology is one of the most important research areas because of an ever increasing global energy need. The conventional solid-state silicon based solar cells, though highly efficient, are yet to become popular for mass applications as they are highly expensive. The necessity for developing low cost devices for harvesting solar energy was, therefore, very much desirable. A new hope was generated in this direction when O'Regan and Gräetzel reported to have achieved an unprecedented high energy conversion efficiency (η) of 7.1% through a dye-sensitized solar cell (DSSC) developed by using nanocrystalline TiO_2 thin film electrode sensitized by a highly efficient Ru(II) polypyridyl complex [1]. This has proven that significantly high light-to-electricity conversion efficiency can be achieved through DSSCs as well. Once this was established, such cells attracted greater attention of the scientists particularly because of two reasons; first, their production cost was expected to be quite low due to ease of their fabrication, and second, they are more environment friendly as compared to conventional solid-state silicon based photovoltaic devices [2]. Being optimistic that DSSCs have the potential to become a commercially viable alternative to expensive silicon solar cells, extensive studies have been conducted on such devices during last two decades.

A dye-sensitized solar cell is usually composed of a dye-capped nanocrystalline porous semiconductor electrode, a metal counter electrode, and a redox electrolyte mediating electron transfer processes occurring in the cell. The performance of the cell is primarily dependent on the material and quality of the semiconductor electrode and the sensitizer dye used for the fabrication of the cell. For their application in DSSCs, many wide band-gap metal oxide semiconductors have been studied but most extensively employed semiconductors are TiO_2 and ZnO [3–8]. Titanium dioxide (TiO_2) has several advantages, including long-term thermal and photostability. The essential properties of semiconductor can be changed significantly by using different techniques for their deposition on the substrate [9]. The sensitizer (dye) plays a key role in absorbing light, and in this respect the highest efficiency obtained so for is with Ru (II) polypyridyl complexes [10, 11]. However, the ruthenium complexes are expensive due to the paucity of the Ru metal and the complexity of preparation procedure limiting the production of low cost DSSC. This has stimulated the search for potential alternative metal complex sensitizers. Simultaneously, organic dyes [12, 13] and natural dyes [14–20] extracted from plants were also studied to explore the possibility of their application as photosensitizer. Organic dyes have been reported to meet the efficiency as high as 9.8% [12]. However, these dyes have been fraught with problems, such as complicated synthetic routes and low yields. On the other hand, the natural dyes found in flowers, leaves, and fruits of plants can be extracted by simple procedures and then employed in DSSCs. The advantages of natural dyes, resembling in functionalities to organic dyes, are their easy availability, nontoxicity, complete biodegradability, and temperature compatibility. Several of natural dyes such as tannin [21], carotene [22], anthocyanin [23], betalain [24], and chlorophyll [25, 26] have been extensively investigated as sensitizers in dye-sensitized solar cells [27].

In this paper, we report the performance of four natural dyes extracted from the leaves of teak (Tectona grandis), tamarind (Tamarindus indica), eucalyptus (Eucalyptus globulus), and the flower of crimson bottle brush (Callistemon citrinus). The basic structures of the coloring components found in these extracts are given in Figure 1. Tannin, that is, gallic acid [3,4,5–trihydroxybenzoic acid] and ellagic acid [2,3,7,8-(1)tetrahydroxybenzopyrano(5,4,3-cde)(1) benzopyran-5,10-dione] are the main constituents of these natural dyes along with some minor components [28–30]. Teak extract mainly contains tectoleafquinone, 1,4,5,8-tetrahydroxy-2 isopentadienyl anthraquinone and tannin [28]. To the best of our knowledge, the use of these plant extracts is being reported for the first time as sensitizers for TiO_2 based dye-sensitized solar cells (DSSCs).

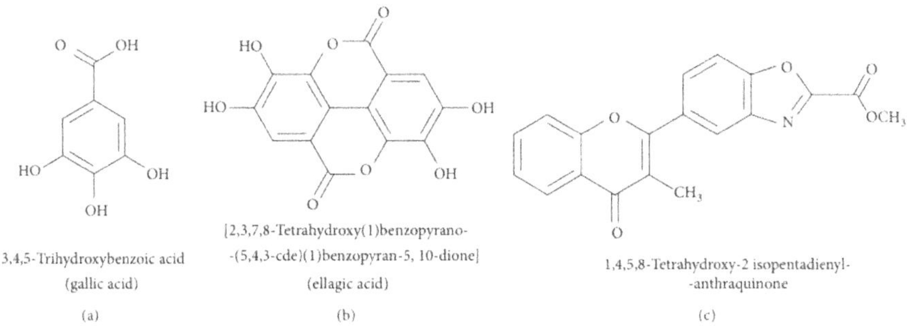

3,4,5-Trihydroxybenzoic acid
(gallic acid)
(a)

[2,3,7,8-Tetrahydroxy(1)benzopyrano-
-(5,4,3-cde)(1)benzopyran-5, 10-dione]
(ellagic acid)
(b)

1,4,5,8-Tetrahydroxy-2 isopentadienyl-
-anthraquinone
(c)

Figure 1: Basic molecular structure for the main components of the extracts.

EXPERIMENTAL

Materials

Ethanol (A.R. grade, 99.9%, Merck) was used for extracting natural dyes from plants. Titanium paste (HT), platinum catalyst (T/SP), and the sealing tape (SX1170–60, 50 μm thick) were obtained from Solaronix. Propylene carbonate (>99%, Merck) was taken as the medium of cell electrolyte. Anhydrous lithium iodide (99.9%, Aldrich) and iodine (G. R. grade, 99.8%, BDH) were used as redox couple in photoelectrochemical (PEC) experiments without any further purification. FTO-coated (Fluorine-doped tin oxide) conductive glass slides (surface resistivity $15/, \Omega/\square$, thickness 2.2 mm) obtained from Pilkington, USA, were used as substrates for preparing TiO_2 thin film electrode and Platinum counter electrode.

Apparatus and Instruments

A bipotentiostat (model number AFRDE 4E, Pine Instrument Company, USA) and e-corder (model 201, eDAQ, Australia) were used for current-potential measurements. For photoelectrochemical (PEC) measurements, a 150 W Xenon arc lamp with lamp housing (model number 66057) and power supply (model number 68752), all from Oriel Corporation, USA, was used as the light source. The semiconductor electrode was illuminated after passing the collimated light beam through a 6-inch long water column (to filter IR part of the light) and condensing it with the help of fused silica lenses (Oriel Corporation, USA). The UV part of this IR-filtered light (referred to as "white light") was cut off by using a long pass filter (model number 51280, Oriel Corporation, USA) and the light obtained this way is mentioned as "visible light." The light was monochromatised, when required, by using a grating monochromator (Oriel model 77250 equipped with model 7798 grating). The width of the exit slit of the monochromator was kept at 0.5 mm. To obtain the action spectrum (J_{photo}-λ) of the dye-sensitized TiO_2 electrode, monochromatic light-induced photocurrent was measured with the help of a digital multimeter (Philips Model number 2525) in combination with the potentiostat. The intensities of light were measured with a digital photometer (Tektronix model J16 with model J 6502 sensor) in combination with neutral density filters (model number 50490-50570, Oriel, USA). The absorption spectrums were recorded on Shimadzu UV-1700 spectrophotometer. The FT-IR spectra were recorded by Varian 3100 FT-IR spectrometer.

Preparation of Natural Dye Solutions (Extracts)

The natural dyes were extracted with ethanol employing the following procedure: fresh leaves of teak (Tectona grandis), tamarind (Tamarindus indica), eucalyptus (Eucalyptus globulus), and the flower of crimson bottle brush (Callistemon citrinus) were washed with water and dried. After crushing them into small pieces in a mortar, these were kept in glass bottles and filled with ethanol; these solutions were kept for one week in the dark at room temperature. Then, the residual (solid) parts were filtered out and the resulting filtrates were used as dye solutions.

Preparation of TiO_2 Electrode (Photo Anode) and Counter Electrode

TiO_2 thin film electrodes (photoanodes) were prepared by spreading highly transparent paste of TiO_2(Titanium-HT) on FTO-coated conductive glass plate by the doctor's blade method. On the conducting side of glass substrate, a U-shaped frame of adhesive tape was applied to control the thickness of the film and to provide noncoated area for electrical contact. After spreading TiO_2 paste, the adhesive tapes were carefully removed and films were annealed at 450°C in air for half an hour in a tubular furnace. This resulted in TiO_2 film of ~6 μm thickness. The dyes were anchored onto the surface of the TiO_2 thin film electrode by immersing it into ethanol solution of natural dye for overnight. The nonadsorbed dye was washed up with anhydrous ethanol. The dye-coated films were air dried and used as photoelectrode in the cell (Figure 2). The platinum counter electrode was prepared on another FTO-coated glass substrate by depositing platinum catalyst (T/SP, Solaronix) using screen printing method and annealing at 400°C for half an hour in air. The electrolyte consisted of 0.2 M lithium iodide and 0.02 M iodine in propylene carbonate.

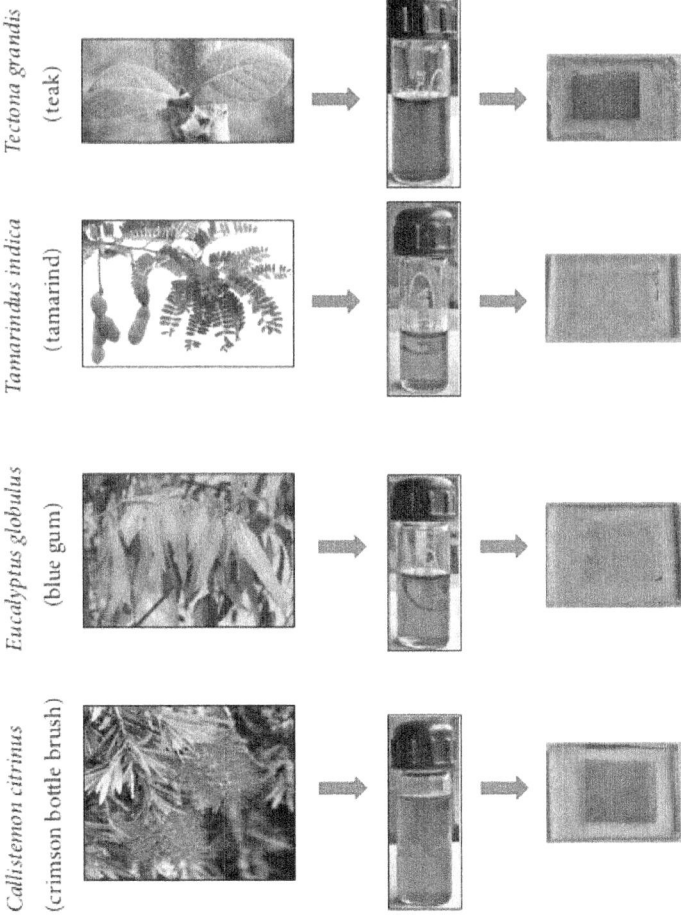

Figure 2: Plants, extracted dyes, and the dye-loaded TiO_2 electrode.

Fabrication of Sandwich Type DSSCs

The photo-electrode (dye-coated TiO_2 film) was put over platinum counter electrode in such a way that the conductive side of both the electrodes faced each other, and the cell was sealed from three sides using spacer/sealing tape (heating it at ~80°C); one side was left open for the injection of electrolyte. The cell electrolyte was injected through open side and was drawn into the space between the electrodes by capillary action. Thereafter, the open side of the cell assembly was sealed properly with Araldite and the contacts were made by copper wires using silver paste (Figure 3).

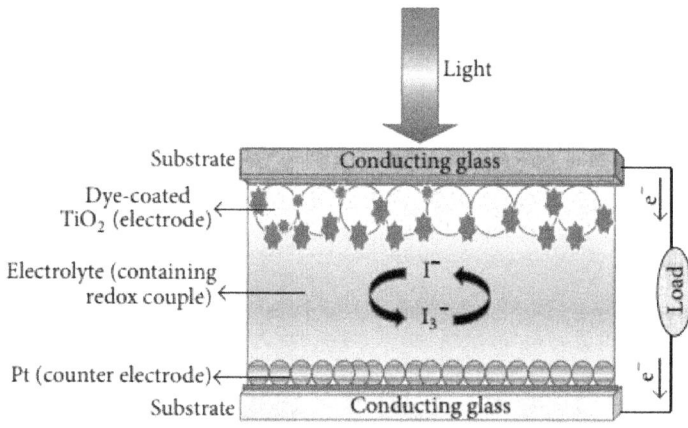

Figure 3: Schematic diagram of dye-sensitized solar cell (DSSC) assembly.

RESULTS AND DISCUSSION

Absorption Spectra of Natural Dyes

Figure 4 shows the absorption spectra of the ethanol extracts of Tectona grandis, Tamarindus indica, Eucalyptus globulus, and Callistemon citrinus. From this figure, it is evident that these natural extracts absorb in the visible region of light spectrum and hence fulfill the primary criterion for their use as sensitizers in DSSCs. To be more specific, Tectona grandis exhibited broad absorption band in the range 425–550 nm besides showing a sharp absorption peak at 662 nm. Tamarindus indica and Eucalyptus globulus have absorption peaks at 410 nm and 472 nm, respectively. Each of them has a common peak at 663 nm which is consistent with the characteristic absorption band of chlorophyll [25, 26]. Callistemon citrinus absorbs in the wide range of 410–600 nm with an absorption peak at 450 nm. The differences and variations in the absorption characteristics of dyes can be attributed to the different colors of the extracts due to respective pigments present in them.

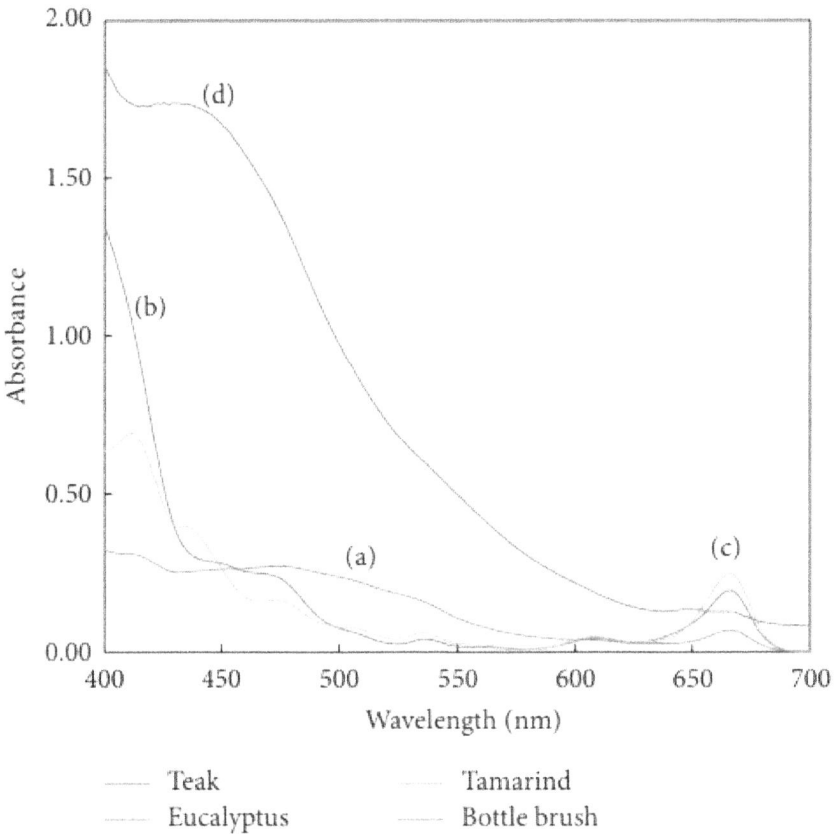

Figure 4: Absorption spectra of ethanol solution of natural dyes extracted from (a) teak, (b) eucalyptus, (c) tamarind, and (d) bottle brush, respectively.

FTIR Spectra

The infrared spectra of these four natural extracts were obtained by pressing them in pellets with KBr. The respective FTIR spectra were recorded in the range from 4000 to 400 cm^{-1} and shown in Figure 5. An examination of the spectra reveals that they exhibit broad absorption in the range 3000–3700 cm^{-1} with a wide and strong band at 3407 cm^{-1} which is attributed to the –OH stretching and due to the wide variety of hydrogen bonding between OH. In these spectrums, a sharp peak at around 2927 and a small shoulder at 2855 cm^{-1} associated with the symmetric and antisymmetric –C–H– stretching vibrations of CH$_2$ and

CH$_3$groups, respectively, is observed. Also, the signal characteristics bands of C=O (carbonyl) stretching vibration at 1730–1705 cm^{-1} and C–O at 1100–1300 cm^{-1} can be observed due to presence of some aromatic esters. The bands observed in the range 1669–1400 cm^{-1} are due to aromatic ring vibrations, while the ones at 1190 and 1052 cm^{-1} are due to ester linkage. The band at around 751 cm^{-1} is assigned to aromatic C–H bending vibration. Hence, the IR spectra of extracts contain bands that can be assigned to the coloring components found in these extracts as given in Figure 1 Tannin, that is, gallic acid, ellagic acid, and tectoleafquinone, 1,4,5,8-tetrahydroxy-2 isopentadienyl anthraquinone.

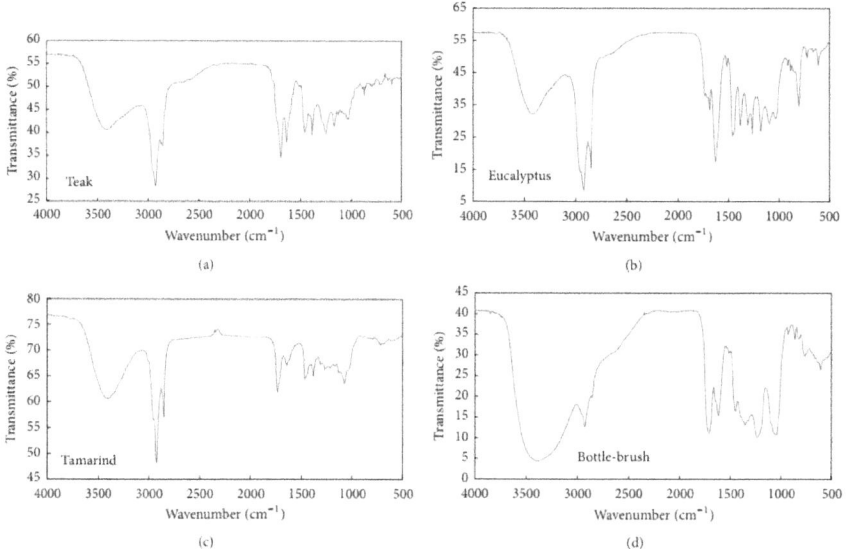

Figure 5: Infra-red spectra of extracts obtained from (a) teaks (b) tamarinds (c) eucalyptuss and (d) bottle brush.

Photoelectrochemical Studies

Current-Potential (J-V) Curves

The photovoltaic performances of DSSCs using natural dyes as photosensitizer (TiO$_2$-dye/electrolyte containing I$^-$,I$_3^-$/Pt counter electrode) were determined by recording the current-potential (J-V) curves under visible light illumination and displayed in Figure 6. The similar curve for the cell using bare TiO$_2$ electrode determined under identical experimental conditions is also shown in the figure (curve (e)). Almost insignificant current is observed in this case as expected, since visible light is incapable of exciting wide band-gap TiO_2. The values of photovoltaic parameters derived from these curves are given in Table 1.

Table 1: The cell output of DSSCs sensitized by four kinds of natural dyes: (a) teak, (b) tamarind, (c) eucalyptus, and (d) bottle brush under visible light (256 mW/cm²) illumination

Natural extract	Peak wavelength λ (nm)	J_{sc} (mA/cm²)	V_{oc} (mV)	IPCE (%)	P_{max} (mW/cm²)	FF
Teak (*Tectona grandis*)	470, 662	0.29	460	37	0.105	79
Tamarind (*Tamarindus indica*)	410, 663	0.18	610	33	0.061	56
Eucalyptus (*Eucalyptus globulus*)	472, 663	0.15	500	12	0.070	93
Bottle brush (*Callistemon citrinus*)	450	0.11	430	34	0.030	63

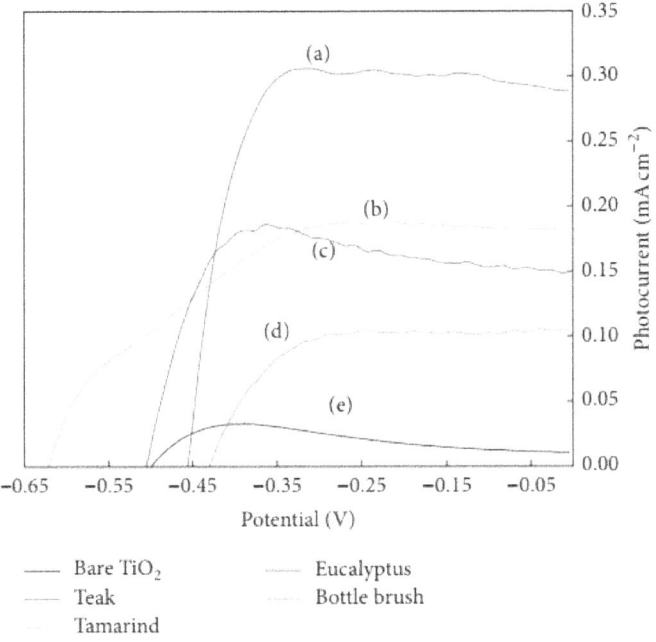

Figure 6: Photocurrent-voltage (J-V) curves for the DSSCs sensitized by four kinds of natural dyes: (a) teak, (b) tamarind, (c) eucalyptus, and (d) bottle brush under visible light illumination of intensity 256 mW/cm² (electrolyte composition: 0.2 M LiI, 0.02 M I₂ in propylene carbonate). Curve (e) is the same for bare TiO_2 electrode.

With DSSCs using these dyes, open circuit voltage (V_{oc}) from 0.430 to 0.610 V and the short circuit photocurrent densities (J_{sc}) in the range of 0.11–0.29 mA/cm² could be achieved. The highest V_{oc} (0.610 V) was obtained with tamarind extract-sensitized DSSC, whereas maximum J_{sc} (0.29 mA/cm²) was obtained with the DSSC sensitized by teak extract.

Transient Photocurrent-Time (J_{photo}-t) Profile

The transient current-time profiles were recorded to know the sustainability of the photocurrent observed initially on illumination of the DSSCs with desired intensity of light. For such an assessment, initially the dark current was monitored for a few seconds; then the semiconductor electrode was illuminated and the short circuit photocurrent was monitored as a function of time. The photocurrent-time (J_{photo}-t) profile obtained under visible light (256 mW/cm²) illumination of natural dye sensitized DSSCs are shown in Figure 7.

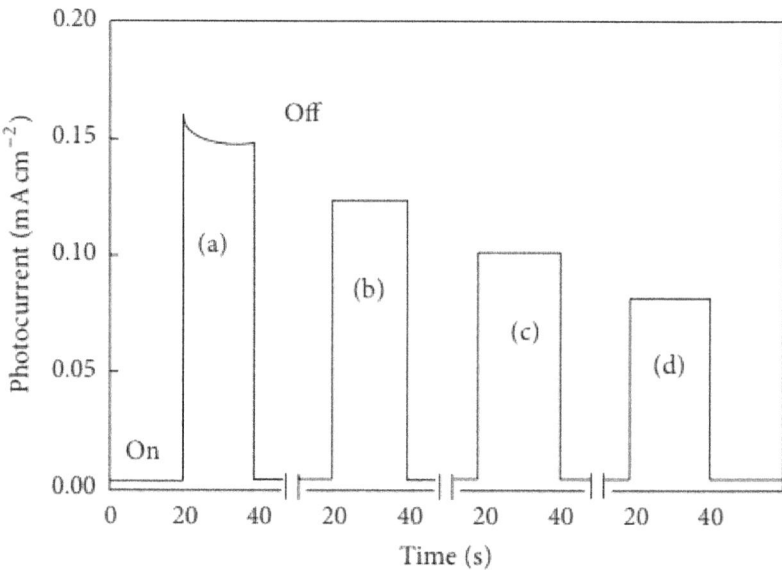

Figure 7: Transient current-time (J_{photo}-t) profiles obtained under visible light illumina(intensity 256 mW/cm²) for the four DSSCs sensitized by (a) teak, (b) tamarind, (c) eucalyptus, and (d) bottle brush, respectively. Electrolyte composition and intensity are the same as in Figure 6.

Except for the curve (a), in all the other cases, ideal behavior (no decay in photocurrent) was observed. In case of curve (a), initially the photocurrent reached maximum, but the same was not sustained and it decayed to ~93% of its initial value before getting stabilized. This may be the result of slowness of dye regeneration process as compared to rate of charge carriers' injection by the excited dye molecule.

Photocurrent Action Spectrum (IPCE)

In order to conclusively ascertain the sensitization of photocurrent by the dyes under investigation, the short-circuit photocurrent (J_{photo}) spectra of dye modified TiO_2 electrodes were determined. From the values of J_{photo} and the intensity of the corresponding monochromatic light (I_{inc}), the incident photon-to-current conversion efficiency (IPCE) was calculated at each excitation wavelength (λ) using the following relation:

$$IPCE\,(\%) = \frac{1240 J_{photo}\left(A/cm^2\right)}{\lambda\,(nm) \cdot I_{inc}\left(W/cm^2\right)} \times 100. \tag{1}$$

The IPCE versus wavelength (λ) curves for different cases (the natural dyes) are shown in Figure 8. It is clearly seen from this figure that there is close resemblance of the nature of IPCE curve with the absorption spectrum of the respective dye providing clear evidence of the sensitization of photocurrent by dye. The IPCE values observed at the characteristic wavelengths of the dyes ranged from 12% to 37%, decreasing in the order Tectona grandis > Callistemon citrinus > Tamarindus indica > Eucalyptus globules. The variation in IPCE values for different natural dyes could be due to the varied amount of dye loaded onto the TiO_2 thin film, different degree of charge carrier's recombination, different energy levels of excited dye molecule, and the quenching of excited state.

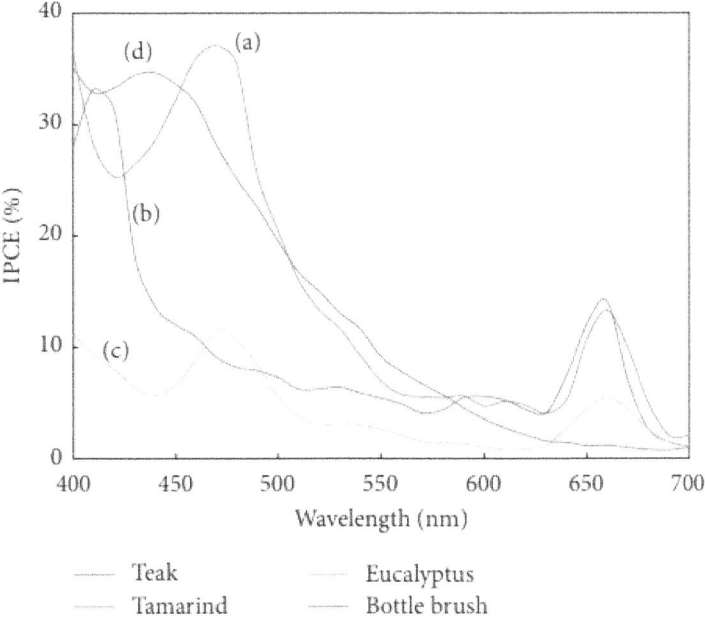

Figure 8: Action spectra of solar cell sensitized by the extracts (a) teak, (b) tamarind, (c) eucalyptus, and (d) bottle brush.

Power Conversion Efficiency (η) and Fill Factor (FF)

The power conversion efficiency and the fill factor of dye-sensitized solar cells were determined from the (J-V) curve of the respective cell under illumination by visible light. From the experimentally determined J-Vcurves (Figure 6), the values of fill factor (FF) and power conversion efficiency (η) were evaluated using the following relations:

$$FF = \frac{P_{max}}{P_{ideal}} = \frac{J_{max}\left(A/cm^2\right) \times V_{max}(V)}{J_{sc}\left(A/cm^2\right) \times V_{oc}(V)},$$

$$\eta(\%) = \frac{J_{max}\left(A/cm^2\right) \times V_{max}(V)}{I_{inc}\left(W/cm^2\right)} \times 100.$$

$$(2)$$

Here J_{sc}, V_{oc}, and I_{inc} are short-circuit photocurrent, open-circuit potential, and intensity of incident light, respectively. With the use of these dyes power conversion efficiency follows the order (Tectona grandis >Eucalyptus globulus > Tamarindus indica > Callistemon citrinus), while fill factor is obtained as (Eucalyptus globulus > Tectona grandis > Callistemon citrinus > Tamarindus indica).

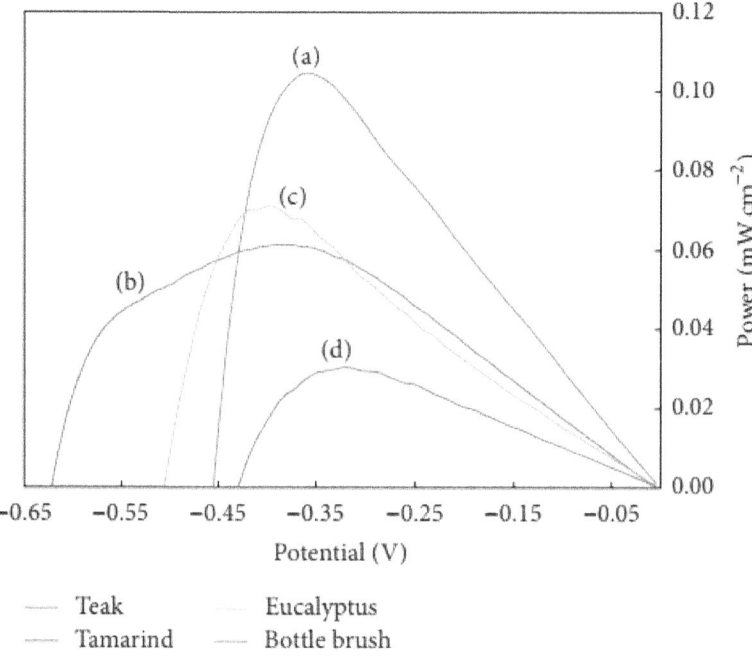

Figure 9: Power versus voltage curves of the DSSCs using the natural dyes extracted from the (a) teak, (b) tamarind, (c) eucalyptus, and (d) bottle brush.

The maximum output power (P_{max}) is obtained by choosing a point on experimentally determined (J-V) curve corresponding to which the product of current (J_{max}) and potential (V_{max}) gives the maximum value. Figure 9 shows the (power versus potential) curves for the natural dye(s)-sensitized solar cells, and the corresponding powers (p_{max}) obtained from various extracts are revealed in Table 1. The maximum photopower was obtained in the case of teak leaf extract, however low conversion responses may be due to poor interaction of sensitizers with the semiconductor electrode that restricts the transport of electrons from the excited dye molecules to the TiO_2 film.

CONCLUSIONS

Four natural dyes extracted from the leaves or flowers of the plants were used as sensitizer and their photovoltaic characteristics were studied. The extracted dyes contain tannins as the major coloring component along with some other minor components. Chlorophyll is the common component present in all the dyes extracted from the leaves. Tectoleafquinone is the key component present in the teak leaf extract. The chemical adsorption of these dyes becomes possible because of the condensation of hydroxyl and methoxy protons with the hydroxyl groups on the surface of nanostructured TiO_2. The DSSCs made using the extracted dyes showed the open circuit voltages (v_{oc}) varying between 0.430 and 0.610 V, and the short circuit photocurrent densities (J_{sc}) ranged from 0.11 to 0.29 mAcm^{-2}. The incident photo-to-current conversion efficiencies (IPCEs) varied from 12 to 37%. Among the four dyes studied, the extract obtained from teak has shown the best photosensitization effects in terms of the cell output as against the expectation arising from the apparent matching profile of the bottle brush extract with the solar spectrum. The natural dye extracts are, generally, a mixture of several pigments. Therefore, the possible reason for the observed differences in sensitization actions of dyes is their varied abilities towards adsorption onto the semiconductor surface. The impact of the different rates of electron transfer from the dye molecule to the conduction band of semiconductor electrode (energy levels alignments) is also reflected. Sometimes, a complication such as dye aggregation on semiconductor film produces absorptivity that results in either the nonelectron injection or the steric hindrance preventing the dye molecules from effectively arraying on the semiconductor film. This leads to the weaker binding and greater resistance, resulting in the low output of cells. Addition of appropriate additives for improving V_{oc} without causing dye degradation might result in further enhancement of the cell performances. Hence, though photocurrent densities, photovoltages, and IPCE obtained with these dyes are somewhat low, they are quite useful for their nontoxicity, greater availability, and very low cost of production opening up a perspective of feasibility for inexpensive and environmentally friendly dye cells.

ACKNOWLEDGMENTS

Reena kushwaha acknowledges the financial support received from the University Grant Commission, New Delhi, and the Ministry of New and Renewable Energy (MNRE), New Delhi, for this work.

REFERENCES

1. B. O'Regan and M. Gräetzel, "A low-cost, high-efficiency solar cell based on dye-sensitized colloidal TiO_2 films," Nature, vol. 353, pp. 737–740, 1991.

2. M. Grätzel, "Dye-sensitized solar cell," Journal of Photochemistry and Photobiology C, vol. 4, pp. 145–153, 2003.

3. M. Grätzel, "Sol-gel processed TiO_2 films for photovoltaic applications," Journal of Sol-Gel Science and Technology, vol. 22, no. 1-2, pp. 7–13, 2001.

4. J. Jiu, S. Isoda, M. Adachi, and F. Wang, "Preparation of TiO_2 nanocrystalline with 3–5 nm and application for dye-sensitized solar cell," Journal of Photochemistry and Photobiology A, vol. 189, no. 2-3, pp. 314–321, 2007.

5. C.-S. Chou, F.-C. Chou, and J.-Y. Kang, "Preparation of ZnO-coated TiO_2 electrodes using dip coating and their applications in dye-sensitized solar cells," Powder Technology, vol. 215-216, pp. 38–45, 2012.

6. T. S. Senthil, N. Muthukumarasamy, D. Velauthapillai, S. Agilan, M. Thambidurai, and R. Balasundaraprabhu, "Natural dye (cyanidin 3-O-glucoside) sensitized nanocrystalline TiO_2 solar cell fabricated using liquid electrolyte/quasi-solid-state polymer electrolyte," Renewable Energy, vol. 36, no. 9, pp. 2484–2488, 2011.

7. S. Kushwaha and L. Bahadur, "Characterization of synthetic Ni(II)-xylenol complex as a photosensitizer for wide-band gap ZnO semiconductor electrodes," International Journal of Photoenergy, vol. 2011, Article ID 980560, 9 pages, 2011.

8. S. S. Kanmani and K. Ramachandran, "Synthesis and characterization of TiO_2/ZnO core/shell nanomaterials for solar cell applications," Renewable Energy, vol. 43, pp. 149–156, 2012.

9. F. C. Krebs, "Fabrication and processing of polymer solar cells: a review of printing and coating techniques," Solar Energy Materials and Solar Cells, vol. 93, no. 4, pp. 394–412, 2009.

10. Y. Chiba, A. Islam, Y. Watanabe, R. Komiya, N. Koide, and L. Han, "Dye-sensitized solar cells with conversion efficiency of 11.1%," Japanese Journal of Applied Physics, vol. 45, no. 24–28, pp. L638–L640, 2006.

11. R. Buscaino, C. Baiocchi, C. Barolo et al., "A mass spectrometric analysis of sensitizer solution used for dye-sensitized solar cell," Inorganica Chimica Acta, vol. 361, no. 3, pp. 798–805, 2008.

12. G. Zhang, H. Bala, Y. Cheng et al., "High efficiency and stable dye-sensitized solar cells with an organic chromophore featuring a binary π-conjugated spacer," Chemical Communications, no. 16, pp. 2198–2200, 2009.

13. P. Srivastava and L. Bahadur, "Dye-sensitized solar cell based on nanocrystalline ZnO thin film electrodes combined with a novel light absorbing dye Coomassie Brilliant Blue in acetonitrile solution,"International Journal of Hydrogen Energy, vol. 37, no. 6, pp. 4863–4870, 2012.

14. S. Hao, J. Wu, Y. Huang, and J. Lin, "Natural dyes as photosensitizers for dye-sensitized solar cell," Solar Energy, vol. 80, no. 2, pp. 209–216, 2006.

15. K. Wongcharee, V. Meeyoo, and S. Chavadej, "Dye-sensitized solar cell using natural dyes extracted from rosella and blue pea flowers," Solar Energy Materials and Solar Cells, vol. 91, no. 7, pp. 566–571, 2007.

16. G. Calogero and G. D. Marco, "Red Sicilian orange and purple eggplant fruits as natural sensitizers for dye-sensitized solar cells," Solar Energy Materials and Solar Cells, vol. 92, no. 11, pp. 1341–1346, 2008.

17. K. E. Jasim, S. Al-Dallal, and A. M. Hassan, "Natural dye-sensitised photovoltaic cell based on nanoporous TiO_2," International Journal of Nanoparticles, vol. 4, no. 4, pp. 359–368, 2011.

18. C. Sandquist and J. L. McHale, "Improved efficiency of betanin-based dye-sensitized solar cells,"Journal of Photochemistry and Photobiology A, vol. 221, no. 1, pp. 90–97, 2011.

19. S. Sönmezoğlu, C. Akyürek, and S. Akin, "High-efficiency dye-sensitized solar cells using ferrocene-based electrolytes and natural photosensitizers," Journal of Physics D, vol. 45, Article ID 425101, 2012.

20. L. U. Okoli, J. O. Ozuomba, A. J. Ekpunobi, and P. I. Ekwo, "Anthocyanin-dyed TiO_2 electrode and its performance on dye-sensitized solar cell," Research Journal of Recent Sciences, vol. 1, pp. 22–27, 2012.

21. R. Espinosa, I. Zumeta, J. L. Santana et al., "Nanocrystalline TiO$_2$ photosensitized with natural polymers with enhanced efficiency from 400 to 600 nm," Solar Energy Materials and Solar Cells, vol. 85, no. 3, pp. 359–369, 2005.

22. E. Yamazaki, M. Murayama, N. Nishikawa, N. Hashimoto, M. Shoyama, and O. Kurita, "Utilization of natural carotenoids as photosensitizers for dye-sensitized solar cells," Solar Energy, vol. 81, no. 4, pp. 512–516, 2007.

23. H. Zhu, H. Zeng, V. Subramanian, C. Masarapu, K.-H. Hung, and B. Wei, "Anthocyanin-sensitized solar cells using carbon nanotube films as counter electrodes," Nanotechnology, vol. 19, no. 46, Article ID 465204, 2008.

24. D. Zhang, S. M. Lanier, J. A. Downing, J. L. Avent, J. Lum, and J. L. McHale, "Betalain pigments for dye-sensitized solar cells," Journal of Photochemistry and Photobiology A, vol. 195, no. 1, pp. 72–80, 2008.

25. W. H. Lai, Y. H. Su, L. G. Teoh, and M. H. Hon, "Commercial and natural dyes as photosensitizers for a water-based dye-sensitized solar cell loaded with gold nanoparticles," Journal of Photochemistry and Photobiology A, vol. 195, no. 2-3, pp. 307–313, 2008.

26. R. Hernández-Martínez, M. Estevez, S. Vargas, F. Quintanilla, and R. Rodríguez, "Natural pigment based dye sensitized solar cells," Journal of Applied Research and Technology, vol. 10, pp. 38–47, 2012.

27. M. R. Narayan, "Review: dye sensitized solar cells based on natural photosensitizers," Renewable and Sustainable Energy Reviews, vol. 16, no. 1, pp. 208–215, 2012.

28. R. Aradhana, K. N. V. Rao, D. Banji, and R. K. Chaithanya, "A review on Tectona grandis.linn: chemistry and medicinal uses," Herbal Tech Industry, vol. 6, no. 11, 2010.

29. R. Mongkholrattanasit, J. Kryštůfek, J. Wiener, and J. Studničková, "Natural dye from Eucalyptus leaves and application for wool fabric dyeing by using padding techniques," in Natural Dyes, E. A. Kumbasar, Ed., chapter 4, 2011.

30. E. D. Caluwé, K. Halamová, and P. V. Damme, "Tamarindus indica L.: a review of traditional uses, phytochemistry and pharmacology," Afrika Focus, vol. 23, pp. 53–83, 2010.

Chapter 9

PIGMENT PRODUCTION AND GROWTH OF ALTERNANTHERAPLANTS CULTURED IN VITRO IN THE PRESENCE OF TYROSINE

Alítcia Moraes Kleinowski; Isabel Rodrigues Brandão; Andersom Milech Einhardt; Márcia Vaz Ribeiro; José Antonio Peters; Eugenia Jacira Bolacel Braga

Laboratório de Cultura de Tecidos de Plantas; Departamento de Botânica; Instituto de Biologia;Universidade Federal de Pelotas; Capão do Leão - RS - Brasil

ABSTRACT

The aim of the present study was to investigate the influence of tyrosine on the in vitro growth and the production of the betacyanin pigment in*Alternanthera philoxeroides* and *A. tenella*. Nodal segments were inoculated in MS medium containing different concentrations of tyrosine (0, 25, 50 and 75 µM), and the number of sprouts and buds, height, root length, fresh matter of shoots and roots and betacyanin content were evaluated. In *A. philoxeroides* , the highest production of betacyanin (51.30 mg 100 g^{-1} FM) was in the stems with the addition of approximately 45 µM tyrosine, while the increase in the leaves was proportional to the tyrosine concentration, and the best average was obtained with a tyrosine concentration of 75 µM (15.32 mg 100 g^{-1} FM). Higher tyrosine concentrations were deleterious to the growth of *A. tenella* plants, and a concentration of 75 µM was considered toxic. However, a tyrosine concentration of 50 µM benefitted betacyanin production, which reached 36.5 mg 100 g^{-1} FM in the plant shoots. These results showed the positive effect of tyrosine on the production of betacyanin in both species; however, application at high concentrations hampered the growth of *Alternanthera* plants.

INTRODUCTION

The Caryophyllales order in the family Amaranthaceae comprises 65 genera and approximately 1,000 described species that originate from the tropical, subtropical and temperate zones of Africa, South America and Southeast Asia. The genus *Alternanthera* is a prominent member of this family and consists of 80 species (30 of which have been described in Brazil) (Siqueira 1995). *A. philoxeroides* (Mart). Griseb (alligator weed) and *A. tenella* Colla (joy weed) are two species that deserve special attention due to their medicinal and economic importance. The former, a herbaceous and perennial plant, is considered a vigorous invader in many regions of the world due to its ability to adapt to different ecosystems (Gunasekera and Bonila 2001). *A. philoxeroides* contains flavonoid glycosides and betalains (Blunden et al. 1999; Rattanathongkom et al. 2009), which are known for their antitumor and antiviral properties (Fang et al. 2007) in addition to anti-inflammatory and immunomodulatory activities (Salvador and Dias 2004). *A. tenella* has been used to treat the infections, fevers, bruises and itches and also has diuretic and anti-inflammatory properties (Vendruscolo and Mentz 2006). Studies with the water extract of joy weed demonstrated immunomodulatory and antitumor activities on the rats (Moraes et al. 1994; Guerra et al. 2003) and *in vitro* antibiotic activity against Gram-positive and Gram-negative bacteria (Silveira and Olea 2009).

The major *A. tenella* components that have been studied are tannins, saponins, flavonoid glycosides and heterosides, such as isorhamnetin, quercetin and kaempferol (Biella et al. 2008), as well as other flavonoids, including vitexin and acacetin (Salvador et al. 2006). Studies on the chemical composition of species of *Alternanthera* showed the presence of betalains, betacyanins, betaxanthins, chromo-alkaloids and flavonoids (Brochado et al. 2003; Cai et al. 2005; Salvador et al. 2006). Betalains are N-heterocyclic natural pigments that are derived from the amino acid tyrosine. Under the action of tyrosine hydroxylase, tyrosine forms an intermediate DOPA (4,5 dihydroxyphenylalanine), which is oxidized into cDOPA. This compound leads to the formation of the class of betalains (Tanaka et al. 2008), which are classified into two groups: betaxanthins (yellow) and betacyanins (red). The betacyanins may also be chemically classified into four types: betanin, amaranthine, gonferine and bougainvillein (Volp et al. 2009).

Several authors have studied the bioactivities of these pigments, including their antiviral and antibiotic effects (Strack et al. 2003). Their antioxidant properties have been shown in a wide range of tests (Kanner et al. 2001; Gentile et al. 2004; Tesoriere et al. 2008). For example, Tesoriere et al. (2003) reported that enrichment of human low-density lipoproteins with betalains increased the oxidation resistance, slowing cell aging. Muntha et al. (2005) also documented that natural pigments, such as betanin, could inhibit cell proliferation of a wide variety of human tumor cells (Muntha et al. 2005). Several authors have suggested *in vitro* culturing of betacyanin-producing plants to optimize the large-scale production of these pigments (Santos-Diaz et al. 2005; Savitha et al. 2006; Georgiev et al. 2008; Pavokovi et al. 2009). Chemical synthesis of these molecules does not seem feasible due to a lack of clarification of several steps involved in this synthetic process (Pavokovi and Krsnik-Rasol 2011). Precursor amino acids, which are considered chemical elicitors, have been used successfully both in the cultured plant cells and in intact plants to improve the pigment production (Berlin et al. 1986; Silva et al. 2005).

Given the medicinal importance of betacyanin and species of the genus *Alternanthera*, the aim of the present study was to evaluate the influence of tyrosine on the morphological characteristics of and betacyanin production in *A. philoxeroides* and *A. tenella* cultured *in vitro*.

MATERIALS AND METHODS

The taxonomy of *A. philoxeroides* (alligatorweed) plants from the municipality of Rio Grande, state of Rio Grande do Sul (RS), Brazil was confirmed by the Amaranthaceae identification key, and the plants were cataloged in the PEL Herbarium under the number 24.53. *A. tenella* (joy weed) plants from the municipality of Pelotas, RS, Brazil had their taxonomy confirmed and were cataloged in the PEL Herbarium under the number 25.26.

Nodal segments of new stems containing one, or two axillary buds of the plants kept for 15 days in the greenhouse were used for *in vitro* establishment of both the species. The segments were washed in tap water and distilled water under mechanical stirring for 15 min. The material was immersed in 70% ethanol for 20 seconds, followed by immersion in 1% sodium hypochlorite with three drops of Tween (20 min). All the procedures were intercalated with autoclaved water baths.

Basic MS medium (Murashige and Skoog 1962) with no growth regulators, pH adjusted to 5.8 and 7.0 g L^{-1} of agar was used for *in vitro* culturing of *Alternathera* plants. Bottles containing 40 mL of culture medium were autoclaved at 121°C for 20 min. After autoclaving, tyrosine (0, 25, 50 and 75 µM) filtered and solubilized in dimethyl sulfoxide (DMSO) was added to the medium. Experiments with both the species were performed separately as follows.

Experiment 1

Explants of *A. philoxeroides* were inoculated in the culture medium in a laminar flow hood under aseptic conditions. Bottles with the explants were placed in a growth room kept under a photoperiod of 16 h, photon flux density of 48 μmoL m^{-2} s^{-1} and temperature of $23 \pm 2°C$. After 40 days of experimental implantation, the average number of auxiliary buds and stems, height (cm), stem fresh matter (mg), root length (cm), root fresh matter (mg) and betacyanin quantification (amaranthine, mg 100 g^{-1} FM) were evaluated. To analyze the betacyanin content, leaves and stems were macerated separately in 0.5 mL of distilled water and then centrifuged at 13.632 g at 4°C for 25 min. The supernatant was used to quantify the betacyanins by monitoring the absorbance at 536 nm and 650 nm with an Ultrospec 2100 Pro spectrophotometer (Amersham Biosciences®). Betacyanin concentration was determined using the molar extraction coefficient of amaranthine (5.66×10^4) according to the method described by Cai et al. (1998).

The experimental design was completely randomized with four concentrations of tyrosine and five replicates. Each experimental unit consisted of one bottle containing five explants. The results were subjected to analysis of variance (ANOVA) and polynomial regression using WinStat statistical software (Machado and Conceição 2002).

Experiment 2

The *in vitro* culturing and growth evaluations of *A. tenella* followed the same method described above for *A. philoxeroides*. However, for betacyanin quantification, stems and leaves were added together and considered as shoots to obtain the amount of fresh matter required for analysis. The experimental design used was completely randomized with four treatments (concentrations of tyrosine) and four replicates. The experimental unit consisted of one bottle containing four explants. The results were subjected to ANOVA, and the means were compared by a Tukey's post-test with a 5% error probability using WinStat statistical software (Machado and Conceição 2002).

RESULTS AND DISCUSSION

In vitro culturing of *A. philoxeroides*

The concentration of tyrosine negatively influenced the plant height, with the highest average (6.91 cm) found in the control plants (Fig. 1 and 2A). Although nitrogen was an essential nutrient for plant growth and development, excess organic nitrogen, due to the presence of tyrosine, might have acidified the culture medium, which hampered the growth of *A. philoxeroides* shoots.

Figure 1: Plants of Alternanthera philoxeroides grown for 40 days in MS medium with different tyrosine concentrations. Control (A); 25 pM tyrosine (13): 50 pM tyrosine (C); 75 pM tyrosine (D). Scale 1 (cm).

The length and fresh matter of roots decreased as the tyrosine concentration increased and no rooting was detected at a concentration of 75 μM. In a study with lentil (*Lens culinaris* Medik), Sarker et al. (2003) found that 20 μM tyrosine combined with other growth regulators showed successful regeneration and *in vitro* rooting. However, in the studies with *A. brasiliana,* Silva et al. (2005) showed that the plants cultured in MS medium with 10 μM tyrosine under white light showed no difference in the root length compared to the control, with averages of 7.40 cm and 7.05 cm, respectively.

In the present study, the treatment with 25 μM tyrosine led to a significant difference in the root formation of *A. philoxeroides*. As the concentration of tyrosine increased in the culture medium, the length and fresh matter of the roots were significantly reduced (Fig. 2B and 2C). In a study with soy plants (*Glycine max* L.) treated with hydroxylated tyrosine, Soares (2006) found that the average total length and number of roots were 20.9 and 76.7% smaller than the controls for treatments with 25 μM and 100 μM, respectively, indicating that this amino acid might have an adverse effect on rooting. According to Oliveira et al. (2009), direct amino acid absorption by the roots offers advantages to plants because they do not need to metabolize the mineral nitrogen (nitrate and ammonium), thus directing more energy for rooting. However, excess amino acids can cause substrate acidification, leading to toxicity.

Betacyanin production in the stems peaked at approximately 45 µM (51.30 mg 100 g^{-1} FM), while the increase in the leaves was proportional to the tyrosine concentration, with the highest average level obtained with a tyrosine concentration of 75 µM (15.32 mg 100 g^{-1} FM) (Fig. 2D). Taha et al. (2008) found that the addition of amino acid precursors, such as tryptophan and glutamine, in *Catharanthus roseus* cultured in MS medium, increased the production of vinblastine and vincristine in the treated cells by up to 75%. In callus tissue cultures of velvet bean (*Mucuna pruriens*), Desai et al. (2010) found that the application of 140 µM of various amino acid precursors, including tyrosine, increased the production of L-DOPA by up to 2%. These authors noted that the enzyme triggering the formation of betacyanin, tyrosine hydroxylase, required copper and that copper was a micronutrient in the formulation of MS medium.

Thus, the addition of the substrate tyrosine might have facilitated the action of the enzyme and triggered reactions that increased the content of betacyanin. Similar to what happened in *A. brasiliana* cultured in MS medium containing tyrosine, Silva et al. (2005) found that the accumulation of betacyanin was higher than in the control.

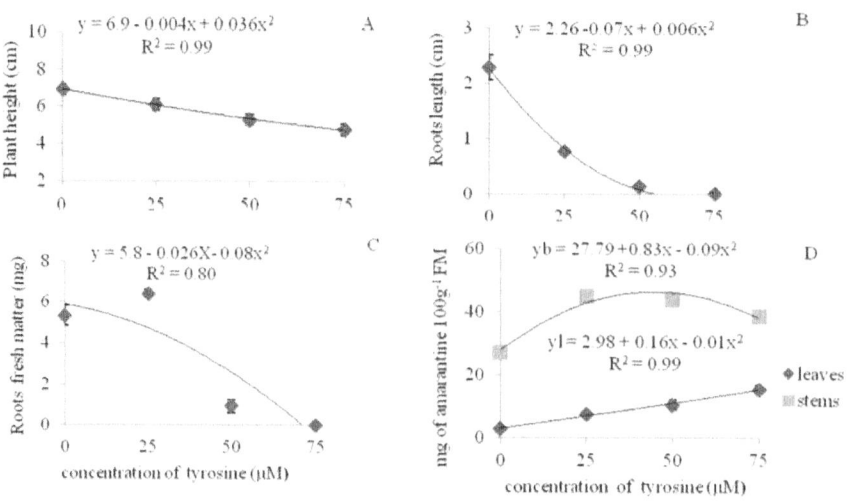

Figure 2: Plant height (A), roots length (B), root fresh matter (C) and betacyanin production (D) of Alternanthera philoxeroides grown in vitro for 40 days in MS culture medium with different of tyrosine concentrations. Vertical bars represent the standard error for the average of five repetitions.

Tyrosine was used in the present study because it was the precursor amino acid in betacyanin synthesis. According to Georgiev et al. (2008), tyrosine addition should stimulate the synthesis pathway of its corresponding secondary metabolite. This was observed in the leaves and stems of *A. philoxeroides*, except with the highest concentration used in the stem. Certain factors must be considered, such as the cellular capacity for the accumulation of the compound and whether the metabolite of interest is an end product of the biosynthetic pathway (Verpoorte and Maraschi 2001). The target metabolite is not a final product of the biosynthetic pathway; the rate of its catabolism may become ineffective with the addition of the precursor. Moreover, the cellular capacity for accumulation of this secondary metabolite is currently not well understood. However, limited accumulation of these compounds, which may influence the yield and accumulation of this pigment in *A. philoxeroides*, is expected.

In vitro Culturing of *A. Tenella*

The presence of the tyrosine in the culture media had a negative influence on the morphological characteristics of *A. tenella*. The highest concentration (75 μM) was completely deleterious, preventing the development of the explants. In the present study, the high levels of organic nitrogen in the culture media might have hindered nitrogen assimilation in the form of nitrate, which was preferentially used by the plants (Sen and Batra 2011) and was also present in the MS culture medium formulation. Difficulty in assimilation could account for the deficit in the growth of shoots and roots and the death of nodal segments of *A. tenella* in the presence of 75 μM tyrosine as nitrogen was essential for the growth and development of plants (Schröder et al. 2000).

The formation of new buds was also negatively influenced by the increase in the amino acid concentration in the culture medium and the smallest average was obtained with 50 μM tyrosine. Nitrogen is a constituent of amino acids, nucleotides and coenzymes (Kanashiro et al. 2007); therefore, there is a correlation between the availability of nitrogen and the formation of new organs. The presence of aromatic amino acids with hydroxyl groups, such as Ser and Tyr, in the nutrient solution could have affected the magnesium transporter activity (Garcia et al. 2011), and thus, the development off the plant and the formation of new sprouts and buds (Fig. 3A and Fig. 3B).

The increase in the concentration of organic nitrogen in the medium might have increased the osmotic potential and thereof, hindered the uptake of water by the explants of *A. tenella*. This might have affected their growth in height (Fig. 3C), as previously described for the plants of Brazilian ginseng (*Pfaffia glomerata* Spreng.) where growth was higher at the typical concentration of nitrogen in MS medium and decreased as the organic nitrogen concentration increased (Russowski and Nicoloso 2003).

The al deficiency of *A. tenella* with the treatments used in the present study influenced the fresh biomass of these plants. The biomass decreased with increasing tyrosine, and the lowest average (615 mg) was obtained with 50 μM tyrosine (Fig. 3D). These results corroborated those of Silva et al. (2005), who tested other factors besides tyrosine in the *in vitro* culture of *A. brasiliana*. Regardless of the treatment used, these authors found that the presence of tyrosine in the culture medium led to the lowest values of dry matter. Berlin et al. (1986) also found this negative effect in *Chenopodium rubrum* L. (red goosefoot) cell culture using 15 μM tyrosine.

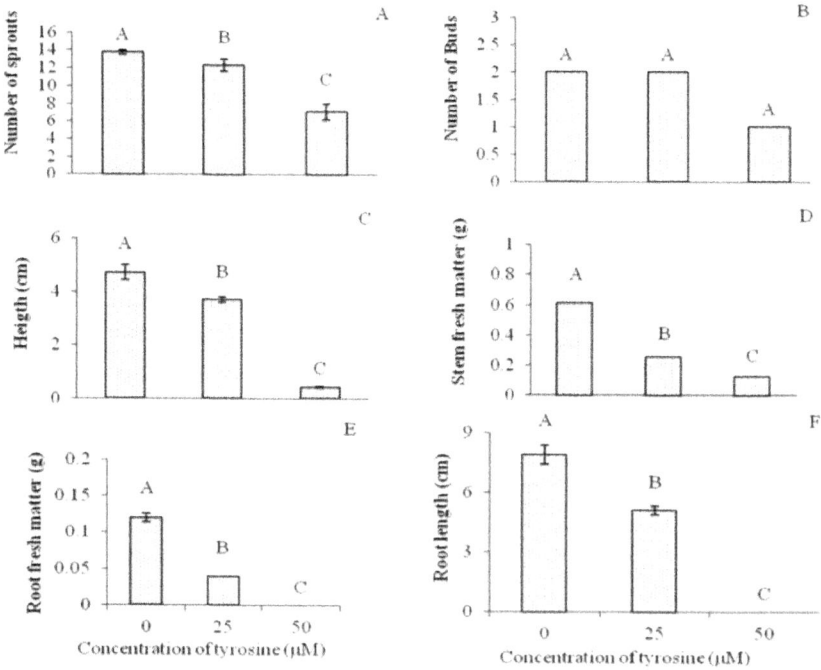

Figure 3: Number of sprouts (A), buds number (B), height (C), stem fresh matter (D), root fresh matter (E), root length, (F) of Alternanthera tenella grown in vitro for 40 days in MS medium with different concentrations of tyrosine. Values followed by different letters are significantly different (p ≤ 0.05; Tukey test). Vertical bars represent the standard error for the average of four repetitions.

Their results revealed that high amounts of this amino acid might be toxic to cell proliferation and growth of this species. The rhizogenesis of *A. tenella* was significantly different with the presence of tyrosine in the culture medium. The highest values of fresh root matter and main root length were found in the plants grown in tyrosine-free media (Fig. 3E and Fig. 3F).

However, after 40 days of culture, there was a significant increase in the betacyanin content with 50 µM tyrosine, and a value of 36.95 mg of amaranthine of 100 g FM⁻¹ was achieved (Fig. 5).

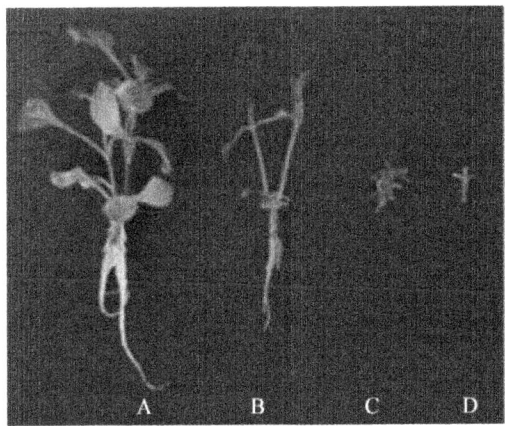

Figure 4: Plants of Alternanthera tenella grown for 40 days in MS medium with different tyrosine concentrations. Control (A), 25 i.tM tyrosine (B), 50 pM tyrosine (C) and 75 pM tyrosine (D). Scale 1 (cm).

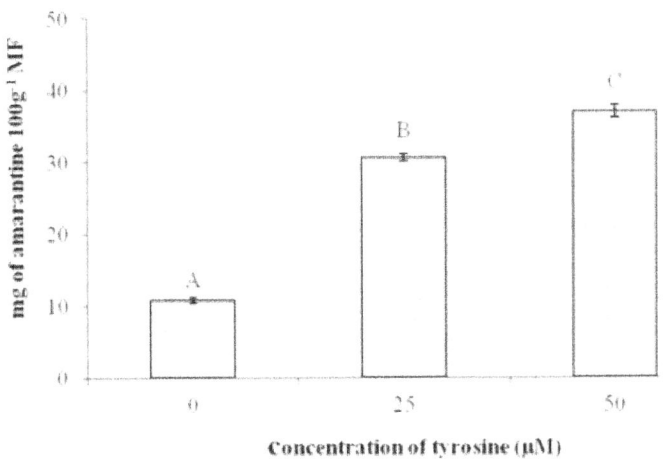

Figure 5: Betacyanin production in plants of Alternanthera Amelia in vitro for 40 days in MS medium with different concentrations of tyrosine. Values followed by different letters are significantly different (pS 0.05; Tukey test). Vertical bars represent the standard error for the average of four repetitions.

Tyrosine has also been tested as an elicitor agent for the production of various secondary metabolites in several plant species. In *Cereus peruvianu* callus tissue culture, an increase in the production of tyramine and hordenine alkaloids previously identified in this species was found with the addition of 1.1 M tyrosine (Rocha et al. 2005). Tyrosine and other amino acids were tested to promote protoberberine alkaloid production in the *in vitro* culture of *Thalictrum minusa*. However, Urmantseva et al. (2005) found that the addition of tyrosine hindered the formation of alkaloids compared to the control treatment.

The increase in betacyanin production due to the elicitor action of tyrosine has been reported. Berlin et al. (1986) used 15 μM tyrosine in the culture of *Chenopodium rubrum* callus tissue and demonstrated an increase in the production of betalains of 50 to 100% after 28 days of culture. Silva et al. (2005) found that the production of betalains increased compared to the control in the *in vitro* culture of *A. brasiliana* for 45 days in MS medium supplemented with 10 μM tyrosine. Other elicitors have also been used to increase the production of betalains in the plants. Savitha et al. (2006) tested seven different metal ions in concentrations up to ten times higher than the ones present in MS medium and found that calcium could induce up to a 47% higher betalain production. Using polyamines as elicitor agents, Bais et al. (2000) doubled the productivity of betalains in beet (*Beta vulgaris*)crops. Suresh et al. (2004) found similar results using putrescine and spermidine in bioreactors, increasing the content of betalains in beet cultures by 1.3 times.

In the present study an increased production of betacyanin was found with the addition of tyrosine. Such a response could be related both to the stress caused by the presence of the elicitor in the medium and mainly due to the increased availability of tyrosine, which was the initial substrate for tyrosine hydroxylase (TOH), the enzyme that converted tyrosine to DOPA and led to the formation of betacyanins. By interacting with membrane receptors, the presence of elicitors in the culture medium can trigger responses in the secondary metabolism of the plants. Savitha et al. (2006) noted that stressors would also be able to activate specific genes in the enzymatic machinery involved in the biosynthesis of secondary metabolites such as betacyanins, key pigments in the plants due to antioxidant and free radical-scavenging properties (Gandía-Herrero et al. 2009).

CONCLUSION

High concentrations of tyrosine had a negative effect for the *in vitro* growth of *Alternanthera* species affecting the formation of shoots and roots, but the addition of this amino acid in the culture medium increased the biosynthesis of betacyanin in the studied two species. The involvement of tyrosine in the metabolism of natural compounds such as alkaloids and other nitrogen products such as betacyanin has been well established in the literature. But to-date little is known about its mechanism of control and regulation, and the role in secondary metabolism of the two medicinal plant species tested. This study demonstrated that the presence of tyrosine, as an elicitor in the culture medium could cause a considerable increase in the production of amaranthine, a natural derivative important to the nutraceutical industry.

REFERENCES

1. Berlin J, Sieg S, Strack D, Bokern M, Harms H. Production of betalains by suspension cultures of *Chenopodium rubrum* L. *Plant Cell Tiss Org.* 1986; 5: 163-174.

2. Biella CA, Salvador MJ, Dias DA, Baruffi DM, Crott LSP. Evaluation of immunomodulatory and anti-inflammatory effects and phytochemical screening of *Alternanthera tenella* Colla (Amaranthaceae) aqueous extracts. *Mem Inst Oswaldo Cruz.* 2008; 103: 569-577.

3. Blunden G, Yang M, Janicsák G, Máthé I, Cabarot-Cuervo C. Betaine distribution in the Amaranthaceae. *Biochem system ecol.* 1999; 27: 87-92.

4. Brochado CO, Almeida AP, Barreto BP, Costa LP, Ribeiro LS, Pereira RLC, Gonçalves-Koat ZVL, Costa SS. Flavonol robinobiosides and rutinosides from *Alternanthera brasiliana* (Amaranthaceae) and their effects on lymphocyte proliferation *in vitro*. *J Braz Chem Soc.* 2003; 14: 449-451.

5. Cai Y, Sun M, Corke H. Characterization and application of betalain pigments from plants of the Amaranthaceae.*Trends Food Sci Tech.* 2005; 16: 370-376.

6. Desai M, Madhuri S, Sharan M. Effect of Culture Conditions on L-Dopa Accumulation in Callus Culture of *Mucuna pruriens*. *J Chem Soc.* 2010; 2: 134-146.

7. Fang JB, Duan HQ, Zhang YW, Yoshihisa T. Cytotoxic triterpene saponins from *Alternanthera philoxeroides*. *J Asian Nat Prod Res.* 2007; 11: 261-266.

8. Gandía-Herrero F, Jiménez-Atiénzar M, Cabanes J, Escribano J, García-Carmona F. Fluorescence detection of tyrosinase activity on dopamine-betaxanthin purified from *Portulaca oleracea* (common purslane) flowers. *J Agric Food Chem.* 2009; 57:2523-2528.

9.   Garcia AL, Madrid R, GimenoV, Rodriguez-Ortega WM, Nicolas N, Garcia-Sanchez F. The effects of amino acids fertilization incorporated to the nutrient solution on mineral composition and growth in tomato seedlings. *Span J Agric Res.* 2011; 852 -861.

10. Gentile C, Tessoriere L, Allegra M, Livrea MA, Alessio PD. Antioxidant betalins from Cactus pear (*Ficus-indica*) inhibit endothelial ICAM-1 expression. *Ann N Y Acad Sc.* 2004; 1028 : 481-486.

11. Georgiev V, Mladenka I, Bleyt B, Pavlov A. Betalain production in plant in vitro systems. *Acta Physiol Plant.* 2008; 30: 581-593.

12. Guerra RNM, Pereira AW, Silveira LMS, Olea RSG. Immunomodulatory properties of *Alternanthera tenella* Colla aqueous extracts in mice. *Braz J Med Bio Res.* 2003; 36: 1215-1219.

13. Gunasekera L, Bonila J. Alligator weed: tasty vegetable in Australian backyards. *J Aquat Plant Manage.* 2001; 39: 17-20.

14. Kanashiro S, Ribeiro RCS, Gonçalves NA, Dias CTS, Jocys T. Efeitos de diferentes concentrações de nitrogênio no crescimento de *Aechmea blanchetiana* (Baker) cultivada *in vitro. Hoehnea.* 2007; 34: 59-66.

15. Kanner J, Harel S, Granit R. Betalains – a new class of dietary cationized antioxidants. *J Agric Food Chem.* 2001; 49: 5178-5185.

16. Moraes VLG, Santos LFM, Castro SB, Loureiro LH, Lima AO, Souza ML, et al. Inhibition of lymphocyte activation by extracts and fractions of *Kalanchoe, Alternanthera, Paullinia* and *Mikani*a species. *Phytomedicine.* 1994; 1: 199-204.

17. Muntha RK, Ruby L, Lino A, Muraleedharan G. Relative inhibition of lipid peroxidation, cyclooxygenase enzymes and human tumor cell proliferation by natural food colors. *J Agric Food Chem.* 2005; 53: 9268-9273.

18. Murashige T, Skoog FA. Revised medium for rapid growth and bioassays with tabacco tissue cultures. *Physiol Plant.* 1962; 15: 473-497.

19. Nicoloso FT, Ferrão GE, Castro GY. pH do meio de cultivo e crescimento de plântulas de ginseng brasileiro cultivadas *in vitro.Cienc Rural.* 2008; 38: 2059-2062.

20. Oliveira MC, Neto JV, Pio R, Oliveira AF, Ramos JD. Enraizamento de estacas de oliveira submetidas à aplicação de fertilizantes orgânicos e AIB. *Ciênc Agrotec.* 2009; 34: 337-344.

21. Pavokovi D, Rusak G, Besendorfer V, Krsnik-Rasol M. Light-dependent betanin production by transformed cells of sugar beet. *Food Technol and Biotechnol.* 2009; 47: 153-158.

22. Perotti JC, Rodrigues ICS, Kleinowski AM, Ribeiro MV, Einhart AM, Peters JA, Bacarin MA, Braga EJB. Produção de betacianina em erva-de-jacaré cultivada *in vitro* com diferentes concentrações de sulfato de cobre. *Cienc Rural*. 2010; 40:1874-1880.

23. Rattanathongkom A, Lee JB, Hayashi K, Sripanidkulchai BO, Kanchanapoom T, Hayashi T. Evaluation of chikusetsusaponin IVa isolated from *Alternanthera philoxeroides* for its potency against viral replication. *Plant Med*. 2009; 75:829-835.

24. Rocha KL, Oliveira AJB, Mangolin AC, Machado MFPS. Effect of different culture medium components on production of alkaloid in callus tissues of *Cerus Peruvians*. *Acta Sci*. 2005; 27: 37-41.

25. Russowski D, Nicoloso FT. Nitrogênio e fósforo no crescimento de plantas de ginseng brasileiro [*Pfaffia glomerata*(Spreng.) Pedersen] cultivadas *in vitro*. *Cienc Rural*. 2003; 33: 57-63.

26. Salvador MJ, Dias DA. Flavone C-glycosides from *Alternanthera maritima* (Mart.). St. Hil (Amaranthaceae).*Biochem System Ecol*. 2004; 32:107-110.

27. Salvador MJ, Ferreira EO, Mertens-Talcott SU, Castro WV, Butterweck V, Derendorf H, Dias DA. Isolation and HPLC quantitative analysis of antioxidant flavonoids from *Alternanthera tenella* Colla. *Z Naturforsch*. 2006; 61: 19-25.

28. Santos-Diaz MS, Velasquez-Garcia Y, Gonzalez-Chavez, MM. Pigment production by callus of *Mammillaria candida*(Cactaceae). *Agrociencia*. 2005; 39: 619-626.

29. Sarker RH, Barkat MM, Ashapurna B, Shirin M, Mouful NR, Hashem MI. In vitro regeneration in lentil (*Lens culinaris* Medik.). *Plant Tissue Cult*. 2003; 13:155-163.

30. Savitha BC, Thimmaraju RN, Bhagyalakshmi BA, Ravishankar GA. Different biotic and abiotic elicitors influence betalain production in hairy root cultures of *Beta vulgaris* in shake-flask and bioreactor. *Process Biochem*. 2006; 41: 50-60.

31. Schröder JJ, Neeteson JJ, Oenema O, Struik PC. Does the crop or the soil indicate how to save nitrogen in maize production? Reviewing the state of the art. *Field Crop Res*. 2000; 66:151-64.

32. Silva NCB, Macedo AF, Lage CLS, Esquibel MA, Sato A. Developmental effects of additional ultraviolet a radiation growth regulators and tyrosine in *Alternanthera brasiliana* (L.) Kuntze cultured *in vitro*. *Braz Arch Biol Technol*. 2005; 48: 779-786.

33. Silveira LMS, Olea RSG. Isolamento de compostos com atividade antibacteriana em *Alternanthera tenella* Colla (Amaranthaceae). ***Rev Bras Farm.*** 2009; 90: 148-153.

34. Siqueira, JC. Phytogeography of brasilian Amaranthaceae. *Pesquisa Botânica.* 1995; 45: 5-21.

35. Soares AR. Lignificação de raízes de soja sob a ação de L-diidroxifenilalanina (L-Dopa). [Mestrado, Dissertação]. Maringá, Brasil: Universidade Estadual Maringá; 2006.

36. Strack D, Vogt T, Schliemann W. Recent advances in betalain research. *Phytochemistry.* 2003; 62: 247-269.

37. Suresh B, Thimmaraju R, Bhagyalakshmi N, Ravishankar GA. Polyamine and methyl jasmonate-influenced enhancement of betalaine production in hairy root cultures of *Beta vulgaris* grown in a bubble column reactor and studies on efflux of pigments. *Process Biochem.* 2004; 39: 2091-2096.

38. Taha HS, Salah EM, Ola I, El-Hamshary M, Nahla S, Naglaa A, Nazif M, El-Bahr M, Medhat M, El-Nasr S. *In vitro*studies on Egyptian *Catharanthus roseus* (L.). Effects of extra tryptophan decarboxylase and strictosidine synthase genes copies in indole alkaloid production. *J Cell Mol Biol. 2008*; 2: 18-23.

39. Tanaka Y, Sasaki N, Ohmiya A. Biosynthesis of plant pigments: anthocyanins, betalains and carotenoids. *Plant.* 2008; 54: 733-749.

40. Tesoriere L, Butera D, D'arpa D, DI Gaudio F, Allegra M, Gentile C, Livrea MA. Increased resistance to oxidation of betalain-enriched human low density lipoproteins. *Free Radic Res.* 2003; 37: 689-696.

41. Tesoriere L, Fazzari LM, Francesa LM, Gentile AC, Livrea MA. In Vitro Digestion of Betalainic Foods Stability and Bioaccessibility of Betaxanthins and Betacyanins and Antioxidative Potential of Food Digesta. *J Agric Food Chem. 2008;* 56: 10487-10492.

42. Trejo TG, Jiménez AA, Rodríguez MM, Jcobalt D. Cobalt and other microelements on the production of betalains and the growth of growth of suspension cultures of *Beta vulgaris. Plant Cell Tiss Org.* 2001; 67: 19-23.

43. Vendruscolo GS, Mentz, LA. Levantamento etnobotânico das plantas utilizadas como medicinais por moradores do bairro Ponta Grossa. *Iheringia.* 2006; 61: 83-103.

44. Verpoorte R, Maraschi M. Engenharia do metabolismo de plantas medicinais. In: Yunes, R. A., Calixto, J. B. (org.).*Plantas medicinais sob a ótica da química medicinal moderna.* 1° ed Chapecó, Brasil, Argos; 2001, p. 381-432.

45. Volp ACP, Renhe IRT, Stringueta PC. Pigmentos naturais bioativos. *Aliment Nutr.* 2009; 20: 157-166.

Chapter 10

BETALAIN PRODUCTION IS POSSIBLE IN ANTHOCYANINPRODUCING PLANT SPECIES GIVEN THE PRESENCE OF DOPA-DIOXYGENASE AND L-DOPA

Nilangani N Harris[1],[4], John Javellana[1], Kevin M Davies[1], David H Lewis[1], Paula E Jameson[2], Simon C Deroles[1], Kate E Calcott[1],[3], Kevin S Gould[3] and Kathy E Schwinn[1]

[1]New Zealand Institute for Plant & Food Research Limited, Private Bag 11- 600, Palmerston North, New Zealand

[2] School of Biological Sciences, University of Canterbury, Private Bag 4-800, Christchurch, New Zealand

[3] Victoria University of Wellington, Wellington 6140, New Zealand

[4] Commonwealth Scientific and Industrial Research Organization, Ecosystem Sciences, Urrbrea, South Australia 5064, Australia

ABSTRACT

Background

Carotenoids and anthocyanins are the predominant non-chlorophyll pigments in plants. However, certain families within the order Caryophyllales produce another class of pigments, the betalains, instead of anthocyanins. The occurrence of betalains and anthocyanins is mutually exclusive. Betalains are divided into two classes, the betaxanthins and betacyanins, which produce yellow to orange or violet colours, respectively. In this article we show betalain production in species that normally produce anthocyanins, through a combination of genetic modification and substrate feeding.

Results

The biolistic introduction of DNA constructs for transient overexpression of two different dihydroxyphenylalanine (DOPA) dioxygenases (DODs), and feeding of DOD substrate (L-DOPA), was sufficient to induce betalain production in cell cultures of *Solanum tuberosum* (potato) and petals of *Antirrhinum majus*. HPLC analysis showed both betaxanthins and betacyanins were produced. Multi-cell foci with yellow, orange and/or red colours occurred, with either a fungal DOD (from*Amanita muscaria*) or a plant DOD (from *Portulaca grandiflora*), and the yellow/orange foci showed green autofluorescence characteristic of betaxanthins. Stably transformed *Arabidopsis thaliana* (arabidopsis) lines containing *35S: AmDOD*produced yellow colouration in flowers and orange-red colouration in seedlings when fed L-DOPA. These tissues also showed green autofluorescence. HPLC analysis of the transgenic seedlings fed L-DOPA confirmed betaxanthin production.

Conclusions

The fact that the introduction of DOD along with a supply of its substrate (L-DOPA) was sufficient to induce betacyanin production reveals the presence of a background enzyme, possibly a tyrosinase, that can convert L-DOPA to *cyclo*-DOPA (or dopaxanthin to betacyanin) in at least some anthocyanin-producing plants. The plants also demonstrate that betalains can accumulate in anthocyanin-producing species. Thus, introduction of a DOD and an enzyme capable of converting tyrosine to L-DOPA should be sufficient to confer both betaxanthin and betacyanin production to anthocyanin-producing species. The requirement for few novel biosynthetic steps may have assisted in the evolution of the betalain biosynthetic pathway in the Caryophyllales, and facilitated multiple origins of the pathway in this order and in fungi. The stably transformed *35S: AmDOD* arabidopsis plants provide material to study, for the first time, the physiological effects of having both betalains and anthocyanins in the same plant tissues.

BACKGROUND

The variety of colours observed in flowers, fruits and vegetative tissues in plants are due to the presence of chromogenic plant secondary metabolites [1, 2]. These pigments serve diverse functions including photosynthesis and the protection of the photosynthetic machinery, attraction of pollinators and seed dispersers, and protection against biotic and abiotic stresses [1, 3]. In addition to their biological functions, plant pigments are also of much interest regarding their possible beneficial effects on human health, their use as natural colorants and their aesthetic value in ornamental and food crops [4]. Non-chlorophyll plant pigments predominantly belong to two groups: flavonoids and carotenoids. Within the flavonoids, anthocyanins are the most significant type, providing a range of colours including orange, red, pink, mauve, purple and blue. However, in certain families within the order Caryophyllales, another class of pigments, the betalains, replaces the anthocyanins [2, 5, 6]. Betalains are only present in the order Caryophyllales and some fungi. They occur in most families of the Caryophyllales, but species of at least two families accumulate anthocyanin pigments instead [7]. The basis of this differentiation is unknown, but may represent an initial evolution of betalain biosynthesis in an ancestor of the core Caryophyllales and then its subsequent loss on different occasions [7].

No plant has yet been found that produces both betalain and anthocyanin pigments [5–8]. This mutually exclusive nature of the betalain and anthocyanin production in the plant kingdom is a curious phenomenon and the evolutionary and biochemical mechanisms for this restriction are unknown [5–7].

There are two major types of betalains, the red-purple betacyanins and the yellow/orange betaxanthins, both of which accumulate in the vacuole. The betaxanthins also emit green autofluorescence, which is not seen with the betacyanins [9–11]. While the production of flavonoids and carotenoids has been extensively studied and metabolically engineered in a variety of species, betalain biosynthesis has yet to be fully characterised [1, 2]. The betalain biosynthetic pathway is relatively simple with putatively only a few reactions that are enzyme catalysed (Figure 1). The initial biosynthetic step is the hydroxylation of tyrosine to L-3,4-dihydroxyphenylalanine (DOPA), attributed to the activity of a tyrosinase, although the exact role (if any) of tyrosinase in betalain synthesis has yet to be resolved [5, 6, 12, 13] Cleavage of the cyclic ring of L-DOPA by DOPA-4,5-dioxygenase (DOD) forms an unstable *seco*-DOPA intermediate, which is thought to spontaneously convert to betalamic acid. The formation of betaxanthins occurs spontaneously from the condensation of betalamic acid with amines/amino acids [14], probably in the vacuole. The classic model of betacyanin biosynthesis involves the condensation of betalamic acid with (most commonly) *cyclo*-DOPA, which again is a likely spontaneous step. In this model,*cyclo*-DOPA is formed from L-DOPA through an oxidation reaction that also has been attributed to tyrosinase activity [5, 6,13] The conversion proceeds via an unstable dopaquinone intermediate, which spontaneously cyclizes to form *cyclo*-DOPA. Betacyanins are generally *O*-glycosylated (at the C-5 or C-6) and frequently subsequently acylated. The aglycone product of betalamic acid and *cyclo*-DOPA condensation is termed betanidin and the 5-*O*-glucosylated form betanin, as with the anthocyanidin/anthocyanin convention. The timing of the glycosylation, regarding whether it occurs on *cyclo*-DOPA or betanidin, has been debated [15, 16].

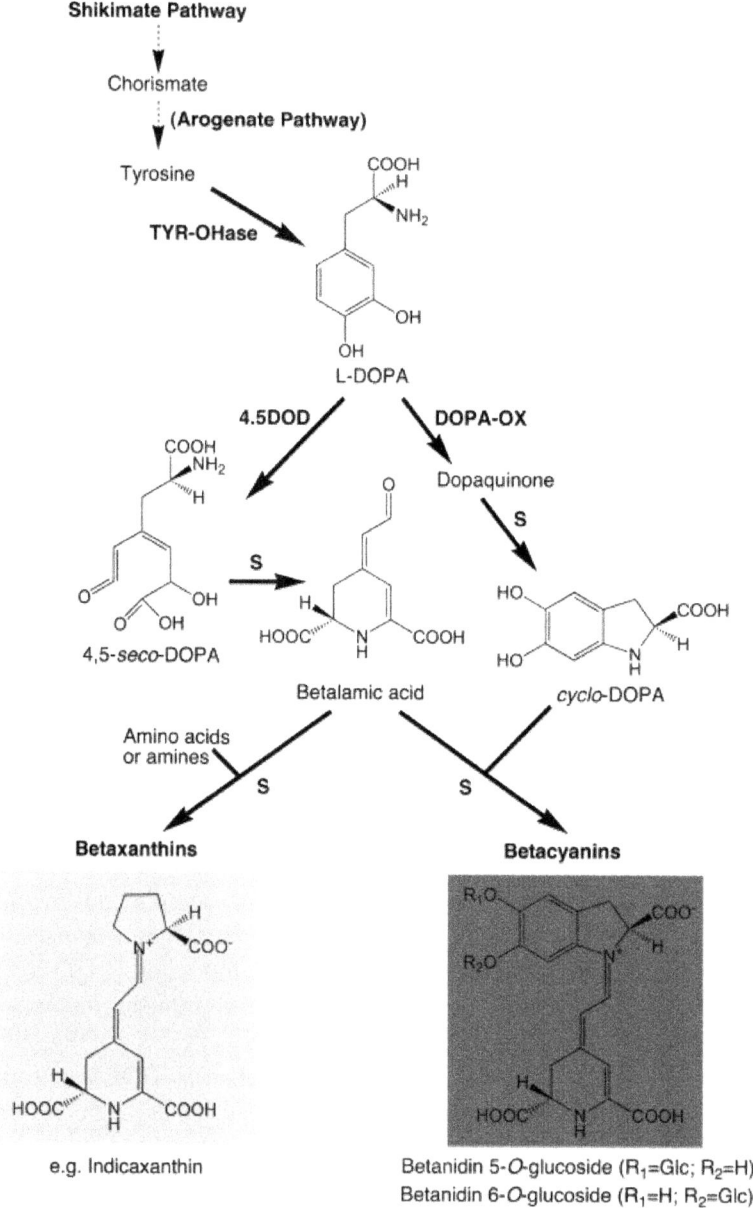

Figure 1: Schematic of the proposed biosynthesis pathway of betalains Abbreviations are; 4,5DOD, DOPA-4,5-dioxygenase; DOPA-OX, DOPA oxidase; S, spontaneous conversion; TY-OHase, tyrosine hydroxylase. The steps for formation of the betacyanin glycosides are not shown.

The only betalain biosynthetic enzymes characterised at the molecular level to date are for DOD and some of the glycosyltransferase activities. The first plant *DOD* gene characterised was from *Portulaca grandiflora*, and it defined a novel plant gene family of non-haem dioxygenases [2, 11]. The plant dioxygenase is phylogenetically unrelated to DOD from the fungi *Amanita muscaria*, although the fungal sequence can complement betalain production in flowers of a *P. grandiflora dod* mutant [17]. The plant DOD carries out only the 4,5-cleavage of DOPA to yield betalamic acid, whereas the fungal enzyme can also conduct 2,3-extradiol cleavage of DOPA, yielding the yellow pigment muscaflavin [18]. Three cDNAs encoding proteins with relevant *O*-glycosylation activity have been identified, two that use both betanidin and flavonoids as substrates [19, 20] and one that uses *cyclo*-DOPA [16].

Studies seeking to understand the genetic basis of the mutual exclusion of betalains and anthocyanins have focused on establishing the extent of the retention of the anthocyanin biosynthetic pathway in betalain producing species [21–23]. Flavonoids are present in betalain producing species, including flavonols and proanthocyanidins, and functional genes have been identified for the flavonoid biosynthetic enzymes chalcone synthase, dihydroflavonol 4-reductase and anthocyanidin synthase [21–23]. This suggests that the lack of anthocyanin production in betalain-producing species may be due to a lack of transcriptional activation of all the necessary biosynthetic genes [7, 23], although a hypothesis based on repressive interaction between anthocyanin and betalain metabolites and the biosynthetic enzymes has also been suggested [24].

Our aim in this study was to determine whether betalain production is possible in anthocyanin-producing species. Using genetic transformation and feeding of pathway intermediates, we have examined what the minimum number is of biosynthetic steps that must be introduced into an anthocyanin-producing species to allow betalain production, and whether betalains can accumulate to significant levels in such species. Stable or transient transgene expression was used with DOD cDNAs from the fungus *A. muscaria* and the plant *P. grandiflora*, introduced into *Arabidopsis thaliana* (arabidopsis) plants, *Solanum tuberosum* (potato) cell cultures and *Antirrhinum majus* (antirrhinum) petals. These are representatives of the asterids and rosids, the two major clades of eudicots.

RESULTS AND DISCUSSION

Betalain Biosynthesis in Potato Cell Cultures by Transient Expression of *DOD*

Potato cell suspension cultures were transformed using particle bombardment with *35S: green fluorescent protein (GFP)* or constructs having either the *P. grandiflora DOD* cDNA (*35S: PgDOD*) or the *A. muscaria DOD* (*35S: AmDOD*) driven by the CaMV35S promoter, and examined for betalain production following feeding with L-DOPA. In cells transformed with *35S: GFP*, GFP was detected in the cells 24 h after biolistic transformation but no pigmentation was apparent (data not shown). In the cells transformed with either *35S: PgDOD* or *35S: AmDOD* and fed L-DOPA, pigmented multi-celled clusters were apparent within 24 h post-bombardment (Figure 2). Both *35S: AmDOD* and *35S: PgDOD* resulted in yellow and orange cell clusters. In addition, *35S: AmDOD* also produced cells with red pigmentation. The yellow and red pigmentation suggests that betaxanthin and betacyanin production, respectively, had been conferred to the cells. Indeed, observation of the yellow pigmented areas under blue light showed the autofluorescence characteristic of betaxanthin pigments (data not shown).

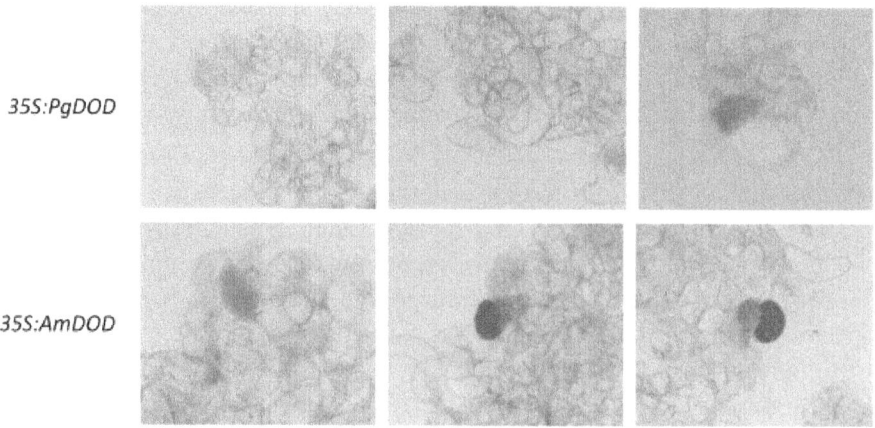

Figure 2: Pigment production in potato cell suspension cultures biolistically trans-formed with*35S: PgDOD*or*35S: AmDOD*and fed 10 mM L-DOPA. Examples are shown of cells with the resulting yellow to orange (*35S: PgDOD*) or yellow to red (*35S: AmDOD*) pigmentation.

Betalain Biosynthesis in Antirrhinum Petals by Transient Expression of *DOD*

The adaxial surface of dorsal petals of antirrhinum was transformed using particle bombardment with *35S: PgDOD* and examined for betalain production with or without infiltration of the petals with L-DOPA. Petals biolistically transformed with*35S: GFP* vector and fed with L-DOPA were used as an additional control. Antirrhinum lines having mutations in flavonoid production were used to provide anthocyanin-free petal backgrounds upon which to observe any pigment production. No pigments were visible in *35S: GFP* shot tissue or in petals transformed with *35S: PgDOD* but not infiltrated with L-DOPA, although positive GFP foci were apparent (Figure 3). The *35S: PgDOD* petals that were infiltrated with L-DOPA had numerous, multi-celled yellow foci 24 h after infiltration (48 h after bombardment) (Figure 3), indicating betaxanthin production. The yellow foci did indeed have the strong green autofluorescence typical of betaxanthins when observed under blue light (Figure 3). When using the *35S: AmDOD* construct, both yellow/orange and pink/red multi-celled foci occurred (Figure 4), although the pink/red foci were present only inconsistently (data not shown). Similarly, pink foci were sometimes present in some replicate experiments using *35S: PgDOD* (data not shown). The pink colouration indicates betacyanin production. The pink regions did not show strong autofluorescence while the central cells did, indicating possible betacyanin accumulation around a central region containing both betacyanins and betaxanthins.

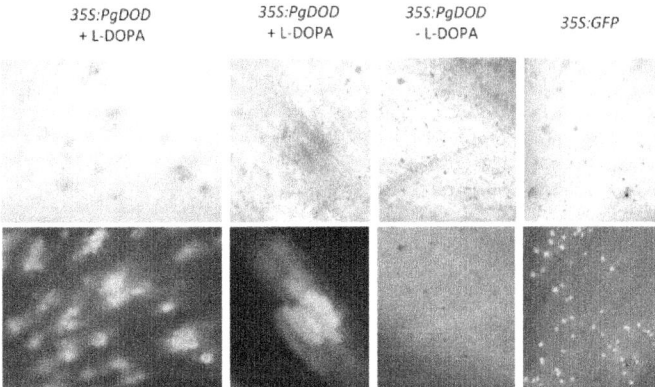

Figure 3: Betaxanthin production in antirrhinum petals transiently transformed with*35S: PgDOD*or*35S: GFP*. Plasmid constructs containing *35S: PgDOD* or *35S: GFP* were introduced into the adaxial epidermis of antirrhinum petals using particle bombardment. At 24 h after bombardment some *35S: PgDOD* bombarded petals were

infiltrated with 10 mM L-DOPA, while the other petals were infiltrated with water, and incubated for a further 24 h before observation. Representative petals are shown as viewed under white light (upper row) or blue light (lower row). The yellow, multi-cell foci for *35S: PgDOD* petals fed with L-DOPA are shown at two magnifications.

35S:AmDOD

White light   Blue light

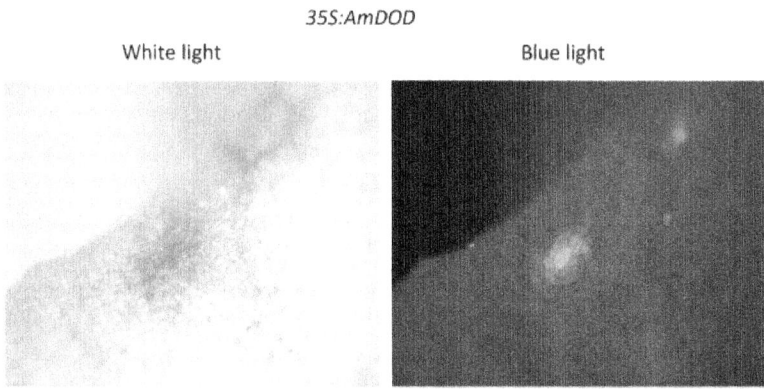

Figure 4: Pigment production in antirrhinum petals transiently transformed with*35S: AmDOD*. A plasmid construct containing *35S: AmDOD* was introduced into the ad-axial epidermis of antirrhinum petals using particle bombardment. At 24 h after bombardment the bombarded petals were infiltrated with 10 mM L-DOPA and incubated for a further 24 h before observation. A representative petal is shown as viewed under white light (left) or blue light (right).

HPLC analysis (LC-DAD) was used to examine the nature of the yellow pigments produced following bombardment with*35S: PgDOD*. The ridge region of the petals was chosen for HPLC analysis, to ensure a similar region was sampled in each case. Betaxanthins and betacyanins have absorbance maxima at around 470 nm and 538 nm, respectively [25], and peak profiles were examined at these two wavelengths (Figure 5). No betalain-related peaks were detected with HPLC analysis of the *35S: GFP* petal tissue or with *35S: PgDOD* petals that were not infiltrated with L-DOPA. Petals transformed with *35S: PgDOD* and fed L-DOPA showed distinct peaks at both 470 nm and 538 nm. The compounds represented by the peaks were putatively identified by comparison of HPLC retention times and spectral data against those of a standard extract (from beetroot, Figure 5A and 5E) and reported spectral data [14, 25, 26]. The beetroot extract showed the expected peaks for the betaxanthin vulgaxanthin I (Peak 1) and the betacyanins betanin (betanidin 5-*O*-glucoside, Peak 2) and isobetanin (Peak 3) (Table 1). The *35S: PgDOD* L-DOPA fed samples showed a small amount of betanin (Peak 2; Figure 5F) and an unknown peak (Peak 5, 21.02 min; Figure 5B) that was possibly dopaxanthin (Table 1).

Generally, the chromatograms at 538 nm showed the same patterns as at 470 nm except that vulgaxanthin I (Peak 1) and the putative dopaxanthin peak (Peak 5) were no longer detected, providing further evidence that Peak 5 is indeed a betaxanthin.

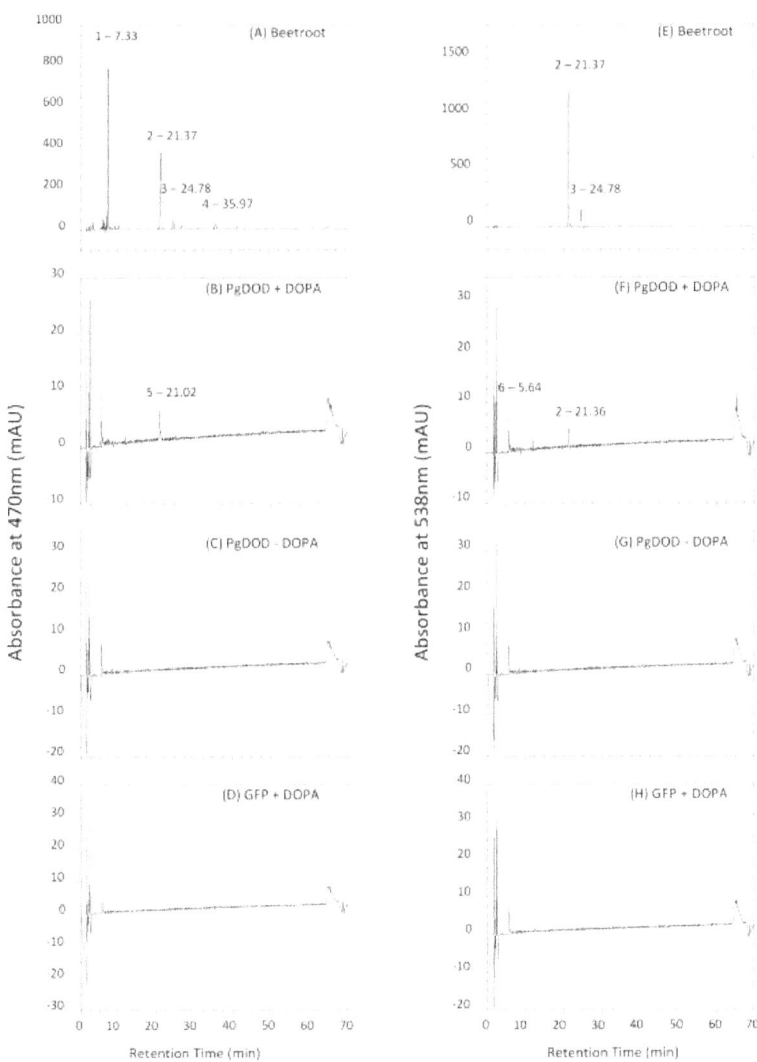

Figure 5: HPLC analysis of pigments produced in antirrhinum petals transiently trans-formed with *35S: PgDOD* or *35S: GFP*. HPLC analysis was conducted on the *35S: Pg-DOD* and *35S: GFP* antirrhinum petal material shown in Figure 3. Absorbance was

monitored at 470 nm for betaxanthins or 538 nm for betacyanins. Chromatograms are for extracts from beetroot root (A and E), *35S: PgDOD*antirrhinum infiltrated with L-DOPA (B and F), *35S: PgDOD* antirrhinum without L-DOPA infiltration (C and G), *35S: GFP* antirrhinum infiltrated with L-DOPA (D and H). The retention times of the major peaks are shown, and these were tentatively identified as per Table 1.

Table 1: Retention times and spectral maxima of the major betalain pigments detected in beetroot tissue (Peaks 1 to 3) or antirrhinum petal tissue bombarded with *35S: Pg-DOD* and infiltrated with L-DOPA (Peaks 4 and 5).

Peak[a]	Betalain	Retention Time (min)	λ_{max} (nm)
1	Vulgaxanthin I	7.33	259, 469
2	Betanin	21.37	269, 534
3	Isobetanin	24.78	269, 534
4	Unknown betaxanthin	35.97	265, 464
5	Unknown betaxanthin[b]	21.02	241, 470

HPLC absorbance traces are shown in Figure 4

[a]The peak present at 5.64 min in antirrhinum tissue was present in all samples, both control and transgenic; its spectral data is not consistent with that of betalain pigments therefore it was excluded from the analysis.

[b]Possibly dopaxanthin

Antirrhinum can produce yellow pigments in the petal face and throat naturally. These pigments are aurones and a product of the flavonoid pathway. In addition to the spectral data (aurones have spectra maxima in the range 390-430 nm), there is other evidence that the yellow pigments observed following *35S: PgDOD* bombardment and L-DOPA feeding are unlikely to be aurones; aurones are not present normally in the dorsal petals of the antirrhinum line used (JI19), and no yellow pigments were observed without infiltration with L-DOPA.

Despite a lack of the red colouration seen following bomardment, the *35S: PgDOD* antirrhinum samples analysed by HPLC did indeed contain small amounts of betacyanin (betanin; Figure 5F). Also, the presence of betacyanin is inferred in the *35S: AmDOD* expressing potato cells, given the red colouration of some foci. How betacyanin production can occur in non-betalain species through the introduced DOD acting on the supplied L-DOPA is not clear. The product of DOD action on L-DOPA is betalamic acid (Figure 1). Betaxanthins can be produced by spontaneous reactions of the DOD reaction end-product, betalamic acid, with amino acids or amines. However, betacyanin formation requires *cyclo*-DOPA, the formation of which from L-DOPA by oxidation has been attributed to the activity of a tyrosinase [6]. As tyrosinases are frequently present in plant cells it is possible that an endogenous enzyme with activity on L-DOPA occurs in most non-betalain producing species. Alternatively, betacyanin could be formed from endogenous tyrosinase activity on dopaxanthin [12]. The betacyanin detected in antirrhinum petals was likely betanin, an *O*-glycosylated betacyanin. This indicates that endogenous glycosyltransferases can act on the novel betacyanin substrates. Two *O*-glucosyltransferases have been characterized with activity on betanidin [19, 20]. Both show sequence similarity to the *O*-glycosyltransferases involved in anthocyanin biosynthesis, and indeed, the betanidin 5-*O*-glucosyltransferase has activity with both betacyanins and flavonoids [20]. Thus, it may be the case that the endogenous flavonoid *O*-glucosyltransferases of antirrhinum are also able to act on betanidin (and/or *cyclo*-DOPA).

Biolistic transformation of both potato and antirrhinum with either *35S: PgDOD* or *35S: AmDOD* resulted in muli-celled foci producing betalains. It is unlikely that this would be due to movement of the DOD enzyme between cells as, similar to GFP (Figure 3), it is too large for passive intercellular movement. As the final betalain pigments accumulate within the vacuole, this suggests that some of the precursors migrate between cells. This is similar to the results of Mueller *et al.* [17] when they biolistically introduced *AmDOD* into different *P. grandiflora* mutant backgrounds. Single coloured cells seen within 18 h after bombardment developed into multi-cell foci by 48 h after bombardment [17]. It was suggested that betalains could have diffused through plasmodesmata to neighbouring cells. However, in contrast to these results, when complementation of the *P. grandiflora dod* mutant was conducted with PgDOD, only single cell foci occurred [11]. Furthermore, cell-specific betalain production is commonly observed in plants, such as in the epidermal cells of petals. In the case of our results, it is possible that high L-DOPA levels produced from tissue feeding allowed the movement of betalamic acid not just to the vacuole but also to neighbouring cells.

Betalain Pigment Production by Stable Transformation of Arabidopsis

Stably transformed arabidopsis plants were produced through *Agrobacterium*-mediated transformation with *35S: AmDOD*. T2 generation seedlings were checked for expression of the *35S: AmDOD* transgene (Additional file 1) and selected lines examined for their potential to produce betalains when fed with L-DOPA. When whole seedlings were fed L-DOPA, novel pigment production was visible within 12 h after feeding (Figure 6A-C), including in the etiolated root tissues and the hypocotyls. The colour ranged from pale yellow through to orange and dark orange-red. Under blue light, the pigments showed the green autofluorescence characteristic of betaxanthins (Figure 6D, E). When detached inflorescences from mature plants were fed L-DOPA pale yellow pigmentation was seen in all tissues 24 h after feeding, including the stem, petals and siliques, and this was accompanied by strong autofluorescence (Figure 7). *35S: AmDOD* seedlings or inflorescences not fed L-DOPA did not produce visible pigments after treatment and did not show significant autofluorescence under blue light (Figure 6F, G and Figure 7).

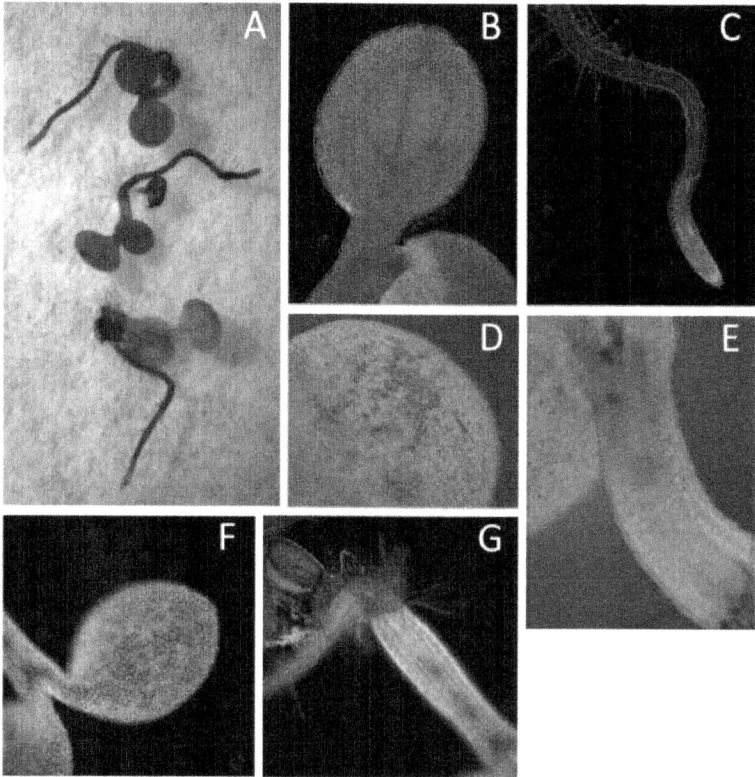

Figure 6: Betaxanthin production in arabidopsis seedlings stably transformed with*35S: AmDOD*. *35S: AmDOD* seedlings were grown on moistened filter disks either with (**A** to **E**) or without (**F** and **G**) 10 mM L-DOPA. Those fed L-DOPA accumulated yellow to orange/red pigments (**A** to **C**) and under blue light showed the autofluorescence typical of betaxanthins, as illustrated for the cotyledon (**D**) and hypocotyl (**E**). The seedlings not fed L-DOPA did not produce coloured pigments in either their cotyledons (**F**) or hypocotyls (**G**).

Water   L-DOPA

Figure 7: Pigment production in inflorescences of arabidopsis plants stably trans-formed with*35S: AmDOD*. Inflorescences from *35S: AmDOD* arabidopsis line 6 were infiltrated with either water or 10 mM L-DOPA for 24 h and examined under white light (upper panel) or blue light (lower panel). Those fed L-DOPA accumulated yellow to orange pigments and under blue light showed the autofluorescence typical of betax-anthins. The inflorescences not fed L-DOPA did not produce visible coloured pigments and did not show autofluorescence.

HPLC analysis was used to examine the nature of the pigments produced following L-DOPA feeding of the *35S: AmDOD* arabidopsis. Seedlings were fed L-DOPA and the entire seedling sampled for chemical analysis. HPLC analysis revealed several peaks present in the L-DOPA fed *35S: AmDOD* tissue but not in the control tissue (Figure 8). Four of these peaks were present in sufficient quantity to confirm that their spectral data and retention times were those characteristic of betaxanthins (Table 2).

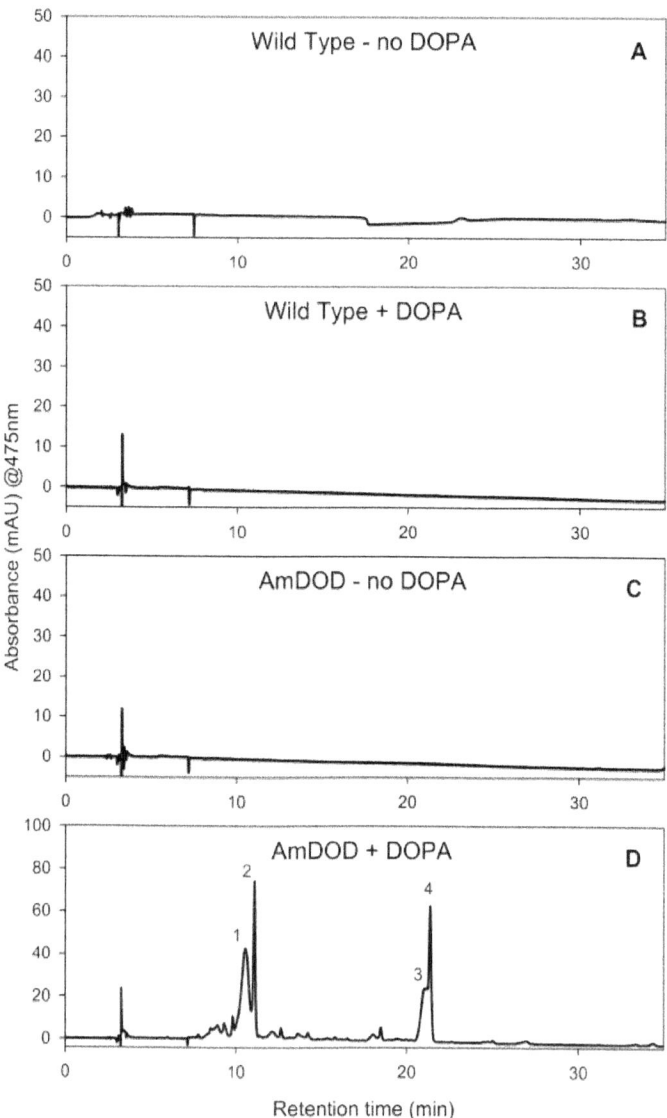

Figure 8: HPLC analysis of pigments produced in arabidopsis seedlings stably transformed with *35S: AmDOD*. Seedlings, representatives of which are shown in Figure 6, were grown on moistened filter paper with or without the addition of 10 mM L-DOPA and the pigments present examined using HPLC. Samples from wild-type (non-transgenic) arabidopsis and from beetroot root were included for comparison. Absorbance was monitored at 475 nm for betaxanthins. The retention times of the major peaks are shown, and these were identified as betaxanthins per Table 2.

Table 2: Retention times and spectral maxima of the major peaks detected by HPLC using samples from *35S: AmDOD* plants fed 10 mM L-DOPA

Peak	Betalain[a]	Retention Time (min)	λ_{max} (nm)
1	Vulgaxanthin I[b]	10.4	268, 468
2	Unknown betaxanthin	11.6	262, 470
3	Unknown betaxanthin	20.4	264, 472
4	Unknown betaxanthin	21.7	262, 468

[a]Spectral data for all four peaks was generally consistent with data reported previously for betaxanthins, but no specific betaxanthin assignments were made.

[b]Peak one has a retention time and spectrum consistent with that of Vulgaxanthin I, as reported for extracts from beetroot [14] and as compared against a beetroot extract run in the same experimental set.

The presence of betaxanthins matched the strong autofluorescence observed in the seedlings under blue light. The failure to detect the presence of betacyanin was a surprise given the orange/red colouration of the seedlings and the ability of *DOD* to confer both betaxanthin and betacyanin production when transiently expressed in antirrhinum petals. In accordance with the HPLC results, while the seedlings did have an orange-red colour that would suggest betacyanin production, the autofluorescence indicated strong betaxanthin presence. Furthermore, the fed inflorescences off the mature plants did not have a red colouration. Thus it appears that a high level of betaxanthin production in the particular background colouration of the seedlings resulted in the orange-red colour. It is unlikely that anthocyanins are contributing to the orange-red colouration of seedlings, as L-DOPA treatment of non-transgenic seedlings did not induce visible pigment production.

Conclusions

We report here methods to enable production of high-levels of betalains in the non-betalain, anthocyanin-producing, species antirrhinum, arabidopsis and potato, which represent both of the major clades of the eudicots. Betalains are not normally observed in anthocyanin-producing species, for unknown reasons.

The introduction of a single enzyme (DOD) and feeding of that enzyme's substrate (L-DOPA) were sufficient to induce both betaxanthin and betacyanin production. Although many steps in betalain production are suggested to be spontaneous, it is somewhat surprising that betacyanin production occurred. It indicates the background presence of an enzyme able to convert L-DOPA to *cyclo*-DOPA, or dopaxanthin to betacyanin/betanidin, and *O*-glycosyltransferases with activity against betanidin. Thus, it is likely that introduction of just two enzymes, DOD and the (currently uncharacterised) enzyme for conversion of tyrosine to L-DOPA, should be sufficient to confer betaxanthin and betacyanin production to anthocyanin-producing species. The requirement for few novel biosynthetic steps may have assisted in the initial evolution of this pathway in the Caryophyllales. A relatively simple mechanism for betalain biosynthesis would also support the multiple origins of the pathway in the Caryophyllales and in some fungi [7].

Given the extensive knowledge available on anthocyanin production and function in arabidopsis, the stably transformed *35S: AmDOD* arabidopsis plants should provide excellent material to study, for the first time, the physiological effects of having both betalains and anthocyanins in the same plant tissues. The large collection of mutant lines in arabidopsis should also allow aspects of betalain biosynthesis, such as vacuolar transport, to be examined.

METHODS

Plant and Cell Culture Material

Antirrhinum majus plants were grown under standard glasshouse conditions in Palmerston North, New Zealand. The glasshouse was heated at 15°C and vented at 25°C, without supplementary lighting. Two antirrhinum lines were used. Transformation with *35S: PgDOD* used line JI19, which produces aurones and flavones but not anthocyanins in the petals, as it is lacking in flavanone 3-hydroxylase activity (homozygous *incolorataII*). 35S: AmDOD transformation used either JI19 or *rosea*[dorsea], which carries a mutation in the R2R3MYB gene *Rosea1* and lacks anthocyanin production in the adaxial epidermis of the petals [27]. *Arabidopsis thaliana* transformation used the Colombia ecotype. Potato cell cultures were obtained by placing 2 cm^2 pieces of potato callus in 50 mL of modified liquid MS media [28] per cotton-plugged 250 mL flask. These cultures were incubated at room temperature (23-27°C) with shaking on an orbital shaker at 100 rpm in the dark. Subculturing was carried out every two weeks with 3 mL of the cell culture used to inoculate a fresh 50 mL of liquid media.

PCR Cloning of *P. Grandiflora* DOD cDNA and Vector Construction

A cDNA for the ORF of *DOD* was RT-PCR amplified from mRNA prepared from *P. grandiflora* betalain pigmented petals using primers designed to the published sequence [11]. The forward primer was 5›-AGTCAGAATCCATGGGTGTTGGGAAGGAA-3›, with the first ATG matching the initiation ATG for DOD, and the reverse primer was 5›-AGTCATCTAGAATCATATGGAAGTGAACT-3›, which incorporated an XbaI site. Standard PCR conditions were used with Taq Polymerase (New England Biolabs, Massachusetts, USA) and cycling parameters of 94°C for 4 min, then 30 cycles of 94°C for 30 s, 60°C for 30 s and 72°C for 1 min, with a final extension phase of 72°C for 7 min. The products were cloned into pGEM-T-Easy (Promega, Wisconsin, USA) and confirmed as *PgDOD* by DNA sequencing. The cDNA was then excised as an EcoRI/XbaI fragment and ligated into EcoRI/XbaI digested *pART7* [29] to from the vector pPN314. The *35S: AmDOD*construct for particle bombardment, containing the *A. muscaria* 4,5-DOPA dioxygenase cDNA driven by the CaMV35S promoter, was *pNcoDod* [17] (courtesy of Dr. Willibald Schliemann, Leibniz-Institute of Plant Biochemistry, Germany). The binary vector *pPN166* for transformation into arabidopsis was constructed by taking the CaMV35S-cDNA region from*pNcoDod* as a PvuII fragment and cloning it into the NotI site (after flushing by end-filling) in the binary vector *pMLBART* (a gift from Dr Bart Janssen, Plant & Food Research). The *35S: GFP* construct was pPN93. Vectors were verified by restriction fragment analysis and/or DNA sequencing.

Biolistic Transformation

Antirrhinum particle bombardment experiments were performed as described in Shang *et al.* [30], with the following variations; the pressure setting was 300 or 400 kPa, the shooting distance 11-12 cm, and petals were bombarded twice. The final DNA concentration for DOD constructs was 1 or 2 µg DNA per mg of 1.0 µm gold particles. Controls included gold particles alone (no DNA) and *35S: GFP* (added at 0.4 µg DNA per mg of 1.0 µM gold particles). Prior to bombardment the petals were surface sterilised by immersion for 15 min in 10% (v/v) bleach containing a few drops of Tween-20, followed by three rinses in sterile water. The adaxial surface of the dorsal petals was bombarded. After bombardment, the plant materials were then cultured on 1/2 MS medium under 20 - 50 µmol m^{-2} s^{-1} light from Osram 36 W grolux fluorescent tubes (16 h photoperiod) at 25°C. At least two flowers were used for each construct per experiment, and each experiment was repeated at least twice.

Potato cell suspension cultures for particle bombardment-mediated transformation were prepared by filtering 3 mL of culture onto sterile filter paper and sub-culturing them on a solid media in tubs for 48 h prior to transformation. The biolistic parameters were the same as those used for the antirrhinum petals with exception that they were only bombarded once. A sterilised metal grid was placed over the cells, on top of the culture tub, to prevent displacement of the cells from the helium in-flow.

Arabidopsis Transformation

The floral spray method developed by Clough and Bent [31] was used for the transformation of arabidopsis. Seeds were harvested from the *Agrobacterium*-inoculated plants and transgenics selected by spraying germinating seedlings with glufosinate herbicide (Basta, Bayer Crop Sciences). Positive transformants were grown through to production of T2 seeds following self-pollination.

L-DOPA Feeding

Infiltration of antirrhinum petals with L-DOPA was carried out 24 h after biolistic transformation. The petals were placed in 10 mM L-DOPA solution and a brief vacuum (30-60 s) applied, until the solution boiled vigorously and the petals became translucent. Following infiltration excess solution was blotted off and the petals incubated on 1/2 MS medium under the same conditions as post-bombardment. Control petals were handled in the same manner but with sterile water substituted for L-DOPA. Petals were observed for betalain production after a further 24 h. Transformed potato cell cultures were fed with 10 mM L-DOPA solution by dispensing 1 mL of the solution on to the filter disks supporting the cells. The L-DOPA solution was applied to half of the samples immediately after transformation.

Germinated arabidopsis seedlings were fed with 10 mM L-DOPA by transfer of the filter paper containing the seedlings on to a plate containing L-DOPA solution. Seedlings were observed for betalain pigment production after a further 12 h. For feeding of inflorescences, six inflorescences were collected from *35S: AmDOD* line 6 and cut to 1 cm long. Three inflorescences were immersed in 10 mM L-DOPA solution and the other three immersed in water, and both sets were left for 2 min. The inflorescences were then placed into 1.5 ml Eppendorf tubes with either 0.5 ml of 10 mM L-DOPA solution or water, and the tubes left open on the bench for 24 h.

PCR Analysis

RT- PCR analysis for *DOD* transgene expression in *35S: AmDOD* arabidopsis plants used total RNA extracted from leaf tissue following the RNeasy protocol of the Qiagen RNeasy mini kit. Five rosette leaves were sampled from each of six individual T2 lines, as well as a non-transgenic wild type line. Following DNAse I treatment 500 ng of RNA from each sample was used to generate cDNA using the Roche Transcriptor First Strand cDNA Synthesis Kit. PCR used Taq polymerase (Roche, New Zealand) and the following primers: DODF1 5›-CATACTACCATGTCCACCAAG-3›, DODF2 5›-AGCACTGCTTCTATATCGTC-3›, Act2S 5›-TCCCTCAGCACA TTCCAGCAGAT-3›, Act2AS 5›-AACGATTCCTGGACCTGCCTCATC-3›. The Actin primers correspond to the arabidopsis *Actin2* gene (AT3G18780). The thermocycling conditions were 94°C for 2 min and 25 cycles of 94°C for 20 s, 55°C for 30 s and 72°C for 30 s. The PCR products were separated on a 1% (w/v) TBE agarose gel containing ethidium bromide and visualised using UV-illumination. PCR products were also cloned and sequenced to confirm that the target PCR product was being amplified.

HPLC Analysis

For arabidopsis analysis whole seedlings were extracted. For antirrhinum analysis, to ensure that similar petal regions were being compared, the ridge region of the biolistically transformed petals was excised away from the remainder of the petals for chemical analysis. The ridge is a distinct raised area that divides the lobes and throat of the dorsal petals. Ridge tissue from between three and six petals was pooled to give total sample fresh weights of 60 to 300 mg. Absorbance at 470 nm and 538 nm was used for detection of betaxanthins or betacyanins, respectively. Beetroot (20 mg freeze dried sample of red root) was used as a standard source of betalains, and strong peaks with the expected retention times and spectra data were observed.

Each sample was extracted three times in 1 ml of 80% (v/v) methanol containing 50 mM sodium ascorbate, as described in Schliemann *et al.* [14], with one overnight extraction at 4°C. The samples were centrifuged for 4 min at 10,000 rpm, the supernatant removed and the pellet re-extracted in the next 1 ml of 80% methanol with ascorbate. The supernatants once removed, were combined to give the crude extract. The extract was dried in vacuo on a Savant SC210 Speedvac to near dryness, then freeze dried overnight to complete dryness. Extracts were resuspended in MilliQ water and made up to a final volume of 500 μl. The extracts were syringe filtered through a 0.45 um nylon filter as per Svenson *et al.* [26] and the pigments analysed by high performance liquid chromatography (HPLC). The analysis of the antirrhinum samples used a Dionex 3000 Ultimate solvent delivery system with a Phenomenex Luna (5 μm, 150 × 4.6 mm) RP-18 endcapped column (column temperature 30°C) and a Dionex 3000 Diode Array Detector (DAD). Elution (1 ml min⁻¹) was performed using a solvent system comprising solvent A [1.0% formic acid in water] and solvent B [80% acetonitrile in water] and a linear gradient starting with 100% A, decreasing to 80% A at 62 mins, and then a linear gradient to 100% B at 67 mins, remaining at 100% B for a further 3 min, then returning to initial conditions. Betaxanthins were detected at 470 nm and betacyanins at 538 nm [25]. The analysis of the arabidopsis samples used the solvent system of Schliemann *et al.* [14]. It was conducted on a Hewlett Packard HP 1100 with two Merck Chromolith analytical columns and a C-18 guard column.

ACKNOWLEDGEMENTS

We thank Ian King and Julie Ryan for care of plants, Steve Arathoon and John Harris for technical assistance, Willibald Schliemann for the *A. muscaria* 4,5-DOPA dioxygenase cDNA vector, Andrew Gleave for *pART7*, Bart Janssen for *pMLBART*, Simon Coupe for *pPN93*, and the New Zealand Foundation for Research, Science, and Technology contract C02X0805 for supporting KES, SCD and KMD. NNH thanks The Todd Foundation, The Freemasons, NZ Federation of Graduate Women and the J. Skipworth Scholarship for their financial support. KEC thanks Victoria University for her PhD scholarship grant.

Electronic Supplementary Material

12870_2011_995_MOESM1_ESM.ppt Additional file 1: PCR analysis for *DOD* transgene expression in *35S: AmDOD* arabidopsis plants. Total RNA was extracted from six lines of *35S: AmDOD* arabidopsis, as well as a non-transgenic wild type line, and analysed for *DOD* transcript levels using RT-PCR. PCR primers for an endogenous actin gene were used as a positive control for RNA/cDNA integrity. PCR products were separated on a 1% (w/v) agarose gel containing ethidium bromide and visualised using UV-illumination. (PPT 356 kb). (PPT 356 KB)

Authors' Contributions

NNH conducted the AmDOD biolistic experiments in antirrhinum, produced and contributed to the analysis of the arabidopsis plants, and contributed to writing of the manuscript; JJ made pPN314 and conducted the biolistic experiments in potato; KMD conducted the PgDOD biolistic experiments in antirrhinum and wrote the main manuscript draft; DHL conducted the HPLC analysis of biolistically transformed material and contributed to the manuscript draft; PEJ contributed to supervision of JJ and NNH and project design; SCD assisted with cell culture experiments and supervision of JJ; KEC conducted PCR analysis and inflorescence feeding for the arabidopsis plants; KSG contributed to supervision of KEC. KES conceived and coordinated the study, supervised NNH and JJ, and contributed to analysis of the results and writing of the manuscript. All authors read and approved the final manuscript.

REFERENCES

1. Grotewold E: The genetics and biochemistry of floral pigments. Ann Rev Plant Biol 2006, 57:761–780.

2. Tanaka Y, Sasaki N, Ohmiya A: Biosynthesis of plant pigments: anthocyanins, betalains and carotenoids. Plant J 2008, 54:733–749.

3. Gould KS, Lister C: Flavonoid functions in plants. In Flavonoids: Chemistry, Biochemistry and Applications. Edited by: Andersen ØM, Markham KR. CRC Press, Boca Raton; 2006:397–441.

4. Davies KM: An introduction to plant pigments in biology and commerce. In Plant Pigments and their Manipulation. Edited by: Davies KM. Blackwell Publishing, Oxford; 2004:1–22.

5. Strack D, Vogt T, Schliemann W: Recent advances in betalain research. Phytochemistry 2003, 62:247–269.

6. Zrÿd J-P, Christinet L: Betalains. In Plant Pigments and their Manipulation. Edited by: Davies KM. Oxford: Blackwell Publishing; 2004:185–213.

7. Brockington SF, Walker RH, Glover BJ, Soltis PS, Soltis DE: Complex pigment evolution in the Caryophyllales. New Phytol 2011,190:854–864.

8. Mabry T: The betacyanins, a new class of red violet pigments, and their phylogenetic significance. New York: Roland Press; 1964.

9. Gandía-Herrero F, García-Carmona F, Escribano J: Fluorescent pigments: new perspectives in betalain research and applications.Food Res Intl 2005, 38:879–884.

10. Gandía-Herrero F, Jiménez-Atiénzar M, Cabanes J, Escribano J, García-Carmona F: Fluorescence detection of tyrosinase activity on dopamine-betaxanthin purified from Portulaca oleracea (Common Purslane) flowers. J Agric Food Chem 2009, 57:2523–2528.

11. Christinet L, Burdet F, Zaiko M, Hinz U, Zrÿd JP: Characterization and functional identification of a novel plant 4,5-extradiol dioxygenase involved in betalain pigment biosynthesis in Portulaca grandiflora . Plant Physiol 2004, 134:265–274.

12. Gandía-Herrero F, Escribano J, García-Carmona F: Betaxanthins as substrates for tyrosinase An approach to the role of tyrosinase in the biosynthetic pathway of betalains. Plant Physiol 2005, 138:421–432.

13. Steiner U, Schliemann W, Böhm H, Strack D: Tyrosinase involved in betalain biosynthesis of higher plants. Planta 1999, 208:114–124.

14. Schliemann W, Kobayashi N, Strack D: The decisive step in betaxanthin biosynthesis is a spontaneous reaction. Plant Physiol 1999,119:1217–1232.

15. Sasaki N, Adachi T, Koda T, Ozeki Y: Detection of UDP-glucose cyclo -DOPA 5- O -glucosyltransferase activity in four o›clocks (Mirabilis jalapa L.). FEBS Lett 2004, 568:159–162.

16. Sasaki N, Wada K, Koda T, Kasahara K, Adachi T: Isolation and characterization of cDNAs encoding an enzyme with glucosyltransferase activity for cyclo -DOPA from Four O›clocks and Feather Cockscombs. Plant Cell Physiol 2005, 46:666–670.

17. Mueller LA, Hinz U, Uze M, Sautter C, Zrÿd JP: Biochemical complementation of the betalain biosynthetic pathway in Portulaca grandiflora by a fungal 3,4-dihydroxyphenylalanine dioxygenase. Planta 1997, 203:260–263.

18. Mueller LA, Hinz U, Zrÿd JP: The formation of betalamic acid and muscaflavin by recombinant DOPA-dioxygenase from Amanita . Phytochemistry 1997, 44:567–569.

19. Vogt T, Grimm R, Strack D: Cloning and expression of a cDNA encoding betanidin 5-O-glucosyltransferase, a betanidin- and flavonoid-specific enzyme with high homology to inducible glucosyltransferases from the Solanaceae. Plant J 1999, 19:509–519.

20. Vogt T, Zimmermann E, Grimm R, Meyer M, Strack D: Are the characteristics of betanidin glucosyltransferases from cell-suspension cultures of Dorotheanthus bellidiformis indicative of their phylogenetic relationship with flavonoid glucosyltransferases? Planta1997, 203:349–361.

21. Shimada S, Inoue YT, Sakuta M: Anthocyanidin synthase in non-anthocyanin-producing Caryophyllales species. Plant J 2005, 58:950–959.

22. Shimada S, Otsuki H, Sakuta M: Transcriptional control of anthocyanin biosynthetic genes in the Caryophyllales. J Exp Bot 2007,58:957–967.

23. Shimada S, Takahashi K, Sato Y, Sakuta M: Dihydroflavonol 4-reductase cDNA from non-anthocyanin producing species in the Caryophyllales. Plant Cell Physiol 2004, 45:1290–1298.

24. Stafford HA: Anthocyanins and betalains: evolution of the mutually exclusive pathways. Plant Sci 1994, 101:91–98.

25. Kugler F, Stintzing FC, Carle R: Identification of betalains from petioles of differently colored Swiss Chard (Beta vulgaris L. ssp. cicla [L.] Alef. Cv. Bright Lights) by High-Performance Liquid Chromatography-Electrospray Ionization Mass Spectrometry. J Agr Food Chem 2004, 52:2975–2981.

26. Svenson J, Smallfield BM, Joyce NI, Sansom CE, Perry NB: Betalains in red and yellow varieties of the andean tuber crop ulluco (Ullucus tuberosus). J Agr Food Chem 2008, 56:7730–7737.

27. Schwinn K, Venail J, Shang Y, Mackay S, Alm V, Butelli E, Oyama R, Bailey P, Davies K, Martin C: A small family of MYB -regulatory genes controls floral pigmentation intensity and patterning in the genus Antirrhinum . Plant Cell 2006, 18:831–851.

28. Murashige T, Skoog F: A revised medium for rapid growth and bioassays with tobacco tissue cultures. Physiol Plantarum 1962,15:743–747.

29. Gleave AP: A versatile binary vector system with a T-DNA organizational-structure conducive to efficient integration of cloned DNA into the plant genome. Plant Mol Biol 1992, 20:1203–1207.

30. Shang Y, Schwinn KE, Hunter DA, Waugh TL, Bennett MJ, Pathirana NN, Brummell DA, Jameson PE, Davies KM: Methods for transient assay of gene function in floral tissues. Plant Methods 2007, 3:1.

31. Clough SJ, Bent AF: Floral dip: a simplified method for Agrobacterium -mediated transformation of Arabidopsis thaliana . Plant J 1998, 16:735–743.

Chapter 11

AN INTRODUCTION TO PLANT PIGMENTS IN BIOLOGY AND COMMERCE

INTRODUCTION

This introductory chapter presents a general overview of plant pigmentation, together with some general functional and economic aspects not covered in detail in the chapters on specific pigment groups.

Plant Pigmentation

The Physical Basis of Pigmentation

Plant pigmentation is generated by the electronic structure of the pigment interacting with sunlight to alter the wavelengths that are either transmitted or reflected by the plant tissue. The specific colour perceived will depend on the abilities of the observer. Humans without colour blindness can detect wavelengths between approximately 380 and 730 nm, representing the visible spectrum of red, orange, yellow, green, blue, indigo and violet. So chlorophyll with peak absorbancies at 430 and 680 nm will leave wavelengths forming a green colour Of course, often the colours are the result of a mix of residual wavelengths; for example, anthocyanins absorbing yellow-green light wavelengths of 520–530 nm will generate mauve colours formed by the reflection of a mix of orange, red and blue wavelengths. Thus the pigments can be described in two ways: the wavelength of maximum absorbance (λ_{max}) and the colour perceived by humans. Further details of the generation of colours and the behaviour of light in plant tissues can be found in Hendry (1996) and Chapter 10 of this volume. The names of many common pigments convey little information to the general reader, as they tend to reflect historical discoveries rather than a set naming system. For example, carotene was first isolated from Daucus carota (carrot), violaxanthin from Viola tricolor (pansy), and the common anthocyanidins, pelargonidin, cyanidin, peonidin, delphinidin, petunidin and malvidin from Pelargonium, Centaurea, Paeonia, Delphinium, Petunia and Malva, respectively. However, these trivial names are often well-established, familiar to workers in the field and allow easy flow of text. Thus, the trivial names are used extensively in this book. More complete names, giving details of constitution and stereochemistry, have been developed for many compounds to meet the standards of the International Union of Pure and Applied Chemistry (IUPAC) and International Union of Biochemistry (IUB) (Weedon & Moss, 1995). The details of such nomenclature, and lists of the IUPAC semi-systematic names, are available for some plant pigment groups in more specialised publications (Pfander, 1987; Harborne, 1988, 1994; Kull & Pfander, 1995; Weedon & Moss, 1995). Bohm (1998) gives a guide to relating the trivial flavonoid names to those in the Chemical Abstracts. In the same book he also provides a table listing the meanings of the trivial names often used for flavonoid di- and trisaccharides (e.g. sophoroside). The trivial names in this book are often accompanied by structural diagrams of the compounds, providing much of the information that would come from the full name. These diagrams commonly include basic representations of the stereochemistry, with a solid triangle for a bond representing above the plane of the paper and a dashed triangle for below the plane of the paper.

Structural Variation of Plant Pigments

Plant pigments exist in many varied forms, some with highly complex and large structures. For example, over 600 naturally occurring carotenoid structures have been identified (Britton et al., 1995) and over 7000 flavonoids, including over 500 anthocyanins (Chapter 10). The complexity of some pigments is well illustrated by the anthocyanin Ternatin A1, which consists of the base 15-carbon anthocyanin modified with seven molecules of glucose, four molecules of 4-coumaric acid and one molecule of malonic acid, corresponding to $C_{96}H_{107}O_{53}^-$ 53 (Terahara et al., 1990). In this book we have grouped plant pigments on a common structural and biosynthetic basis into four major groups (Table 1.1), and Chapters 2–6 focus on these pigments. In addition to these major groups there is a great array of pigments that are of limited taxonomic occurrence, and often poorly characterised. Some of the more notable of these are covered in Chapter 7.

Table 1: Major pigments of plants and their occurrence in other organisms

Pigment	Common types	Occurrence
Betalains	Betacyanins	The Caryophyllales and some fungi
	Betaxanthins	
Carotenoids	Carotenes	Photosynthetic plants and bacteria
	Xanthophylls	Retained from the diet by some birds, fish, and crustaceans
Chlorophylls	Chlorophyll	All photosynthetic plants
Flavonoids	Anthocyanins	Widespread and common in plants,
	Aurones	including angiosperms,
	Chalcones	gymnosperms, ferns, fern allies and
	Flavonols	bryophytes. Retained from the diet
	Proanthocyanidins	by some insects

The most obvious and widespread pigments of plants are, of course, the chlorophylls. These are cyclic tetrapyrrole pigments chelated with magnesium, and they share structural features with the haem and bile pigments of animals. Also associated with photosynthesis, but additionally providing bright colours to flowers and fruits, are the carotenoids. Carotenoids are terpenoid pigments present in all photosynthetic plants and they also occur in photosynthetic bacteria such as Erwinia and Rhodobacter. Annual production of carotenoids by plants, algae and dinoflagellates has been estimated at 100 million tons (Britton et al., 1995). The flavonoids are phenylpropanoid compounds of widespread occurrence. There are several major classes of flavonoids; however, only a few of these provide pigments to plants, in particular the anthocyanins and proanthocyanidins (condensed tannins). Reviews of the biosynthesis and function of non-pigmented flavonoids can be found in earlier volumes in this Annual Plant Reviews series (Wink, 1999a, 1999b) and in Bohm (1998). The betalains are nitrogenous pigments that are the most taxonomically restricted of the major plant pigment groups, being found only in a few families of the order Caryophyllales and some fungi. Curiously, their occurrence is mutually exclusive to that of the anthocyanins. Within plants, the major pigment groups show wide occurrence in the different tissues. For example, flavonoids occur in almost all tissues, carotenoids in leaves, roots, tubers, seeds, fruits and flowers, and even chlorophylls occur in flowers and fruits as well as leaves. Within tissues, there is often distinct localisation of pigment types in different cell layers. For example, anthocyanins are typically found in epidermal cells in petals and sub-epidermally in leaves, and chlorophyll in the sub-epidermal photosynthetic cell layers of leaves. The subcellular localisation of the different pigment groups is also generally distinct. The chlorophylls and carotenoids are principally lipid-soluble, plastid-located pigments, although there are examples of water-soluble carotenoids, at least some of which are located in the vacuole via plastid–vacuole interactions (Bouvier et al., 2003a). The betalains are water-soluble and vacuolar-located. While flavonoids occur in many subcellular locations, as well as extracellularly, the coloured flavonoids are principally found in the vacuole (Bohm, 1998). For flavonoids, the subcellular localisation is just one of several factors that determine the behaviour of the pigment molecule in the cell and the colour generated from it (Brouillard & Dangles, 1993). The mechanisms by which pigments such as the flavonoids are directed to the correct subcellular compartment are poorly defined, although some of the steps for anthocyanins have been elucidated (Winefield, 2002). Interactions of plant pigments with other cellular compounds have been well defined for flavonoids and small molecules (Brouillard & Dangles, 1993), and it is known that both flavonoids and carotenoids interact with specific proteins in the cell (Vishnevetsky et al., 1999; Winefield, 2002). Modern phytochemical techniques can enable the rapid identification of the class of pigment present in a plant tissue of interest, and with more extensive analysis the detailed structure of the compound may be elucidated. The techniques required for identification and analysis of pigments in a plant of interest can be daunting to newer researchers in the field. With this in mind, Chapter 10 provides an overview of the modern and most recent techniques used for extraction, separation, identification and quantification of the major pigment classes in plants, providing sufficient detail to support the practical application of such techniques.

The History of Plant Pigment Research

Pigmentation is one of the oldest subjects in formal plant science and has lead to many discoveries with impacts much wider than those in the pigmentation field alone. Early studies included discovery of the purple pigment from Viola as a natural pH indicator (Boyle, 1664), comparisons of the solubility of different pigments (Nehemiah Grew in 1682, quoted extensively in Onslow, 1925 and cited in Bohm, 1998), and Mendel's studies on the genetics of flower and seed colour in Pisum sativum – pea (Bhattacharya & Bhattacharya, 2001). The first publications on carotenoids date from the early nineteenth century, and detail the basis of colours of common food colourants of the time, some of which are still in use today, e.g. saffron and annatto (Eugster, 1995). The term chlorophyll was first used by Pelletier and Caventou (1818), and the different chlorophyll pigments were first separated from each other by Stokes (1864),

Sorby (1873) and Tswett (1906) (cited in Jackson, 1976 and Eugster, 1995). Tswett's research included the development of column chromatography, one of the basic methods of modern biochemistry, and allowed the ground-breaking work of Willsta"tter on the structure of carotenoids. The conspicuousness of the flower colour trait and the easy identification of mutants has established it as a favourite system for geneticists ever since Mendel, from pioneering studies such as those of Darwin (1868), Onslow/Whedale (1925/1907) through to transposon studies and recent breakthroughs such as those on transcriptional regulation in plants featured in Chapter 4. Studies on transposons in pigment genes of maize were part of the work that earned Barbara McClintock a Nobel Prize in 1983, and plant pigments have also featured in research earning Nobel prizes for Willsta"tter (1915), Fischer (1930, the chemistry of blood pigments and chlorophyll), Karrer (1937, work on the structures of plant pigments, in particular carotenoids) and Kuhn (1938, work on the structure of carotenoids and related vitamins). Studies on anthocyanins have been responsible for several important breakthroughs in plant science, including the isolation and identification of a plant transcription factor gene for the first time (Cone et al., 1986; Paz-Ares et al., 1986), the isolation of one of the first cDNAs for a plant cytochrome P450 enzyme (Holton et al., 1993), the first demonstration of antisense RNA technology in a transgenic plant (van der Krol et al., 1988) and the first description of transgene co-suppression (Napoli et al., 1990; van der Krol et al., 1990).

The Biosynthesis of Plant Pigments

The biosynthetic pathways for the major pigments of plants are now well defined at the genetic and enzymatic level. Of course, some knowledge gaps do exist, but these are becoming fewer. In the case of flavonoids, much is also known on the transcriptional regulation of the pathway. However, little is known on gene regulation mechanisms for any of the other major pigmentation pathways. Changes in the transcription rate of pigment biosynthetic genes usually precede pigment production, and it is thought that transcriptional regulation is the major controlling step for the pigmentation pathways studied to date. However, translational and post-translational regulation may also be important in specific cases, particularly for the photosynthetic pigments. Indeed, the general emphasis on transcriptional regulation may in part be due to lack of appropriate studies to identify post-transcriptional control. This is illustrated by the recent use of genomics to show rhythmic expression of many Arabidopsis thaliana genes, including those of the phenylpropanoid and carotenoid pathways, with the potential involvement of changing RNA stability in generating the rhythms (Staiger, 2002). Although the biosynthetic pathways of many plant pigments are now well defined, little is known of the turnover and degradation of most pigments. There is an obvious loss of chlorophylls during the autumn senescence of leaves, and some information is available on chlorophyll degradation. However, anthocyanin levels can also change rapidly, e.g. in flowers of the Yesterday, Today and Tomorrow plant (Brunfelsia calcina), which can turn from white to purple and back to white within three days. Simpson et al. (1976) provided a comprehensive review on the knowledge to that date on the metabolism of the various pigment groups. However, this earlier chemistry and biochemistry has not been followed by extensive findings on the molecular biology of the metabolism of pigments in plants. Indeed, the metabolism is perhaps better understood for the fate of pigments when absorbed from the diets of animals than it is for the plant tissues themselves. One gene has been identified that has a direct impact on colour fading of flowers, the Fading locus of Petunia hybrida (de Vlaming et al., 1982). Fully coloured flowers turn rapidly to white towards the end of flower development when the Fading gene is dominant and the genetic background is appropriate with regard to other genes that affect the type of anthocyanin present and the flower petal pH. Active degradation of the anthocyanins is thought to occur, rather than a change to a colourless form of anthocyanin (Schram et al., 1984).

The Functions of Pigments in Plants

The Function of Pigments in Vegetative Tissues

Chlorophylls and carotenoids are required for photosynthesis, chlorophylls for the capture of light energy and as the primary electron donors and carotenoids as essential structural components of the photosynthetic apparatus, where they protect against photo-oxidation. The roles of the pigments in these processes are covered in Chapters 2 and 3. Plant pigments are also involved in other interactions of plants with light, in particular the response to UV radiation (described in detail in Chapter 9), which is of growing concern with regard to changes in the global environment. Anthocyanins also frequently occur in vegetative tissues. The most spectacular example is their contribution to autumn colours in leaves of many deciduous species, which they generate in combination with the retention of carotenoids and loss of chlorophyll (Matile, 2000; Hoch et al., 2001; Lee, 2002). In non-senescing tissues their occurrence is more sporadic. Some species accumulate them in significant amounts in healthy leaves, providing red or purple colours to the foliage. In other cases anthocyanin production is induced in leaves in response to stresses such as cold, high light levels, pest and pathogen attack or deficiency of nutrients such as phosphate and nitrogen. Anthocyanin colouration in leaves can vary with season, environment, between individuals of a population and between different leaves on a single plant. It is commonly thought that anthocyanins have a role in protecting the photosynthetic apparatus from damage in many of these situations, and those tissues that show more anthocyanin accumulation are often at greater photoinhibitory risk, e.g. during nutrient reabsorption in senescing leaves or in cold temperatures (Hoch et al., 2001). However, the details of how anthocyanins achieve this are not determined. One hypothesis is that anthocyanins help attenuate the light levels, modifying the quantity and quality of light incident on the chloroplasts and thus reducing excitation pressure (Gould et al., 2002a; Steyn et al., 2002). However, this would not account for their accumulation in other stress situations. An alternative is that anthocyanins are acting as both direct light screens under high light stress and general antioxidants against harmful reactive oxygen species in the various other stress situations in which they are prevalent (Gould et al., 2002a, 2002b, 2002c). Supporting data include observations that red-leafed morphs of some shade species have a significant antioxidant advantage over green morphs, that anthocyanins can enhance oxidative protection in species more directly exposed to the sun, and that anthocyanins can reduce photoinhibition and photobleaching of chlorophyll under light stress conditions (Gould et al., 2002c; Neill et al., 2002; Steyn et al., 2002). Recently, the role of anthocyanins in improving foliar nutrient reabsorption during senescence, through the shielding of the photosynthetic apparatus from excess light, was tested using wild-type and anthocyanin-deficient mutants of three deciduous woody species under varying environmental conditions (Hoch et al., 2003). Nitrogen reabsorption efficiencies of the mutants were significantly lower than the wild-type counterparts, supporting the protection hypothesis of anthocyanins in senescing leaves. There are likely to be many other functions of pigments in plants that have not been determined as yet. A recent example of a new function for anthocyanins is that of protecting light-sensitive phototoxic plant defence compounds from degradation, which was described by Page and Towers (2002). Thiarubrines are phototoxic plant pigments that decompose to thiophenes when exposed to sunlight. In Ambrosia chamissonis, they occur in laticifers that are surrounded by anthocyanin-containing cells. Page and Towers (2002) were able to show that the anthocyanins around the laticifers functioned to photoprotect these defence compounds.

The Function of Pigments in Reproductive Tissues

The most obvious function of plant pigments, with the exception of chlorophyll, is to provide colour to flowers and fruit for attraction of pollinators and seeddispersal agents. These colours arise predominantly from flavonoid and carotenoid pigments, and a short guide to the likely pigments producing specific colours in flowers and fruits of plants is given in Table 1.2. There is, however, also a range of less common pigments that generate colours in specific species. For many angiosperms, colour is key to attracting pollinators, whether they are bees, butterflies, other insects or birds, although it is frequently one of a number of factors, including fragrance, floral shape and nectar reward, which combine to determine pollinator choice. Flavonoids are the most common flower colour pigments and it is on these that most research has been done. The role of flavonoids in pollination was the subject of an extensive review by Harborne and Grayer (1994), in which the authors were able to identify many general trends with regard to pollinator preference for different colours. However, they also noted the shortage of detailed studies on specific pigments and pollinators. This has changed greatly in the last decade, with many studies determining pollinator preference with regard to individual colours (Melendez-Ackerman et al., 1997; Gumbert et al., 1999; Oberrath & Bohning-Gaese, 1999; Gigord et al., 2001; Johnson & Midgley, 2001; Jones & Reithel, 2001; Landeck, 2002), fragrances (Odell et al., 1999; Raguso & Willis, 2002) and even petal epidermal cell shape (Glover & Martin, 1998; Comba et al., 2000). These studies cover a mix of approaches, including observational field studies, laboratory studies using model flowers and studies of the frequency of natural colour variants or white-flowered mutants within a population. Table 1.3 presents some of the general colour preferences identified for different pollinators. Most beetle-pollinated flowers are cream, white or green, and it was thought that beetles were generally insensitive to anthocyanin colours, such as red and blue (Harborne & Grayer, 1994). However, beetle selection of specific cyanic colours has now been demonstrated in some studies, including that of Johnson and Midgley (2001), who found that monkey beetles (Scarabaeidae: Hopliini) preferred orange-coloured model flowers to red-, yellow- or blue-coloured ones. For pollinators such as bees that can detect light in the UV spectrum, UVabsorbing pigments also influence flower selection. The main contributors to the UV absorbance of the flower are the chalcone- and flavonol-type flavonoids. Flavonols are very common in flowers, often being in greater abundance than the coloured pigments. The flavonoids may form UV-visible patterning in petals, often in combination with UV-reflective carotenoid pigments (Harborne & Grayer, 1994; Bohm, 1998).

Table .2: The most common pigment types associated with flower and fruit colours in plants. (The terms pelargonidin, cyanidin and delphinidin are used to refer to pelargonidin, cyanidin or delphinidin derived anthocyanins.)

Colour	Specific pigment type	Pigment group	Examples
Cream	Flavonols or flavones	Flavonoid	Most cream flowers
Pink to red		Carotenoid	Some red flowers and fruit, e.g. *Lycopersicon esculentum* (tomato) fruit
	Pelargonidin and/or cyanidin	Flavonoid	Most pink flowers and some fruit, e.g. *Eustoma grandiflorum* (lisianthus) flowers
	Pelargonidin and/or cyanidin	Flavonoid	Most red flowers and some fruit, e.g. *Malus* (apple) fruit
	Anthocyanin and carotenoid mix	Flavonoid and carotenoid	A few examples, e.g. *Tulipa* flowers
	Betacyanin	Betalain	A few examples in the Caryophyllales, e.g. *Bougainvillea* flowers
Orange		Carotenoid	Most orange flowers and fruit, e.g. *Tagetes erecta* (marigold) flowers
	Pelargonidin alone	Flavonoid	A few examples, e.g. *Pelargonium* flowers
	Anthocyanin and aurone mix	Flavonoid	Rare occurrence, e.g. *Antirrhinum majus* (snapdragon) flowers
	Anthocyanin and chalcone mix	Flavonoid	Rare occurrence, e.g. *Dianthus* (carnation) flowers
	Betacyanin	Betalain	A few examples in the Caryophyllales, e.g. *Portulaca* (purslane) flowers
Yellow		Carotenoid	Most yellow flowers and fruit
	Aurone	Flavonoid	Rare occurrence, e.g. *Antirrhinum majus* flowers
	Chalcone	Flavonoid	Rare occurrence, e.g. *Dianthus* flowers
	Flavonol	Flavonoid	Rare occurrence, e.g. *Gossypium* (cotton) flowers
	Betaxanthin	Betalain	A few examples in the Caryophyllales, e.g. *Portulaca* flowers
Green		Chlorophyll	All green flowers and fruit
Blue	Delphinidin	Flavonoid	Most blue flowers and fruit
	Cyanidin	Flavonoid	Rare occurrence, e.g. *Ipomoea* (morning glory) flowers
Purple		Carotenoid	Rare occurrence, e.g. *Capsicum* (pepper) fruit
	Cyanidin and/or delphinidin	Flavonoid	Most mauve flowers, e.g. *Petunia* and some purple fruit, e.g. *Solanum melongena* (eggplant)
		Flavonoid and carotenoid mix	Some flowers, e.g. *Cymbidium* orchids
Black	Delphinidin	Flavonoid and carotenoid mix	Some black flowers, e.g. *Viola* (pansy)

Table .3: Flower colour preferences of some pollinators, presented in terms of colour perceived by humans

Pollinator	Flower colour preference
Bees	Blue, yellow and UV-absorbing pigments
Birds	Bright red and scarlet
Beetles	White, cream, green and occasionally orange and red
Butterflies	Strong pinks, reds and mauves
Flies	White, green, dark brown and purple
Moths	White, cream and occasionally red
Wasps	Purple and blue

In some cases, colour combinations and floral patterning may help attract a range of pollinators, including organisms as varied as hummingbirds and bumble bees, or provide more specific signals within the flower (discussed extensively in Harborne & Grayer, 1994 and particularly Bohm, 1998). For example, distinctive spots on the flower lip or pigment lines in the flower tube may act as a nectar guide to bees. This is well illustrated by the yellow face and two yellow throat stripes of aurone pigment in Antirrhinum majus (snapdragon) flowers. There are a few cases of elegant plant–animal co-evolution in which colour patterning has been shown to be part of floral mimicry. In particular, flowers of the orchid genus Ophrys use scent, shape and colour to mimic female bees, causing the male bee to attempt copulation, thus achieving pollination (Schiestl et al., 1999; Paxton & Tengo, 2001; Ayasse et al., 2003). The relationship between different orchid species and different bee species can be highly specific, and it is likely based on the pattern of scent compounds produced by the flower (Schiestl et al., 1999). There have been recent breakthroughs in understanding the genetic basis of colour patterning in flowers, discussed briefly in Chapter 4. Change in flower colour during the later stages of flower development or in response to pollination has been recognised for many years. Indeed, as early as the late nineteenth century, Mu"ller described the colour change of Lantana flowers from yellow to purple over an ageing period of three days, and the preference of pollinators for the younger, yellow flowers (see Weiss, 1991). That the yellow flowers were fertile, offered nectar and pollen that older flowers did not, and were preferred by the butterfly pollinators was later confirmed for Lantana species by Weiss (1991). The same study also offered a possible reason for maintenance of the older flowers – larger inflorescences were more successful in attracting butterflies. In most cases, a change in colour is likely to be associated with a change in nectar and pollen availability (Weiss, 1991, 1995; Harborne & Grayer, 1994; Bohm, 1998; Oberrath & Bohning-Gaese, 1999). At least 200 other plant genera contain species that show colour change during flower development and interact with a wide range of pollinator species, making it a common occurrence in plant reproduction (Weiss, 1991).

The specific studies on individual plant and pollinator species are supported at a higher level by general evolutionary and ecological trends (Harborne & Grayer, 1994; Gumbert et al., 1999). In particular, blue flower colours are more common in the temperate ecosystems, in which bees are key pollinators, while bright red colours are more prevalent in tropical ecosystems, in which other insects and birds are more important. Carotenoids and flavonoids also commonly colour pollen, although their functions in pollen are not well elucidated. They have been shown to have a role in signalling to pollinators (Lunau, 2000), and it is possible they also have protective activities against various stresses. Colourless flavonoids are known to be involved in plant fertility in some species (Taylor & Jorgensen, 1992; Jorgensen et al., 2002), but this has not been shown for coloured flavonoids. The role of pigments in fruit is an obvious one, of signalling the ripeness of the fruit to seed-dispersal agents. Both carotenoids and flavonoids commonly provide fruit colours. Flavonoids and carotenoids can also colour seeds, e.g. the yellow carotenoids and purple flavonoids of maize kernels. Such pigmentation may be related to timing of seed germination or plant defence. In some cases the pigments may reach very high levels. In the resinous seed coating of Bixa orellana (Fig. 1.1), which is the source of the commercial food colourant annatto, levels of the apocarotenoid bixin can reach 10% dry weight (Britton, 1996).

The Roles of Plant Pigments in Non-Plant Organisms

Many organisms absorb plant pigments from their diet, and may sequester them until they reach high levels. Flavonoid uptake has been demonstrated for a range of insects. In particular, butterflies and grasshoppers have been shown to take up the colourless or weakly pigmented flavonols and flavones, reaching levels of 2% dry weight of the wings of some butterflies. The sequestered compounds possibly act as visual attractants to mates (Harborne & Grayer, 1994; Bohm, 1998). Uptake of the more strongly pigmented flavonoids, such as anthocyanins, to the level of providing pigmentation to the insect has been implied by observation of insect colour on different plant food sources but it has not been characterised in detail. Flavonoids are taken up into the bloodstream in much smaller amounts by mammals, including humans, and their role in human health is a subject of much current research, which is covered in detail in Chapter 8. The polyphenolic tannins, reviewed in Chapter 5, also impact on animal health through the amelioration of bloat in ruminant animals.

Plant-derived carotenoids feature as key pigments in such familiar animal tissues as flamingo feathers (principally ketocarotenoids), shrimp and lobster shells, beetle shells, egg yolks and fish flesh (such as astaxanthin in goldfish) (Britton et al., 1995; Britton, 1996). In marine invertebrates, the carotenoids can occur as 'carotenoprotein' complexes. These can generate vivid colours, including blues, greens, reds and purples. Two examples of varying carotenoid–protein interactions in lobster are the blue astaxanthin-protein compound crustacyanin of

Figure .1: The red-coloured flower, seed pods and seed (see inset) of the tropical bush Bixa orellana, which is the source of the natural food colourant annatto. (Photograph by the author.)

the carapace and the green astaxanthin-lipoglycoprotein compound ovoverdin of the ovaries (Zagalsky, 1995; Britton et al., 1997). The effect of the protein on the carotenoid colour in invertebrates can often be observed upon cooking of the animal, as the heat can lead to the breakdown of the complex and a dramatic colour change (Britton et al., 1995). Carotenoid occurrence in bird plumage has also been extensively studied in relation to the specific chemical structures that occur, which can be modified by the bird's metabolism (McGraw et al., 2002), and the impact of the colours on mating behaviour (Olson & Owens, 1998). Like flavonoids, carotenoids are also of much interest with regard to their healthpromoting effects in the human diet, a subject also discussed in Chapter 8. An extreme example of the use of ingested pigments by animals, in this case from algae in the diet, is shown by photosynthetic molluscan sea slugs, in particular Elysia chlortica (Rumpho et al., 2000). In these organisms, chloroplasts are taken up intact from the food source and are maintained intracellularly in specific cells that line the highly branched digestive system. The chloroplasts remain functional for at least nine months, providing both camouflage for the mollusc and a supply of carbon. The mollusc has been shown in the laboratory to be able to sustain itself in the absence of food by photoautotrophic CO_2 fixation. The uptake of chloroplasts is widespread in the molluscan order Ascoglossa, but in some cases this may represent use only for camouflage and not the remarkable interaction described for E. chlortica.

Economic Aspects of Plant Pigments

Human use of plant pigment extracts dates back as long as recorded history. The most obvious use has been the application of pigments such as henna for tattooing and carthamin, indigo and other pigments to generate bright colours in clothing. For example, anthraquinone, indigoid and flavonoid pigments have been identified in the fourth century ad Egyptian textiles (Orska-Gawrys et al., 2003). Further back than this the use of carthamin extract (from safflower, Carthamus tinctorius) to dye the wrappings of mummies has been reported, and there is written evidence from 4600 years ago documenting human use of indigo (Gilbert & Cooke, 2001). Furthermore, a key part of garment manufacture since the time of early human societies has been the tanning of animal leathers with polymeric phenols, with record use dating back 10 000 years (Bohm, 1998). Today there are still tanneries carrying out leather treatment and dyeing with plant pigments in much the same manner as they have been doing for centuries (Fig. 1.2). From an economic perspective, putting to one side the vital role of chlorophylls and carotenoids in photosynthesis, the most obvious contribution of plant pigments to agriculture is with regard to consumer choice of fresh fruit, vegetables and floriculture products. However, the impact of the non-photosynthetic pigments is often much wider. For example, they are also of economic importance as

Figure. 2: Traditional tanning and dyeing of leather, including the use of many plant-derived compounds, in the Moroccan city of Fez. (Photograph by the author, taken in 2002.)

flavour and colour components of teas, wine and other beverages, as natural food colourants, for the health of ruminant animals, as plant defence agents and for amelioration of damaging UV light. The great range of non-coloured flavonoids and alkaloids produced by plants is often key to plant defence, and the biosynthesis of these compounds, and their importance to plant defence, agriculture and the pharmaceutical industries, have been extensively reviewed in previous Annual Plant Reviews (Wink, 1999a, 1999b).

Natural Food Colourants

Food quality is first assessed by its visual characteristics such as colour. Fresh food is often highly coloured by the major plant pigment groups, e.g. carotenoids and anthocyanins in fruit and chlorophylls in green vegetables. However, for processed foods the pigmentation is often lost during manufacturing, and the visual appeal of the final product is enhanced using added colourants. Until the discovery of synthetic dyes in the mid-nineteenth century, the food industry was solely reliant on natural food colourants. Although their use in many applications was superseded by synthetic dyes, in recent years there has been a return to the use of natural colourants, and an increased interest in new sources and improving their performance in food applications. Four plant pigment types are widely used as food colourants: annatto, anthocyanins, betalains (beetroot pigment) and curcumin (the main pigment of turmeric spice). Together with the insect-derived pigment cochineal, they account for over 90% of the market for natural food colourants (Hendry, 1996). Use of chlorophyll as a food colourant is very limited in comparison to these pigments, principally because of its poor stability during food processing or in response to light or acid conditions in the final food product. Plant pigments used as colourants in smaller amounts include the carotenoids xanthophyll and lutein, carthamus yellow from C. tinctorius petals, iridoids derived from Gardenia fruit and the carotenoid derivatives crocin and crocetin from Crocus sativus (saffron). One of the most important natural colourants is caramel, which is used extensively in both the food and beverage industries, particularly in soft drinks. However, as it is not a true plant pigment, being derived directly from sugar by processing, it is not considered here. Plant pigments are also key components of spices sold in large amounts, in particular paprika (containing a mix of carotenoids), turmeric and saffron. In addition to their use as food colourants, there is extensive use of pigment extracts as animal feed supplements. For example, carotenoids such as lutein are used in poultry/ egg production and aquaculture. Furthermore, large amounts of nature-identical synthetic pigments are used, in particular b-carotene. The types of pigments used in different food applications are determined by their solubility, and for the water-soluble pigments, their behaviour in response to pH. The anthocyanins and betalains are water-soluble, and the chlorophylls, curcumin and carotenoids typically oil-soluble. Annatto is one of the oldest known dyes used for foods, textiles and cosmetics. It is extracted from the resinous coating on the seeds of the tropical bush B. orellana. The species occurs in the wild in tropical North America and was used by Native Americans in pre-Columbian times as a source of pigment. Today, around 7000 tonnes of seed are processed annually by the pigment industry. The main pigment of annatto is cis-bixin, a monomethyl ester of the diapocarotenoic acid norbixin (Fig. 1.3), and supplies yellow to orange colours. It is sparingly soluble in oil and is principally used in dairy and fat-based foods. Present in smaller quantities in the pigment extract is a water-soluble carotenoid, cis-norbixin, which can also be generated by alkaline treatment of bixin (Britton, 1996). Curcumin (Fig. 1.3) is the principal pigment in the spice turmeric, which is extracted from the rhizomes of Curcuma longa, a perennial member of the ginger family (Zingiberaceae) that has been cultivated in Asia for many centuries. It supplies strong yellow colours and is generally oil-soluble. Turmeric has traditionally been used for colouring and flavouring meals; it is still used in large quantities for this purpose as well as extensively in a wide range of processed foods. Around 300 000 tonnes are produced annually in India, mostly for spice with a small amount for preparation of pure curcumin (Francis, 1996). Anthocyanins are widely used as food colourants (Jackman & Smith, 1996). First described as pH indicators, the colours of anthocyanin vary greatly on the pH of the food, but generally are used only in acidic foods and provide red to blue

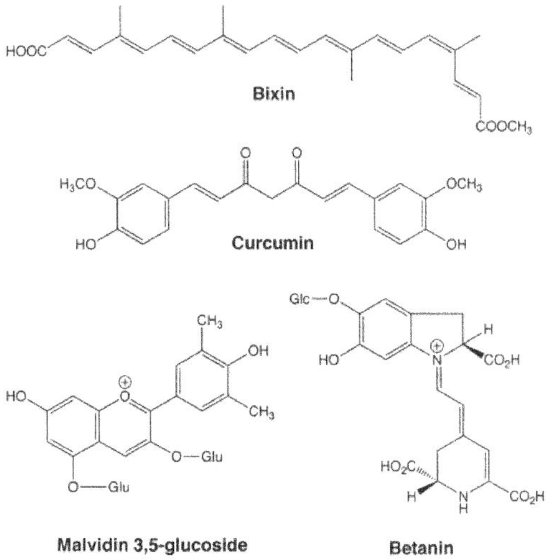

Figure .3: Diagrammatic representation of the major coloured constituents of common plant-based food colourants. Bixin is the main pigment of annatto, curcumin the

main pigment in turmeric, malvidin 3,5-O-glucoside a representative anthocyanin from grape extract and betanin the major betalain of beetroot extract.

colours. Although they are present in many sources, commercial anthocyanin extracts are predominantly prepared from Vitis species (grape)/berry skin or leas (barrel sediment) from the wine industry, which is available cheaply and in large amounts. Around 10 000 tonnes of grape skin are processed annually in Europe yielding about 50 tonnes of anthocyanin. However, although around 20 different anthocyanins have been reported for grapes, the common commercial preparations contain principally only relatively simple 3- and 3,5-diglucosides of cyanidin, delphinidin and malvidin (Fig. 1.3). These have limited colour stability with regard to pH, and are therefore limited in their food applications. This has led to research into sources of more complex, and thus more stable, anthocyanins to widen their food applications (Jackman & Smith, 1996; Giusti & Wrolstad, 2003; Schwarz et al., 2003). Commercial extracts are now available from species such as Ribes nigrum (black currant), Sambucus nigra (elderberry) and Brassica oleracea (red cabbage). Furthermore, extracts from Raphanus sativus (red radish) and Solanum tuberosum (red potatoes) are potential sources of acylated anthocyanins to replace the synthetic dye Allura Red (Giusti & Wrolstad, 2003). This interest in new anthocyanin sources has been intensified by the growing body of evidence on the health benefits of anthocyanins. Beetroot is the main source of betalain food colourants, principally the red betanin (Fig. 1.3) and lesser quantities of yellow betaxanthins. In Europe 20 000 tonnes of beetroot are processed annually for juice and pigment extraction, and betalains can account for 2% of beet-soluble solids (Jackman & Smith, 1996). Betalains are water-soluble pigments, and as they are unstable to heat and light, are used principally in foods with short shelf lives that do not need high heat treatment. At present, most plant-derived food colourants are sourced directly from plant material generated through traditional plant breeding. However, the amount of most plant pigment types, and the structural variation within each type, is affected by seasonal and environmental factors, making availability of defined source material variable. In addition, current source material is limited in the type of pigment produced, particularly for anthocyanins. Thus, there is much interest in applying plant biotechnology to colourant production, both for modifying the type of pigment produced and for developing alternative production routes. To date, most research has been on tissue culture methodologies, and genetic modification (GM) has not been applied to the same extent as it has for food crops and floriculture. Microbial fermentation has been extensively applied to food colourant production, including GM-developed strains. It is mostly based on endogenous microbial pigments, e.g. production of b-carotene in the algae Dunaliella salina, the blue pigment phycocyanin in the filamentous blue-green algae Spirulina platensis, and astaxanthin in the algae Haematococcus pluvialis and the yeast Phaffia rhodozyma (O'Callaghan, 1996). Recently, Bouvier et al. (2003b) have reported the production of the apocarotenoid bixin in Escherichia coli. Production of simple flavonoids in microbial systems has also been achieved (Hwang et al., 2003), but to date this has not been extended to the coloured flavonoids. The first successful biotechnology system for plant-derived pigments was plant cell culture production of the naphthoquinone pigment shikonin. Plant cell culture systems have also been developed for betalains (Strack et al., 2003) and anthocyanins (Do"renburg & Knorr, 1996). However, commercial production has not occurred for either of these pigment groups, principally due to the economics of biotechnology production compared to extraction of field crops. Application of GM techniques may improve the economics by allowing production of higher-value pigments with structures designed for specific food applications. Negative aspects of use of GM techniques include the potential need for additional regulatory approval and the uncertainty of public acceptance of GM products.

Modification of Pigment Biosynthesis in Transgenic Plants

The availability of the biosynthetic genes for several plant pigment pathways, and regulatory genes for flavonoids, has allowed genetic modification approaches for altering pigment production in transgenic plants. This book addresses both the biosynthesis and function of the plant pigments and the prospects for directed modification of their biosynthetic pathways. Modification of plant pigment biosynthesis using genetic technologies can be broken down into a few major approaches: abolishing or reducing flux into sections of a pathway by targeting biosynthetic genes with anti-sense/sense-suppression constructs; increasing levels of specific pathway compounds by overproducing endogenous biosynthetic activities; redirecting a pathway to produce novel compounds by the introduction of new enzyme activities; and changing regulation of a pathway using gene constructs for regulatory factors. All these approaches have been successfully used for modifying flavonoid biosynthesis, and some have been successful in the modification of other pigment pathways, in particular carotenoids and shikonin. To date, one transgenic crop with modified pigment production is commercially available – carnations containing transgenes that allow accumulation of delphinidin-derived anthocyanins for novel mauve flower colours (Lu et al., 2003). This is likely to be the first of many such ornamental products, as suggested by at least a dozen field trials of GM ornamentals with modified pigmentation that have been conducted recently by various biotechnology companies, universities and research institutes. The targets are cut flower crops, pot plants, bedding plants and garden species. There are also opportunities for altering plant colour using non-plant genes that code for biosynthetic activities for novel pigments, some of which are proteinbased. There are only a few published examples of these approaches to date, and as the source pigments are not of plant origin, they are only mentioned briefly here rather than in the main chapters of this book. Green fluorescent protein (GFP) from the jellyfish Aequorea victoria has been used extensively as a molecular marker (Tsien, 1998; Stewart, 2001), and there has been a study on producing GFP flowers as ornamental products (Mercuri et al., 2001). Mutagenesis studies have generated GFP variants with new colours and improved fluorescence in plants (Stewart, 2001), and searches for related sequences have led to a red fluorescent protein (Campbell et al., 2002). Indeed, there is a wide range of protein-based GFP-like chromophores in aquatic organisms of the phylum Cnidaria, which includes the class Hydrozoa containing A. victoria. Perhaps the most notable group of protein chromophores of the phylum is that providing the brilliant colours in coral, in particular the pocilloporin pigments (Dove et al., 2001). Brugliera et al. (2002) have identified GFP-like proteins in a range of coral reef organisms as possible sources for genes that could impart colours to plants, although reports on their performance in transgenics have yet to be published. Preferably, these pigments would readily provide colour in visible light without the need for special light sources. Currently there is widespread interest in applying GM techniques to improving health attributes of crops by changing plant pigment content. Several approaches have been taken to modifying carotenoid biosynthesis in food plants, and these are reviewed in Chapter 3. Of special note to date is the 'Golden Rice' line developed to have higher levels of b-carotene, the precursor of vitamin A (Hoa et al., 2003). This research featured widely in the international media, with the project leader, Dr Ingo Potrykus, appearing on the cover of Time magazine. It is planned that fullscale production of the GM rice will occur within five years, and it is likely to be followed by related products, such as high b-carotene mustard seed oil (Fraley, 2003; Potrykus, 2003) and zeaxanthin-rich potatoes (Ro¨mer et al., 2002). With respect to flavonoids, although there have been many successes altering their biosynthesis in ornamental crops, there are few examples in crop plants. Bovy et al. (2002) introduced into tomato transgenes for two transcription factors (LC and C1) that are involved in the upregulation of flavonoid biosynthesis in maize. Normally, tomato fruit contain only small amounts of flavonoids in their peel, and none in their flesh. However, the transgenic plants accumulated colourless flavonoids in the flesh of the fruit and increased levels of anthocyanins in the leaves. Nine major flavonoids could be detected in the flesh, principally flavonol and dihydroflavonol glycosides (Le

Gall et al., 2003). The total flavonoid content of the whole fruit (flesh and peel together) was tenfold higher for some transgenic lines compared to non-transgenic controls. This experiment is a good illustration of the promise that genes for transcription factors offer for increasing flavonoid levels in food crops. Chapter 4 reviews the knowledge of flavonoid transcription factors from studies of anthocyanin biosynthesis in flowers. Flavonoid biosynthetic genes have also been used successfully to increase the flavonoid levels of tomato fruit (Verhoeyen et al., 2002). A 78-fold increase in total flavonoid levels was achieved in some transgenic lines overexpressing a gene for the flavonoid biosynthetic enzyme chalcone isomerase. Furthermore, transgenes for two other flavonoid biosynthetic enzymes, chalcone synthase and flavonol synthase, were able to increase further the flavonoid levels in the transgenic lines overexpressing chalcone isomerase. Given the key roles of plant pigments in so many important plant processes, it is likely that they will remain the focus of major biotechnology programmes. The recent successes in elucidating the biosynthesis of many plant pigment types, as featured in this book, will provide strong underpinnings for such studies.

REFERENCES

1. Ayasse, M., Schiestl, F. P., Paulus, H. F., Ibarra, F. & Francke, W. (2003) Pollinator attraction in a sexually deceptive orchid by means of unconventional chemicals. Proc. Royal Soc. London Series B – Biol. Sci., 270, 517–522.

2. Bhattacharya, C. & Bhattacharya, N. (2001) Mendel's experiment regarding anthocyanin inheritance in the light of 100 years after its rediscovery. Crop Res., 21, 181–187.

3. Bohm, B. A. (1998) Introduction to Flavonoids. Harwood Academic Publishers, Amsterdam, The Netherlands.

4. Bouvier, F., Suire, C., Mutterer, J. & Camara, B. (2003a) Oxidative remodeling of chromoplast carotenoids: identification of the carotenoid dioxygenase CsCCD and CsZCD genes involved in crocus secondary metabolite biogenesis. Plant Cell, 15, 47–62.

5. Bouvier, F., Dogbo, O. & Camara, B. (2003b) Biosynthesis of the food and cosmetic plant pigment bixin (Annatto). Science, 300, 2089–2092.

6. Bovy, A., de Vos, R., Kemper, M., Schijlen, E., Pertejo, M.A., Muir, S., Collins, G., Robinson, S., Verhoeyen, M., Hughes, S., Santos-Buelga, C. & van Tunen, A. (2002) High-flavonol tomatoes resulting from the heterologous expression of the maize transcription factor genes LC and C1. Plant Cell, 14, 2509–2526.

7.   Boyle, R. (1664) Experiments and Considerations Touching Colours: First Occasionally Written Among Some Other Essays to a Friend and Now Suffered to Come Abroad as the Beginning of an Experimental History of Colours. Henry Herringman, London, UK. Available on microfilm at Early English Books Online, http://www.lib.umich.edu/eebo/

8.   Britton, G. (1996) Carotenoids, in Natural Food Colorants (eds G. A. F. Hendry & J. D. Houghton), hapman& Hall, London, UK, pp. 197–243.

9.   Britton, G., Liaaen-Jensen, S. & Pfander, H. (1995) Carotenoids today and challenges for the future, in Carotenoids, Vol. 1A: Isolation and Analysis (eds G. Britton, S. Liaaen-Jensen, & H. Pfander), Birkha"user Verlag, Basel, Switzerland, pp.13–26.

10.   Britton, G., Weesie, R. J., Askin, D., Warburton, J. D., GallardoGuerrero, L., Jansen, F. J., de Groot, H. J. M., Lugtenburg, J., Cornard, J. P. & Merlin, J. C. (1997) Carotenoid blues: structural studies on arotenoproteins. Pure Appl. Chem., 69, 2075–2084.

11.   Brouillard, R. & Dangles, O. (1993) Flavonoids and flower colour, in The Flavonoids: Advances in Research Since 1986 (ed. J. B. Harborne), Chapman & Hall, London, UK, pp. 565–587.

12.   Brugliera, F., Karan, M., Prescott, M., Mason, J., Dove, S. G., Hoegh-Guldberg, I. O. & Jones, E. L. (2002) Cell visual characteristic-modifying sequences. Patent application, No. WO 02/070703 A2.

13.   Campbell, R. E., Tour, O., Palmer, A. E., Steinbach, P. A., Baird, G. S., Zacharias, D. A. & Tsien, R. Y. (2002) A monomeric red fluorescent protein. Proc. Natl. Acad. Sci. USA, 99, 7877–7882.

14.   Comba, L., Corbet, S. A., Hunt, H., Outram, S., Parker, J. S. & Glover, B. J. (2000) The role of genes influencing the corolla in pollination of Antirrhinum majus. Plant Cell Environ., 23, 639–647.

15.   Cone, K. C., Burr, F. A. & Burr, B. (1986) Molecular analysis of the maize anthocyanin regulatory locus C1. Proc. Nat. Acad. Sci. USA, 83, 9631–9635.

16.   Darwin, C. R. (1868) The variation of animals and plants under domestication. John Murray, London, UK.

17.   de Valming, P., van Eekeres, J. E. M. & Wiering, H. (1982) A gene for flower colour fading in Petunia hybrida. Theor. Appl. Genet., 61, 41–46.

18.   Do"renburg, H. & Knorr, D. (1996) Generation of colors and flavors in plant cell and tissue cultures. Crit. Rev. Plant Sci., 15, 141–168.

19.   Dove, S. G., Hoegh-Guldberg & O., Ranganathan, S. (2001) Major colour patterns of reef-building corals are due to a family of GFP-like proteins.

Coral Reefs, 19, 197–204.

20. Eugster, C. H. (1995) History: 175 years of carotenoid chemistry, in Carotenoids, Vol. 1A: Isolation and Analysis (eds G. Britton, S. Liaaen-Jensen, & H. Pfander), Birkha¨user Verlag, Basel, Switzerland, pp. 1–12.

21. Fraley, R. T. (2003) Improving the nutritional quality of plants, in Plant Biotechnology 2002 and Beyond (ed. I. K. Vasil), Kluwer Academic Publishers, Dordrecht, The Netherlands, pp. 61–67.

22. Francis, F. J. (1996) Less common natural colourants, in Natural Food Colorants (eds G. A. F. Hendry & J. D. Houghton), Chapman & Hall, London, UK, pp. 112–132.

23. Gigord, L. D. B., Macnair, M. R. & Smithson, A. (2001) Negative frequency-dependent selection maintains a dramatic flower color polymorphism in the rewardless orchid Dactylorhiza sambucina (L.) Soo. Proc.Natl. Acad. Sci. USA, 98, 6253–6255.

24. Gilbert, K. G. & Cooke, D. T. (2001) Dyes from plants: past usage, present understanding and potential. Plant Growth Reg., 34, 57–69.

25. Giusti, M. M. & Wrolstad, R. E. (2003) Acylated anthocyanins from edible sources and their application in food systems. Biochem. Eng. J., 14, 217–225.

26. Glover, B. J. & Martin, C. R. (1998) The role of petal cell shape and pigmentation in pollination success in Antirrhinum majus. Heredity, 80, 778–784.

27. Gould, K. S., Vogelmann, T. C., Han, T. & Clearwater, M. J. (2002a) Profiles of photosynthesis within red and green leaves of Quintinia serrata A. Cunn. Physiol. Plant., 116, 127–133.

28. Gould, K. S., Neill, S. O. & Vogelmann, T. C. (2002b) A unified explanation for anthocyanins in leaves? Adv. Bot. Res., 37, 167–192.

29. Gould, K. S., Mckelvie, J. & Markham, K. R. (2002c) Do anthocyanins function as antioxidants in leaves? Imaging of H_2O_2 in red and green leaves after mechanical injury. Plant Cell Environ., 25, 1261–1269.

30. Gumbert, A., Kunze, J. & Chittka, L. (1999) Floral diversity in plant communities, bee colour space and a null model. Proc. Royal Soc. Lon. Series B – Biol. Sci., 266, 1711–1716.

31. Harborne, J. B. (ed.) (1988) The Flavonoids, Advances in Research Since 1980, Chapman & Hall, London,UK.

32. Harborne, J. B. (ed.) (1994) The Flavonoids, Advances in Research Since 1986, Chapman & Hall, London,UK.

33. Harborne, J. B. & Grayer, R. J. (1994) Flavonoids and insects, in The

Flavonoids: Advances in Research Since 1986 (ed. J. B. Harborne), Chapman & Hall, London, UK, pp. 589–618.

34. Hendry, B. S. (1996) Natural food colours, in Natural Food Colorants (eds G. A. F. Hendry & J. D.Houghton), Chapman & Hall, London, UK, pp. 40–79.

35. Hoa, T. T. C., Al-Babili, S., Schaub, P., Potrykus, I. & Beyer, P. (2003) Golden indica and japonica rice lines amenable to deregulation. Plant Physiol., 133, 161–169.

36. Hoch, W. A., Zeldin, E. L. & McCown, B. H. (2001) Physiological significance of anthocyanins during autumnal leaf senescence. Tree Physiol., 21, 1–8.

37. Hoch, W. A., Singsaas, E. L. & McCown, B. H. (2003) Resorption protection: anthocyanins facilitate nutrient recovery in autumn by shielding leaves from potentially damaging light levels. Plant Physiol., 133, 1296–1305.

38. Holton, T. A., Brugliera, F., Lester, D. R., Tanaka, Y., Hyland, C. D., Menting, J. G. T., Lu, C-Y., Farcy, E., Stevenson, T. W. & Cornish, E. C. (1993) Cloning and expression of cytochrome P450 genes controlling flower colour. Nature, 366, 276–279.

39. Hwang, E. I., Kaneko, M., Ohnishi, Y. & Horinouchi, S. (2003) Production of plant-specific flavanones by Escherichia coli containing an artificial gene cluster. Appl. Environ. Microbiol., 69, 2699–2706.

40. Jackman, R. L. & Smith, J. L. (1996) Anthocyanins and betalains, in Natural Food Colorants (eds G. A. F. Hendry & J. D. Houghton), Chapman & Hall, London, UK, pp. 244–309.

41. Jackson, A. H. (1976) Structure, properties and distribution of chlorophylls, in Chemistry and Biochemistry of Plant Pigments (ed. T. W. Goodwin), Academic Press, London, UK, pp. 1–63.

42. Johnson, S. D. & Midgley, J. F. (2001) Pollination by monkey beetles (Scarabaeidae: Hopliini): do color and dark centers of flowers influence alighting behavior? Environ. Ent., 30 861–868.

43. Jones, K. N. & Reithel, J. S. (2001) Pollinator-mediated selection on a flower color polymorphism in experimental populations of Antirrhinum (Scrophulariaceae). Am. J. Bot., 88, 447–454.

44. Jorgensen, R. A., Que, Q. D. & Napoli, C. A. (2002) Maternally controlled ovule abortion results from cosuppression of dihydroflavonol-4-reductase or flavonoid-30 ,50 Funct. Plant Biol., 29, 1501–1506.

45. Kull, D. & Pfander, H. (1995) Appendix: list of new carotenoids, in Carotenoids, Vol. 1A: Isolation and Analysis (eds G. Britton, S. Liaaen-Jensen, & H. Pfander), Birkha¨user Verlag, Basel, Switzerland, pp. 295–317.

46. Landeck, I. (2002) Feeding spectrum of the hairy flower wasp Scolia hirta in Lusatia (Central Europe) with special focus on flower colour, morphology of flowers and inflorescences (Hymenoptra: Scoliidae). Ent. Generalis, 26, 107–120.

47. Le Gall, G., DuPont, M. S., Mellon, F. A., Davis, A. L., Collins, G. J., Verhoeyen, M. E. & Colquhoun, I. J.

48. (2003) Characterization and content of flavonoid glycosides in genetically modified tomato (Lycopersicon esculentum) fruits. J. Ag. Food Chem., 51, 2438–2446.

49. Lee, D. W. (2002) Anthocyanins in autumn leaf senescence. Adv. Bot. Res., 37, 147–165.

50. Lu, C., Chandler, S. F., Mason, J. G. & Brugliera, F. (2003) Florigene flowers: from laboratory to market, in Plant Biotechnology 2002 and Beyond (ed. I. K. Vasil), Kluwer Academic Publishers, Dordrecht, The Netherlands, pp. 333–336.

51. Lunau, K. (2000) The ecology and evolution of visual pollen signals. Plant Systematics Evol., 222, 89–111.

52. Matile, P. (2000) Biochemistry of an Indian summer: physiology of autumnal leaf coloration. Exp. erontology, 35, 145–158.

53. McGraw, K. J., Adkins-Regan, E. & Parker, R. S. (2002) Anhydrolutein in the zebra finch: a new, etabolically-derived carotenoid in birds. Comp. Biochem. Physiol. Part B – Biochem. Mol. Biol., 132, 811–818.

54. Melendez-Ackerman, E., Campbell, D. R. & Waser, N. M. (1997) Hummingbird behavior and mechanisms of selection on flower color in Ipomopsis. Ecology, 78, 2532–2541.

55. Mercuri, A., Sacchetti, A., De Benedetti, L., Schiva, T. & Alberti, S. (2001). Green fluorescent flowers. Plant Sci., 161, 961–968.

56. Napoli, C., Lemieux, C. & Jorgensen, R. (1990) Introduction of a chimeric chalcone synthase gene into Petunia results in reversible co-suppression of homologous genes in trans. Plant Cell, 2, 279–289.

57. Neill, S. O., Gould, K. S., Kilmartin, P. A., Mitchell, K. A. & Markham, K. R. (2002) Antioxidant capacities of green and cyanic leaves in the sun species, Quintinia serrata. Funct. Plant Biol., 29, 1437–1443.

58. Oberrath, R. & Bohning-Gaese, K. (1999) Floral color change and the attraction of insect pollinators in lungwort (Pulmonaria collina). Oecologia, 121, 383–391.

59. O'Callaghan, M. C. (1996) Biotechnology in natural food colours. The role of bioprocessing, in Natural Food Colorants (eds G. A. F. Hendry & J. D. Houghton), Chapman & Hall, London, UK, pp. 80–111.

60. Odell, E., Raguso, R. A. & Jones, K. N. (1999) Bumblebee foraging responses to variation in floral scent and color in snapdragons (Antirrhinum: Scrophulariaceae). Am. Mid. Nat., 142, 257–265.

61. Olson, V. A. & Owens, I. P. F. (1998) Costly sexual signals: are carotenoids rare, risky or required? Trends Ecol. Evol., 13, 510–514.

62. Onslow, M. A. (1925) The Anthocyanin Pigments of Plants, 2nd edn, Cambridge University Press, Cambridge, UK.

63. Orska-Gawrys, J., Surowiec, I., Kehl, J., Rejniak, H., Urbaniak-Walczak, K. & Trojanowitcz, M. (2003) Identification of natural dyes in archeological Coptic textiles by liquid chromatography with diode array detection. J. Chromatogr. A., 898, 239–248.

64. Page, J. E. & Towers, G. H. N. (2002) Anthocyanins protect light-sensitive thiarubrine phototoxins. Planta, 215, 478–484.

65. Paxton, R. J. & Tengo, J. (2001) Doubly duped males: the sweet and sour of the orchid's bouquet. Trends Ecol. Evol., 16, 167–169.

66. Paz–Ares, J., Wienand, U., Peterson, P. A. & Saedler, H. (1986) Molecular cloning of the c locus of Zea mays: a locus regulating the anthocyanin pathway. EMBO J., 5, 829–833.

67. Pfander, H. (1987) The Key to Carotenoids, 2nd edn, Birkhäuser Verlag, Basel, Switzerland.

68. Potrykus, I. (2003) Nutritional improvement of rice to reduce malnutrition in developing countries, in Plant

69. Biotechnology 2002 and Beyond (ed. I. K. Vasil), Kluwer Academic Publishers, Dordrecht, The Netherlands, pp. 401–406.

70. Raguso, R. A. & Willis, M. A. (2002) Synergy between visual and olfactory cues in nectar feeding by naïve hawkmoths, Manduca sexta. Animal Behav., 64, 685–695.

71. Römer, S., Lubeck, J., Kauder, F., Steiger, S., Adomat, C. & Sandmann, G. (2002) Genetic engineering of a zeaxanthin-rich potato by antisense inactivation and co-suppression of carotenoid epoxidation. Metab.Eng., 4, 263–272.

72. Rumpho, M. E., Summer, E. J. & Manhart, J. R. (2000) Solar-powered seaslugs. Mollusc/Algal chloroplast symbiosis. Plant Physiol., 123, 29–38.

73. Schiestl, F. P., Ayasse, M., Paulus, H. F., Lofstedt, C., Hansson, B. S., Ibarra, F. & Francke, W. (1999) Orchid pollination by sexual swindle. Nature, 399, 421–422.

74. Schram, A. W., Jonsson, L. M. V. & Bennink, G. J. H. (1984) Biochemistry of flavonoid synthesis in Petunia hybrida, in Monogrpahs on Theoretical and Applied Genetics 9: Petunia (ed. K. C. Sink), SpringerVerlag, Berlin, pp. 68–76.

75. Schwarz, M., Hillebrand, S., Habben, S., Degenhardt, A. & Winterhalter, P. (2003) Application of high-speed countercurrent chromatography to the large-scale isolation of anthocyanins. Biochem. Eng. J., 14, 179–189.

76. Simpson, K. L., Lee, T-C., Rodriguez, D. B. & Chichester, C. O. (1976) Metabolism in senescent and stored tissues, in Chemistry and Biochemistry of Plant Pigments Vol. 1 (ed. T. W. Goodwin), Academic Press, London, UK, pp. 780–842.

77. Staiger D. (2002) Circadian rhythms in Arabidopsis: time for nuclear proteins. Planta, 214, 334–344.

Chapter 12

SIMPLE EXTRACTION METHODS THAT PREVENT THE ARTIFACTUAL CONVERSION OF CHLOROPHYLL TO CHLOROPHYLLIDE DURING PIGMENT ISOLATION FROM LEAF SAMPLES

Xueyun Hu[1] , Ayumi Tanaka[1,2] and Ryouichi Tanaka[1,2]

[1]Institute of Low Temperature Science, Hokkaido University
[2]CREST/JST, Hokkaido University

ABSTRACT

Background

When conducting plant research, the measurement of photosynthetic pigments can provide basic information on the physiological status of a plant. High-pressure liquid chromatography (HPLC) is becoming widely used for this purpose because it provides an accurate determination of a variety of photosynthetic pigments simultaneously. This technique has a drawback compared with conventional spectroscopic techniques, however, in that it is more prone to structural modification of pigments during extraction, thus potentially generating erroneous results. During pigment extraction procedures with acetone or alcohol, the phytol side chain of chlorophyll is sometimes removed, forming chlorophyllide, which affects chlorophyll measurement using HPLC.

Results

We evaluated the artifactual chlorophyllide production during chlorophyll extraction by comparing different extraction methods with wild-type and mutant Arabidopsis leaves that lack the major isoform of chlorophyllase. Several extraction methods were compared to provide alternatives to researchers who utilize HPLC for the analysis of chlorophyll levels. As a result, the following three methods are recommended. In the first method, leaves are briefly boiled prior to extraction. In the second method, grinding and homogenization of leaves are performed at sub-zero temperatures. In the third method, N, N'-dimethylformamide (DMF) is used for the extraction of pigments. When compared, the first two methods eliminated almost all chlorophyllide-forming activity in *Arabidopsis thaliana, Glebionis coronaria, Pisum sativum* L. and *Prunus sargentii* Rehd. However, DMF effectively suppressed the activity of chlorophyllase only in Arabidopsis leaves.

Conclusion

Chlorophyllide production in leaf extracts is predominantly an artifact. All three methods evaluated in this study reduce the artifactual production of chlorophyllide and are thus suitable for pigment extraction for HPLC analysis. The boiling method would be a practical choice when leaves are not too thick. However, it may convert a small fraction of chlorophyll *a* into pheophytin *a*. Although extraction at sub-zero temperatures is suitable for all plant species examined in this study, this method might be complicated for a large number of samples and it requires liquid nitrogen and equipment for leaf grinding. Using DMF as an extractant is simple and suitable with Arabidopsis samples. However, this solvent cannot completely block the formation of chlorophyllide in thicker leaves.

BACKGROUND

Chlorophyll analysis has been conducted in numerous studies due to the importance of this pigment in the physiology of plants. Chlorophyll is involved in the absorption and transfer of light energy, and electron transfer, all of which are vital processes in photosynthesis. Chlorophyll content can change in response to biotic and abiotic stresses such as pathogen infection [1], and light stress [2, 3]. Thus, quantification of chlorophyll provides important information about the effects of environments on plant growth [4–8].

Historically, spectroscopic methods have been most frequently used for chlorophyll measurement because they provide a quick, accurate and inexpensive estimation of chlorophyll concentration [9–11]. However, conventional spectroscopic methods, where bulk photosynthetic pigments are measured in the same cuvette, have limitations in their ability to simultaneously measure multiple photosynthetic pigments due to the overlapping absorption spectra of these pigments. For this reason, it has become more common to separate photosynthetic pigments by high-pressure liquid chromatography (HPLC) prior to spectrophotometric analysis [12, 13]. When separating pigments by HPLC, extra care must be taken since HPLC analysis is prone to the artifactual modification of pigments. In particular, cleavage of the phytol chain of chlorophyll molecules readily occurs with the use of common extraction solvents such as 80% acetone [14]. The products of chlorophyll hydrolysis are chlorophyllide and free phytol. Since chlorophyllide has the same absorption spectra as chlorophyll in the visible light spectrum and phytol does not, cleavage of the phytol chain does not affect the values obtained using conventional spectroscopic methods of chlorophyll determination when samples are extracted with organic solvents. However, due to the polar nature of chlorophyllide, it is readily separated from chlorophyll with HPLC, and thus the artifactual formation of chlorophyllide can result in erroneous data using HPLC-based determination of chlorophyll concentration.

Conversion of chlorophyll to chlorophyllide induced by the extraction agent reduces the apparent concentration of chlorophyll in samples. It is usually difficult to distinguish whether or not the chlorophyllide detected during HPLC analysis is an artifact or a natural product. In fact, chlorophyllide has been considered a natural product in leaves without examining the basis of its formation [15, 16]. In order to avoid possible misinterpretation of chlorophyll levels, it is essential to employ extraction methods that result in a minimal amount of conversion of chlorophyll to chlorophyllide.

It has been reported that the hydrolase enzyme, chlorophyllase (CLH) catalyzes the formation of chlorophyllide during pigment extraction [17] (Figure 1A). This enzyme is unusually stable in high concentrations of organic solvents such as 50-70% aqueous acetone [18, 19]. Higher plants contain one or two isoforms of this enzyme [20] and Arabidopsis has two CLH isoforms encoded by *CLH1* and *CLH2* genes, respectively [21]. *CLH1* encodes the isoform of CLH that accounts for the majority of CLH in Arabidopsis leaves. *CLH1* gene expression is significantly upregulated by methyl-jasmonate (MeJA), a phytohormone mediating various biotic and abiotic signaling pathways [22]. In contrast, *CLH2* is constitutively expressed and only represents a minor fraction of CLH activity [23]. In the present study, we assessed how much chlorophyllide is formed during pigment extraction compared to the amount that naturally occurs in leaves. In a subsequent analysis, we then examined whether or not CLH is involved in chlorophyllide formation during extraction by comparing its formation in leaves of wild-type and an Arabidopsis mutant which is deficient in CLH activity. Collectively, these experiments indicated that the majority of chlorophyllide detected in extracts obtained using 80% acetone or pure acetone is produced during pigment extraction through the reaction catalyzed by CLH. We also compared three different methods of pigment extraction that were previously reported in literature. Bacon and Holden [17] reported that CLH activity could be suppressed by boiling leaves for a period of 5 min. They also indicated, however, that the boiling treatment also removes Mg^{2+} from chlorophyll [17]. We found that, in the case of Arabidopsis leaves, CLH can be inactivated and Mg^{2+} removal from chlorophyll can be reduced when samples were boiled for only 5 sec. In the method of Schenk et al. [23], leaves were first ground into powder in liquid nitrogen and pigments were subsequently extracted in buffered acetone cooled to −20°C. We found this method is very efficient when processing a relatively small number of samples. Finally, we tested the use of N, N'-dimethylformamide (DMF) as an extraction agent to eliminate the formation of chlorophyllide during sample preparation. Although Moran and Porath [24] reported that chlorophyll is stable in this solvent, they did not characterize the effect of DMF on chlorophyllide formation. In our study, DMF was capable of extracting pigments without enabling the conversion of chlorophyll to chlorophyllide in Arabidopsis, however, for the other species which we have tested in this study, DMF cannot completely suppress the activity of CLH. Collectively, all three methods (boiling leaf sample, freezing leaf samples in liquid nitrogen with the use of pre-cooled acetone, and the use of DMF as an extraction agent) were superior to the methods only using 80% or pure acetone for the extraction of photosynthetic pigments. It is important to understand advantages and disadvantages of each method and choose an appropriate one for each plant species and for the purpose of pigment analysis.

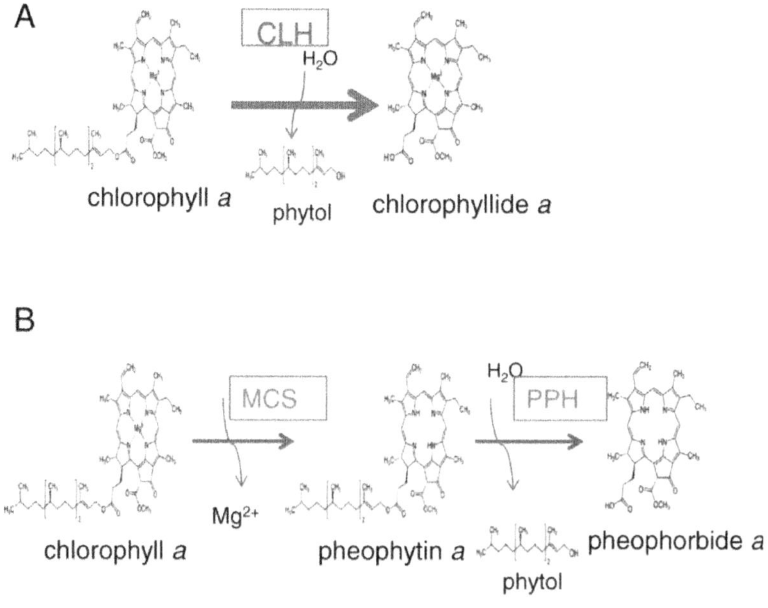

Figure 1: Comparison of the CLH reaction and a proposed *in vivo* degradation pathway of chlorophyll *a*. **A.** CLH catalyzes the hydrolysis of the ester bond of chlo-

rophyll to form chlorophyllide and phytol. **B**. An *in vivo* degradation pathway of chlorophyll *a* proposed by Hörtensteiner and coworkers [29]. MCS denotes magnesium dechelating substances. PPH denotes pheophytinase.

RESULTS

Use of 80% and pure acetone as extraction solvents results in the formation of chlorophyllide It was our objective to provide a simple and reliable method to extract chlorophyll for HPLC analysis that would be free from artifacts. In order to achieve this goal, we started with one of the simplest methods to extract pigments and attempted to improve it. We first compared two conventional methods in which pigments are extracted from Arabidopsis leaf samples by soaking them in 80% or pure acetone for 12 hours at 4°C. Since CLH has been reported to be active in aqueous acetone but precipitated in pure acetone [18, 19], it was expected that chlorophyllide would only be produced in the 80%-acetone extracts. In line with this expectation, we determined that nearly 70% of the combined chlorophyll and chlorophyllide *a* content was composed of chlorophyllide *a* in the 80%-acetone extracts (Figure 2). In contrast, only a small amount of chlorophyllide *a* was produced in pure acetone (Figure 2). Therefore, it is likely that most of the chlorophyllide *a*detected in 80% acetone was formed during extraction or after extraction. Since chlorophyll *a* is much more abundant than chlorophyll *b* and the trend of chlorophyllide *b* formation was similar to that of chlorophyllide *a*, we only describe the results on chlorophyll *a* and chlorophyllide *a* in the present study.

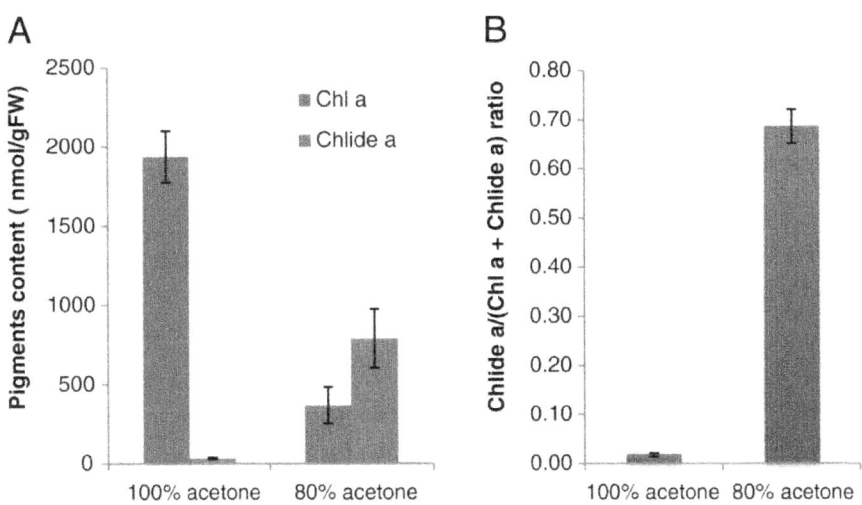

Figure 2: Formation of chlorophyllide *a* by the extraction of chlorophyll with pure or 80% buffered acetone. Frozen leaves were immersed in pure acetone or 80% acetone

containing 20% (v/v) Tris–HCl, pH 8 and then incubated in these solvents for 12 h at 4°C in the dark. Pigments were subsequently extracted by grinding with stainless steel beads as described in the Methods section. **A**. Levels of chlorophyll *a* and chlorophyllide *a* per gram fresh weight of leaves. **B**. Chlorophyllide *a* levels in sample extracts expressed as the ratio of cholorophyllide *a* to the sum of chlorophyll *a* and chlorophyllide *a*. Chl *a*, chlorophyll *a*. Chlide *a*, chlorophyllide *a*. Error bars indicate standard deviations. Sample size, n = 3.

CLH activity increases in leaves that are either senescent [25] or wounded [16], and in response to MeJA treatment [21]. Thus, we examined chlorophyllide formation using naturally-senescent leaves from 8-week-old Arabidopsis plants, those in which senescence was induced by a 4-day dark treatment, and those in which chlorophyll breakdown was induced by MeJA. Eight to ten percent of the chlorophyll content in naturally-senescent, dark-treated, and MeJA-treated leaves was composed of chlorophyllide (Figure 3) even when pure acetone was used. These results indicate that pure acetone does not sufficiently suppress the formation of chlorophyllide during chlorophyll extraction.

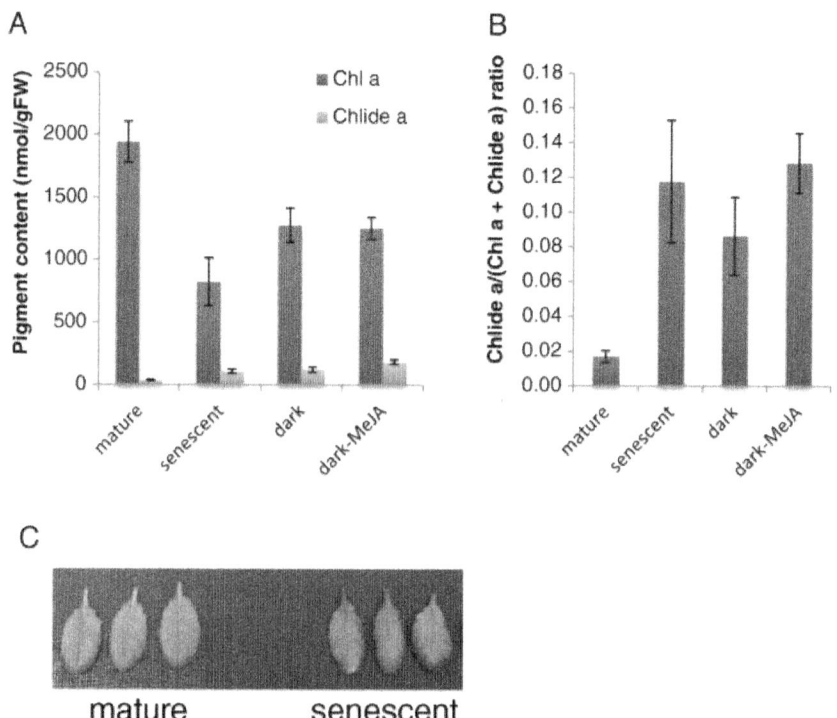

Figure 3: Chlorophyllide formation during pigment extraction from senescent or Me-JA-treated leaves. Pigments extracted from mature leaves (7th to 9th leaves counting

from the bottom of the plant) of wild-type (WT) Arabidopsis plants under four different conditions: leaves collected from 4-week-old plants ("mature"), leaves from 8-week-old plants ("senescent"), leaves from 4-week-old plants where the removed leaves were incubated in complete darkness on filter paper saturated with 3 mM MES buffer ("dark") or the same buffer plus 50 μM MeJA ("dark-MeJA"). Pigments were extracted by immersing the leaves in pure acetone that was cooled to 4°C and the extracts were subsequently analyzed by HPLC (see the Methods section for detail). **A**. Levels of chlorophyll *a* and chlorophyllide *a* per gram fresh weight of leaves. **B**. Chlorophyllide *a* levels in sample extracts expressed as the ratio of chlorophyllide *a* to the sum of chlorophyll *a* and chlorophyllide *a*. Chl *a*, chlorophyll *a*. Chlide *a*, chlorophyllide *a*. Error bars indicate standard deviations. Sample size, n = 3. **C**. The photograph below the bar graphs illustrates the 4-week-old (left) and 8-week-old (right) leaves used in these analyses.

CLH Is Responsible For the Formation of Chlorophyllide during Chlorophyll Extraction

The experiments described above indicated that the majority of chlorophyllide is formed during extraction. In the next series of experiments, we examined the time-course of chlorophyllide formation during acetone extraction. We also tested whether or not CLH was involved in chlorophyllide formation in pure acetone by comparing chlorophyllide formation during extraction using Arabidopsis leaves from wild type (WT) and a *clh1-1* mutant that lacks the major isoform of CLH. After immersing leaves in pure acetone, leaves were incubated in pure acetone for up to 6 min (Figure 4). In this series of experiments, chlorophyllide formation was also compared in leaves from WT plants that had been either treated or not with 50 μM MeJA because the CLH activity is more evident after MeJA treatment (Figure 4).

Production of chlorophyllide was negligible unless incubation was at an ambient temperature (Figure 4). Increasing incubation time at an ambient temperature resulted in elevated amounts of chlorophyllide (Figure 4). MeJA-treated WT leaves yielded a higher amount of chlorophyllide (up to 12% of the total chlorophyll plus chlorophyllide after 6 min of incubation) compared to non-treated WT leaves. In contrast, extracts from leaves from *clh1-1* plants only contained a trace amount of chlorophyllide with or without MeJA treatment (Figure 4). These results are consistent with the report of Schenk et al. [23], in which they detected a much lower level of chlorophyllide in acetone extracts of dark-incubated *clh1-1* leaves compared to extracts from dark-incubated wild-type leaves. These results indicate that the major isoform of CLH is responsible for chlorophyllide formation during extraction.

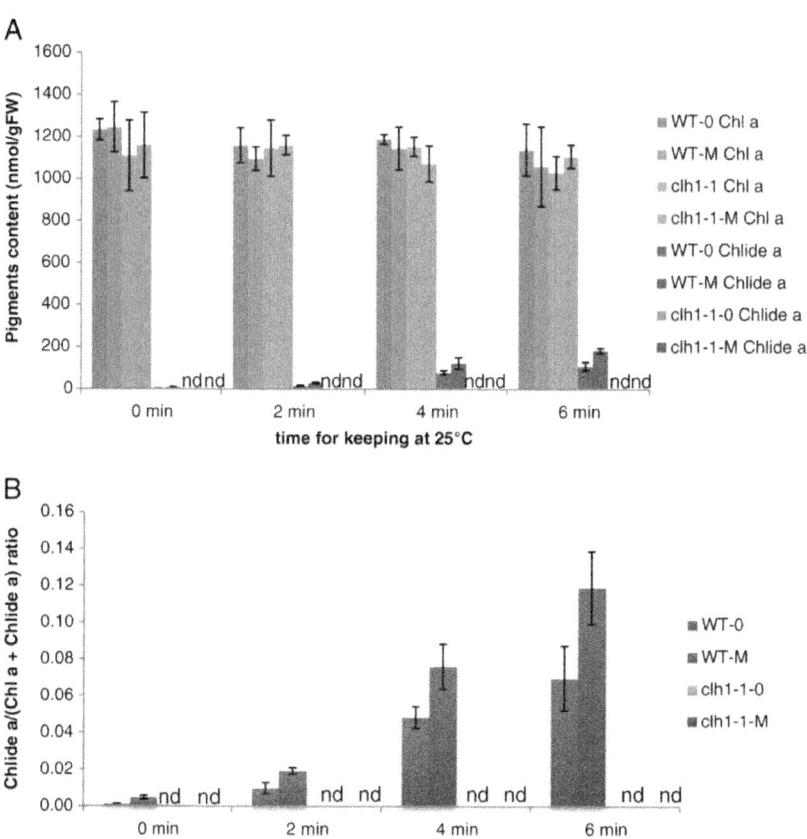

Figure 4: Time course of chlorophyllide *a* formation during pigment extraction. Chlorophyll was extracted by immersing leaves in pure acetone as described in the "Time-course experiments" subsection of the Methods section. **A.** Levels of chlorophyll *a* and chlorophyllide *a* per gram fresh weight of leaves. **B.** Chlorophyllide *a* levels in sample extracts expressed as the ratio of chlorophyllide *a* to the sum of chlorophyll *a* and chlorophyllide *a*. Chl *a*, chlorophyll *a*. Chlide *a*, chlorophyllide *a*. Error bars indicate standard deviations. Sample size, n = 3. n.d. = not detected.

Stability of Chlorophyll and Chlorophyllide in Pure Acetone

In our next line of investigation, we tested the stability of chlorophyll and chlorophyllide in pure acetone. Chlorophyllide production was evident from an extraction of WT leaves in pure acetone at ambient temperatures for 6 min (Figure 4 and Table 1). Specifically, the pigments were extracted by grinding leaves in pure acetone with stainless steel balls and the extracts were separated from cell debris by centrifugation. This procedure yielded approximately

200 nmol/gFW (fresh weight) chlorophyllide in the extract. In contrast, chlorophyllide levels were less than 10 nmol/gFW in the absence of the 2 – 6 min incubation. Extracts were also maintained for one day in darkness at room temperature and their chlorophyll and chlorophyllide levels were measured using HPLC. No significant changes in chlorophyll or chlorophyllide levels were observed in the one-day-old extracts (Table 1). Therefore, it is likely that the majority of chlorophyllide detected in the extracts were formed during the extraction or homogenization procedures.

Table 1: Stability of chlorophyll *a* and chlorophyllide *a* after extraction with pure acetone

	0 day		1 day	
Time	Chlide *a*	Chl *a*	Chlide *a*	Chl *a*
(min)	(nmol/gFW)	(nmol/gFW)	(nmol/gFW)	(nmol/gFW)
0	9	1284	9	1317
	5	1107	7	1167
	8	1338	6	1340
6	211	979	203	983
	191	1249	203	1329
	168	1052	171	1081

Mature Arabidopsis leaves (7th - 9th leaves counting from the bottom of the 4-week-old plants) were treated with 50 μM MeJA and were kept in complete darkness for 3 days. Chlorophyll was extracted by immersing leaves in pure acetone as it is described in the "Time-course experiments" subsection of the Methods section. "Time" indicates the incubation time at room temperature. After removing the residual leaf tissue by centrifugation, the extracts were incubated for 24 hours at room temperature. Each line represents a single experiment with an extract from a single leaf, and experiments were repeated three times in the same conditions. Values represent chlorophyllide *a* or chlorophyll *a* concentrations per gram fresh weight of leaves. Chl *a*, chlorophyll *a*. Chlide *a*, chlorophyllide *a*.

Chlorophyllide Formation Is Suppressed With Rapid Boiling Of Leaves

In order to provide a simple chlorophyll extraction method that is less affected by CLH activity, we assessed if CLH activity could be inactivated by boiling leaves. Bacon and Holden [17] have already reported that CLH activity can be suppressed by boiling leaves for 5 min. However, they also found that this treatment destroys some pigments. Therefore, we examined whether shorter (approximately 5 or 10 sec) periods of boiling can adequately suppress CLH activity while avoiding pigment decomposition. In this experiment, mature Arabidopsis leaves were sprayed with 50 μM MeJA and subsequently harvested. After collection they were dipped in boiling water for 5 or 10 sec, and then soaked in pure acetone. For pigment extraction, leaves were homogenized in acetone using stainless beads or kept in pure acetone overnight at 4°C.

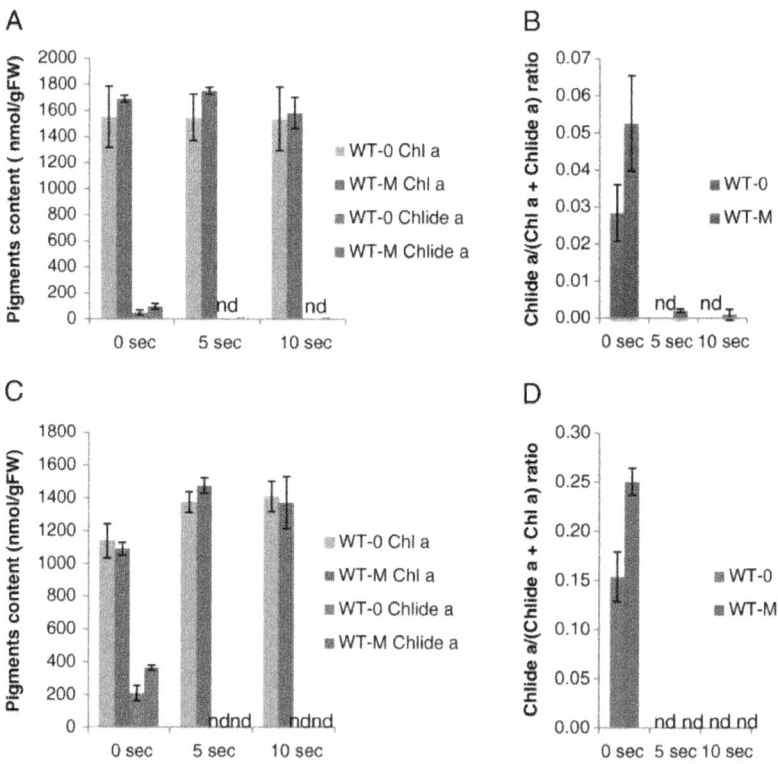

Figure 5: Effect of quick boiling on the formation of chlorophyllide *a* during extraction. Mature leaves (7th to 9th leaves counting from the bottom of the plant) were collected from 4-week-old wild-type plants that had been sprayed with 100 μM MeJA and kept for 3 days at the same growth conditions. The collected leaves were immersed in pure acetone at room temperature directly or were boiled for 5 and 10 sec respectively before they were immersed in pure acetone, then they were ground with stainless steel beads by vigorous shaking (**A** and **B**). Alternatively, pigments were extracted by immersing leaves in pure acetone at 4°C for 12 hours (**C** and**D**). **A** and **C**, Levels of chlorophyll *a* and chlorophyllide *a* per gram fresh weight of leaves. **B** and **D**, Chlorophyllide *a* levels in sample extracts expressed as the ratio of chlorophyllide *a* to the sum of chlorophyll *a* and chlorophyllide *a*. Chl *a*, chlorophyll *a*. Chlide *a*, chlorophyllide *a*. Error bars indicate standard deviations. Sample size, n = 3.

Leaves from WT plants that were homogenized without boiling yielded 94 nmol/gFW and 46 nmol/gFW chlorophyllide *a*when they were treated or not with MeJA, respectively, prior to extraction (Figure 5A). In contrast, chlorophyllide *a* was below a detectable level in leaves that were not treated with MeJA when the leaves were boiled for 5 or 10 sec before chlorophyll extraction. When MeJA-treated WT leaves were boiled for 5 or 10 sec prior to extraction, only 3.6 and 1.4 nmol/gFW chlorophyllide *a* were detected, respectively (Figure 5A). The combined sum of chlorophyll and chlorophyllide were not affected by boiling (Figure 5A and B), indicating that the significant pigment losses, observed when leaves were subjected to 5 min boiling [17], did not occur when the brief boiling procedure was used. The overall profiles of detectable photosynthetic pigments obtained by HPLC were not altered by boiling except for pheophytin *a*. This pigment increased slightly from 30 nmol/gFW to 50 nmol/gFW in boiled samples. These levels represented approximately 0.2% to 0.3% of total chlorophyll *a* levels, and are almost negligible in the HPLC profiles (Figure 6B).

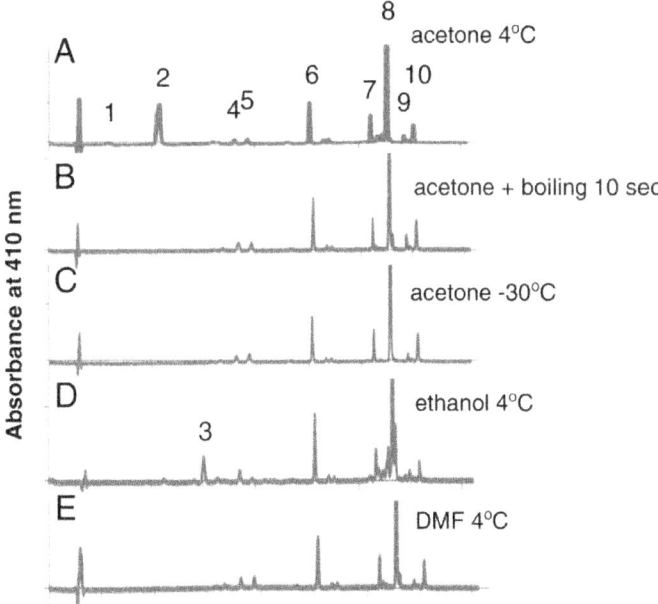

Figure 6: The elution profiles of pigments from extracts separated by HPLC. Representative HPLC profiles from each experiment are shown. **A.** HPLC profile of pigments extracted by immersing wild-type (WT) leaves in pure acetone for 12 hours at 4°C. **B.** An HPLC profile of WT pigments extracted by the quick boiling method. **C.** HPLC profile of pigments extracted from leaves of WT plants by immersing leaves in pure acetone for 96 hours at −30°C. **D.** HPLC profile of pigments extract-

ed from leaves of WT plants by immersing leaves in ethanol at 4°C for 12 hours. **E**. HPLC profile of pigments extracted from leaves of WT plants by immersing leaves in DMF at 4°C for 12 hours. Peak 1, chlorophyllide *b*. Peak 2, chlorophyllide *a,* Peak 3, pheophorbide *a*. Peak 4, neoxanthin. Peak 5, violaxanthin. Peak 6, lutein. Peak 7, chlorophyll *b*. Peak 8, chlorophyll *a*. Peak 9, pheophytin a. Peak 10, β-carotene.

To further simplify the extraction method, boiled leaves were kept overnight in pure acetone at 4°C. Chlorophyllide formation in both MeJA-treated and non-treated boiled WT leaves were negligible after overnight incubation (Figure 5C and5D), indicating that rapid boiling almost completely inactivated CLH activity.

Pure Acetone Extraction at Sub-Zero Temperatures

Schenk et al. [23] have already demonstrated that chlorophyllide formation could be minimized if leaf samples were ground in liquid nitrogen and extracted with acetone cooled to −20°C. In this study, we attempted to simplify their method. We evaluated whether or not chlorophyll could be extracted just by immersing frozen leaves in pure acetone cooled to −30°C. Our data indicated that a substantial amount of chlorophyll remained in leaf tissue after an overnight incubation in −30°C acetone (data not shown). After 4 days of incubation, the majority of chlorophyll had been extracted as evidenced by the white appearance of extracted leaves (data not shown). HPLC analysis showed that only trace amounts of chlorophyllide were detected in samples obtained from WT leaves that had been treated or not with MeJA (Figures 6 and 7), indicating that CLH activity was negligible at −30°C.

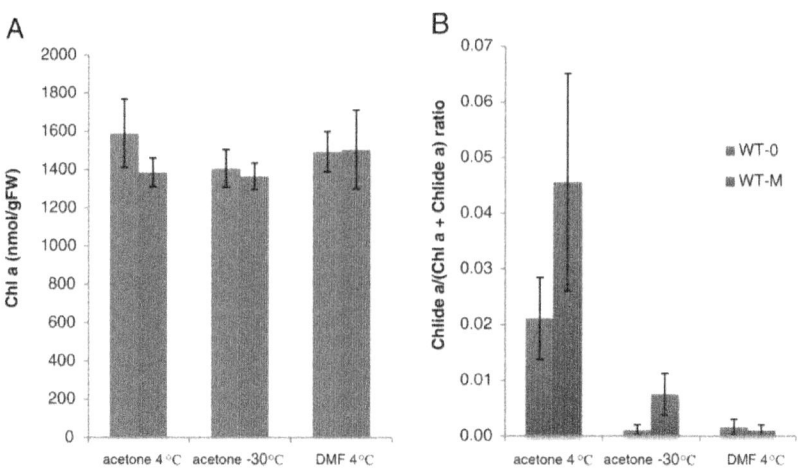

Figure 7: Pigment extraction at −30°C in pure acetone and at 4°C in DMF. Mature Arabidopsis leaves (7th to the 9th leaves counting from the bottom of the plant) that

were treated with or without MeJA (see the Method section for the detail) were harvested and their pigments were extracted by three different methods. In the first two methods, leaves were immersed in pure acetone at 4°C for 12 hours ("acetone 4°C") and at −30°C for 96 hours ("acetone −30°C") respectively. In the third method, leaves were immersed in DMF at 4°C for 12 hours ("DMF 4°C"). **A**. Levels of chlorophyll *a* and chlorophyllide *a* per gram fresh weight of leaves. **B**. Chlorophyllide *a* levels in sample extracts expressed as the ratio of chlorophyllide *a* to the sum of chlorophyll *a* and chlorophyllide *a*. Chl *a*, chlorophyll *a*. Chlide *a*, chlorophyllide *a*. Error bars indicate standard deviations. Sample size, n = 3.

Extraction of Chlorophyll by DMF Or Ethanol

DMF is reported to efficiently extract chlorophyll without the need for homogenization [26]. In the present study, we tested whether or not chlorophyllide formation occurs during extraction with DMF. WT leaves, treated or not with MeJA, were incubated for 12 h in DMF at 4°C. Only a trace amount of chlorophyllide was detected (Figure 7). Modification of pigment structure that impacts the profiles obtained by HPLC separation of major photosynthetic pigments, including chlorophyll *a*, did not occur with the use of DMF (Figure 6). We also examined the ability of ethanol to extract chlorophyll in this study. WT leaves, those were either treated or not with MeJA, were incubated for 12 h in ethanol at 4°C. This solvent did not extract all of the chlorophyll from leaves after a 12-h incubation, as evidence by the leaves retaining some greenish color. Additionally, this solvent induced significant modifications of the pigments (Figure 6).

Comparison of the Chlorophyll Extraction Methods in Different Plant Species

For testing the general utility of the three methods as described above, chlorophyll was extracted from the leaves of three other plant species, namely, *Glebionis coronaria* (garland chrysanthemum), *Pisum sativum* L. (pea) and *Prunus sargentii*Rehd. (North Japanese hill cherry) (Figure 8A-F). Chlorophyllide *a* was detected in all three species when pigments were extracted by homogenizing leaves in pure acetone at room temperature or by immersing leaves in 4°C pure acetone. These results indicate that all three species possess CLH activity. Among these species, the largest accumulation of chlorophyllide was observed with pea leaves when the pigments were extracted by immersing leaves in pure acetone at 4°C, which converted 20% of chlorophyll *a* to chlorophyllide *a* (Figure 8D). In contrast, the *G. coronaria* leaves did not show high CLH activity, which yielded only 2% of chlorophyllide *a* compared to total chlorophyll *a* levels by the acetone immersion method (Figure 8B). The sub-zero temperature extraction yielded negligible amounts of chlorophyllide from all three samples (Figure 8A-F), demonstrating that the chlorophyllide formed during the other extraction methods was predominantly an artifact.

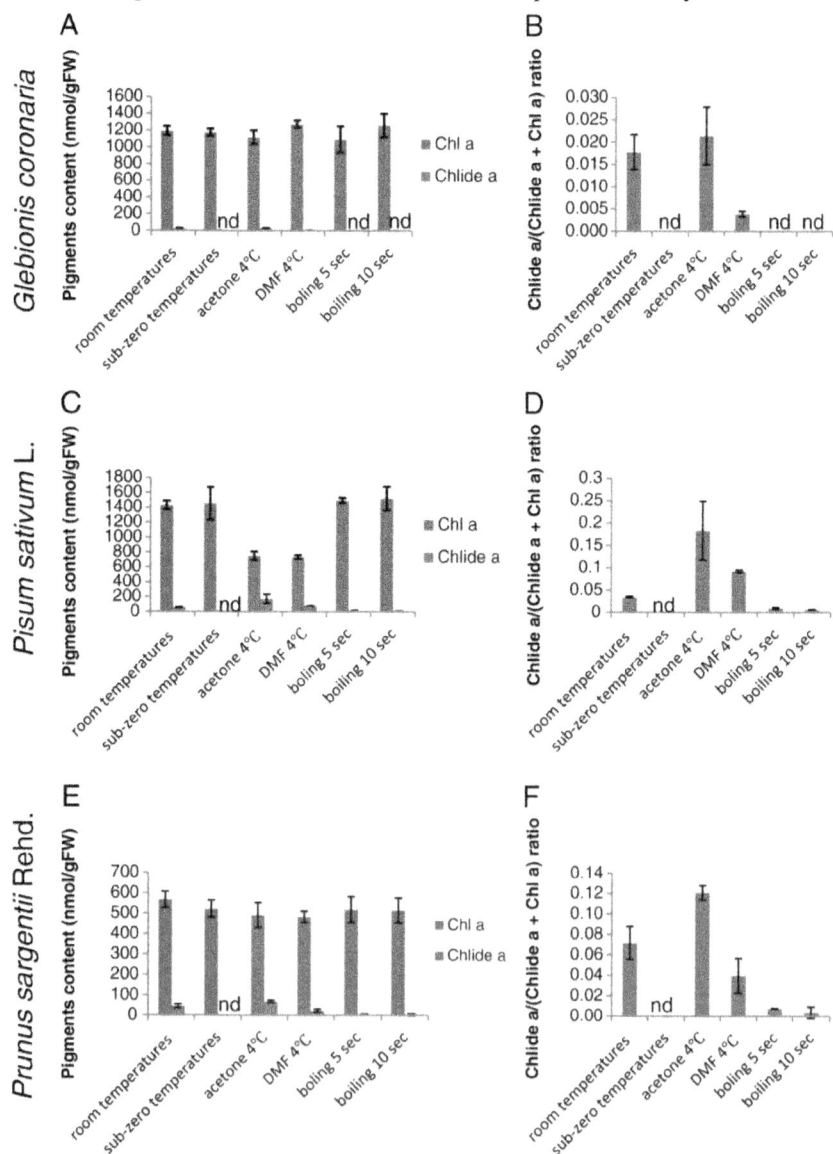

Figure 8: Comparison of extraction methods for chlorophyllide formation with three different species (*Glebionis coronaria* , *Pisum sativum* L. and *Prunus sargentii* Rehd.). Leaves were homogenized in acetone at room temperatures ("room temperatures") or sub-zero temperatures ("sub-zero temperatures"),

respectively, by grinding leaves with steel stainless beads. Alternatively, leaves were immersed in pure acetone at 4°C ("acetone 4°C") or in pure DMF at 4°C ("DMF 4°C"), respectively, without the grinding procedure. In the last pair of experiments, leaves were boiled for 5 ("boiling 5 sec") or 10 sec ("boiling 10 sec"), and were subsequently immersed in acetone at 4°C for chlorophyll extraction. Details of the methods is described in the Method section. **A,C** and **E**. Levels of chlorophyll *a* and chlorophyllide *a* per gram FW of leaves. **B, D** and **F**. Chlorophyllide *a* levels in sample extracts expressed as the ratio of chlorophyllide *a* to the sum of chlorophyll *a* and chlorophyllide *a*. Chl *a*, chlorophyll *a*. Chlide *a*, chlorophyllide *a*. Error bars indicate standard deviations. Sample size, n = 3.

The boiling method worked well with the leaves of *G. coronaria,* which formed negligible amounts of chlorophyllide*a* (Figure 8A and B). This method resulted in the formation of small amounts chlorophyllide *a* with pea and cherry leaves (Figure 8C-F). The chlorophyllide *a* levels in this method were in a similar range with Arabidopsis, which was approximately 1% or less of total chlorophyll *a* (Figure 8D and F). The DMF extraction method did not work as well for *G. coronaria,* pea and cherry leaves as it did for Arabidopsis. This method allowed the formation of chlorophyllide *a* up to 10% of total chlorophyll *a* in pea leaves (Figure 8D). Moreover, immersing leaves in DMF (48 h) and acetone only extracted half of the pigments in comparison to other methods (Figure 8C). Interestingly, a short boiling before immersing leaves in pure acetone drastically improved the extraction efficiency of pigments (Figure 8C).

Discussion

It has been reported that the commonly used method of extraction of photosynthetic pigments with aqueous acetone sometimes results in artifactual chlorophyllide formation [18, 23]. In the present study, we determined the quantity of chlorophyllide formation before, during or after extraction of these pigments using several different methods of extraction. By suppressing CLH activity during extraction, we demonstrated that only trace amounts of chlorophyllide, if any, are present in cells prior to extraction (see Figures 4, 5, 7 and 8). We also showed that both chlorophyll and chlorophyllide are stable in acetone after extraction (Table 1). Therefore, it is unlikely that chlorophyllide is formed in the solvents after the extraction procedure is completed. Based on our collective results, we concluded that chlorophyllide is formed during the extraction process. We speculate that chlorophyllide is formed when acetone infiltrates the tissue, or when the tissue is homogenized in acetone. During these processes, the actual concentration of acetone to which cells are exposed may increase gradually rather than immediately, thus allowing an opportunity for aberrant enzymatic reactions to occur. Although CLH is known to precipitate in pure acetone, it is capable of remaining highly active in lower concentrations of aqueous acetone [18, 27]. Therefore, it is likely that CLH catalyzes the formation of chlorophyllide during extraction until the actual acetone concentration reaches nearly 100%. This hypothesis explains the differential effects of DMF on chlorophyllide formation during chlorophyll extraction from different plant species (Figures 7 and 8). DMF suppressed chlorophyllide formation in Arabidopsis leaves almost perfectly, while it allowed chlorophyllide formation in other plant species whose leaves are thicker than Arabidopsis (Figure 8). These observations can be explained by the assumption that the infiltration of DMF occurs more slowly in thicker leaves as compared to in thinner leaves.

The aforementioned hypothesis raises the question why CLH is only active after the tissue is homogenized with organic solvents or soaked in organic solvents. A possible answer to this question may be that CLH is active in cells but separated from chlorophyll in cells. Schenk et al. [23] used CLH-GFP targeting experiments and confirmed that CLH is localized outside of chloroplasts. If CLH is indeed separated from chlorophyll in intact cells, the homogenization of cells or immersion of tissue in acetone may disrupt cell structures and enable CLH to act on chlorophyll.

Chlorophyllide has been long considered to be an intermediate of both chlorophyll biosynthesis and breakdown [20, 28]. Hörtensteiner and co-workers [23, 29], however, suggested that chlorophyllide is not a true intermediate of chlorophyll breakdown, at least during leaf senescence in Arabidopsis. Instead, they indicated that chlorophyll is degraded via pheophytin (Figure 1B). Our results are consistent with the chlorophyll breakdown model of Hörtensteiner and coworkers [23, 29]. The majority of chlorophyllide detected in acetone extracts of leaf pigments in our experiments were formed by the action of CLH during extraction (see Figures 3 and 4). These results suggest that plants only accumulate a small amount (if any) of chlorophyllide in cells under either normal growth conditions ([23] and this study) or when exposed to MeJA.

We compared three methods that suppress CLH activity with a conventional acetone-extraction method. In the first method, Arabidopsis leaves were boiled for a short time (5 or 10 sec). This procedure almost completely suppressed chlorophyllide formation with Arabidopsis and G. coronaria leaves. Bacon and Holden [17] already reported that a 5 minute period of boiling eliminates chlorophyllide formation. Their boiling time, however, appears to have been too long since they observed extensive decomposition of the pigments [17].

In principle, the boiling time used in this procedure should be optimized for each plant species but we do not suggest boiling leaves for more than 10 sec for most plant species (see Figure 8). Thicker leaves may necessitate a longer boiling time. For example, we found that a 30 sec boiling time worked well to eliminate CLH activity in mulberry leaves in our laboratory (data not shown). This method appears to have another advantage in increasing the extraction efficiency of pigments from thicker leaves such as pea leaves when pigments are extracted by immersing leaves in organic solvents (see Figure 8C). Thus, the boiling method combined with the use of DMF as an extractant would be worth testing when pigments are extracted from thicker leaves. A possible drawback of the boiling method is the potential for additional types of modification to chlorophyll molecules. For instance, we observed a slight increase in pheophytin *a* concentration in our extracts (Figure 6) indicating that 0.1 to 0.2% chlorophyll *a* might be converted to pheophytin *a* by boiling. Thus, the boiling method is recommended in studies where the quantitation of pheophytin *a* is not being considered.

In the second method, frozen leaves were ground at sub-zero temperatures in a metal box that was cooled with liquid nitrogen. The leaves were then homogenized in pure acetone cooled to $-30°C$ using an automatic bead shaker, Shake Master. The use of this shaker facilitates the processing of a relatively large number of samples. It is also possible to use cooled mortar and pestles for grinding leaves at sub-zero temperatures. However, this approach may be laborious and time-consuming when the analysis of a large number of samples is required. In addition, the recovery of a sufficient amount of solvent from a mortar can be problematic when only a small amount of sample tissue is used or available [24]. Therefore, the usage of a mortar and pestle with this method is recommended only when a relatively small number of samples need to be analyzed and when a sufficient amount of tissue is available for each sample. Another limitation of this method will be a requirement of liquid nitrogen, which might not be readily available in field research. Regardless of these limitations, this method is superior to other methods in completely suppressing CLH activity in all plant species tested in this study. This method would be suitable for determining the minimum levels of chlorophyllide formation.

In the third method, pigments were extracted with DMF. This solvent has been previously used for pigment extraction [24,26] but, to the best of our knowledge, was not tested for chlorophyllide formation. This solvent prevents CLH activity even during an overnight incubation of Arabidopsis leaves at 4°C (Figure 7). Therefore, the use of DMF appears to be the best option for extracting photosynthetic pigments from this model organism for downstream analysis using HPLC without introducing artifacts. However, this solvent is not as effective for *G. coronaria,* pea and cherry leaves as it is for Arabidopsis leaves (Figure 8). Moreover, this solvent is a possible liver toxin [30] and all appropriate safety guidelines should be adhered to in its use. Although the volatility of DMF is low, it should be carefully handled in an exterior venting fume hood. In conclusion, the use of DMF might be restricted to Arabidopsis or similar plant species under well-ventilated laboratory conditions.

CONCLUSIONS

We demonstrated that the most-widely used acetone-based procedures for the extraction of photosynthetic pigments from leaf samples potentially results in the rapid, artifactual conversion of chlorophyll to chlorophyllide, especially when pigments are extracted from leaves with high amounts of CLH. This alteration affects HPLC analysis of photosynthetic pigments by decreasing the apparent content of chlorophyll in extracts. The artifactual conversion can be prevented or reduced by adopting one of three simple methods described in this study, namely, short-time boiling of samples prior to extraction with acetone, extraction at sub-zero temperatures, and the use of DMF as a solvent. A researcher may consider one of the three extraction methods depending on the plant material, availability of equipment or liquid nitrogen, and the purposes of pigment analysis.

METHODS

Chemicals

Acetone (HPLC grade, 99.7% purity) was purchased from Wako Pure Chemical Industries, Ltd, Osaka, Japan. DMF (Guaranteed Reagent Grade), ethanol (HPLC grade) and other solvents (Guaranteed Reagent Grade) were purchased from Nacalai Tesque, Inc., Kyoto, Japan.

Plant Materials

Wild-type Arabidopsis (Columbina-0 ecotype) and a T-DNA insertion line (SALK_124978; designated as *clh1-1*,[23]) were primarily used in this study. Plants were grown in soil under long-day conditions (16 h light/8 h dark) in growth chambers under fluorescent light (70–90 μmol photons $m^{-2}s^{-1}$) at 23°C. For pigment extraction, leaves (7th to 9th leaves counting from the bottom of the plant) were harvested either after 4 weeks or after a period of 8 weeks to allow natural senescence. For the MeJA treatment, 4-week-old plants were sprayed with 100 μM MeJA in 0.1% ethanol, 0.01% Tween 20, or solvent control (0.1% ethanol, 0.01% Tween 20), and kept for 3 days at the same growth conditions. For dark-induced senescence, the 7th-9th leaves of 4-week-old plants were detached and placed on wet filter paper (3 mM MES buffer, pH 5.8, with or without 50 μM MeJA) and incubated in complete darkness at 23°C for up to 3 days. In addition to Arabidopsis, three other plant species were tested in this study. *Glebionis coronaria* (garland chrysanthemum) adult plants were purchased from a supermarket, and their mature leaves were used for the experiments. *Pisum sativum* L. (pea) was grown in soil under long-day conditions (16 h light/8 h dark) for seven days in growth chambers under fluorescent light (70–90 μmol photons $m^{-2}s^{-1}$) at 23°C. Then, young leaves were harvested for pigment extraction. Young leaves of *Prunus sargentii* Rehd. (North Japanese hill cherry) were collected from the campus of Hokkaido University, Japan in mid-May.

Chlorophyll Extraction and Analysis

Leaves were harvested and the fresh weight (18–30 mg) of each sample was recorded. Leaves were then frozen in liquid nitrogen and stored at −80°C in a deep freezer. In most experiments described in this study, pigments were extracted by immersing leaves in organic solvents for 10 to 48 hours. Incubation time and the organic solvent were varied from experiment to experiment, which is described in the result section. The procedure described below is common to all extraction methods used in this study unless otherwise noted. Firstly, tubes with leaves were removed from the liquid nitrogen and 1 ml of organic solvents cooled to 4°C or −30°C was immediately added to each tube and incubated at 4°C in the dark for 12 h for Arabidopsis, 20 h for *G. coronaria*, 48 h for pea and 10 h for cherry leaves. The time length of incubation was determined for each plant species by preliminary experiments. For Arabidopsis leaves, the results of longer incubation in acetone at −30°C in the dark for 4 days were also described in the results section. The 80% acetone employed in this study contained 20% (v/v) 0.2 M Tris–HCl pH 8.

In the boiling method, the leaves were dipped into boiling water for 5 or 10 sec. The leaves were then placed on filter paper to absorb excess water and then homogenized in pure acetone at room temperature. Alternatively, four-degree acetone was added to the boiled samples in a 2-ml microtube. Tubes were then kept at 4°C for 12 h for Arabidopsis, 20 h for *G. coronaria*, 48 h for pea and 10 h for cherry leaves. After incubation, extracts were transferred to a glass vial and analyzed using HPLC as described below.

In the second extraction method, acetone was cooled to −30°C prior to its use. An aluminum metal box (BioMedical Science Co. Ltd., Tokyo, Japan) for holding sample tubes during shaking was cooled in liquid nitrogen for 30 min prior to its use. Two-ml microcentrifuge tubes containing leaf samples and homogenization beads were frozen in liquid nitrogen and then stored at −80°C Acetone was added to the tubes containing the frozen samples while they were in the nitrogen-cooled metal box. Homogenization of the tissue was performed immediately by shaking the sample tubes containing the homogenization beads and leaf samples in an automatic bead shaker (Shake Master, BioMedical Science Co. Ltd, Tokyo, Japan).

In the third extraction method, 1 ml of DMF cooled to 4°C was immediately added to frozen leaves. The samples were subsequently incubated at 4°C in the dark for 12 h for Arabidopsis, 20 h for *G. coronaria*, 48 h for pea and 10 h for cherry leaves. Subsequently, the organic solvent was recovered by centrifugation and its pigment composition was determined by HPLC as described below.

HPLC Separation of Photosynthetic Pigments

The microtubes containing the homogenized samples were subjected to centrifugation at 15,000 rpm for 5 min at 4°C and the resulting supernatant was analyzed by HPLC with a Symmetry C8 column (150 mm in length, 4.6 mm in i.d.; Waters, Milford, MA, USA) according to the method of Zapata et al. [31]. Elution profiles were monitored by measuring absorbance at 410 nm. Pigments used as standards were purchased from Juntec Co. Ltd. (Odawara, Japan).

Time Course Experiments

Leaf samples from 4-week-old Arabidopsis plants were immersed in pure acetone in 2.0 mL-microtubes that were held in a metal box that had been pre-cooled with liquid nitrogen. The sample tubes were then transferred to a tube rack and incubated at ambient temperature for the times indicated in Figure 4. After a prescribed time, the tubes were returned to the nitrogen-cooled metal box to terminate the incubation. Pigments were extracted from samples at a sub-zero temperature while tubes were in the cooled metal box by adding stainless beads and shaking the box in a bead shaker (Shake Master).

ACKNOWLEDGEMENTS

The authors thank Prof. Stefan Hörtensteiner (Zürich University) for the gift of the *clh1-1* seeds. This research was supported by the Ministry of Education, Culture, Sports, Science and Technology, Japan, a Grant-in-Aid for Scientific Research, no. 23570042 to R.T XH is supported by a scholarship from China Scholarship Council.

AUTHORS' CONTRIBUTIONS

The experiments were conceived by XH, AT and RT, and performed by XH. The manuscript was written by XH and RT. RT is the principal investigator of the research grant. All authors read and approved the final manuscript.

REFERENCES

1. Mur LAJ, Aubry S, Mondhe M, Kingston-Smith A, Gallagher J, Timms-Taravella E, James C, Papp I, Hörtensteiner S, Thomas H, Ougham H: Accumulation of chlorophyll catabolites photosensitizes the hypersensitive response elicited by Pseudomonas syringae in Arabidopsis. New Phytol. 2010, 188: 161-174. 10.1111/j.1469-8137.2010.03377.x.

2. Brouwer B, Ziolkowska A, Bagard M, Keech O, Gardeström P: The impact of light intensity on shade-induced leaf senescence. Plant Cell Environ. 2012, 35: 1084-1098. 10.1111/j.1365-3040.2011.02474.x.

3. Kitajima K, Hogan KP: Increases of chlorophyll a/b ratios during acclimation of tropical woody seedlings to nitrogen limitation and high light. Plant Cell Environ. 2003, 26: 857-865. 10.1046/j.1365-3040.2003.01017.x.

4. Bianchi TS, Findlay S: Plant pigments as tracers of emergent and submergent macrophytes from the Hudson River. Can J Fish Aquat Sci. 1990, 47: 492-494. 10.1139/f90-054.

5. Rosevear MJ, Young AJ, Johnson GN: Growth conditions are more important than species origin in determining leaf pigment content of British plant species. Funct Ecol. 2001, 15: 474-480. 10.1046/j.0269-8463.2001.00540.x.

6. Oserkowsky J: Quantitative relation between chlorophyll and iron in green and chlorotic pear leaves. Plant Physiol. 1933, 8: 449-468. 10.1104/pp.8.3.449.

7. Cartelat A, Cerovic ZG, Goulas Y, Meyer S, Lelarge C, Prioul JL, Barbottin A, Jeuffroy MH, Gate P, Agati G, Moya I: Optically assessed contents of leaf polyphenolics and chlorophyll as indicators of nitrogen deficiency in wheat (Triticum aestivum L.). Field Crops Res. 2005, 91: 35-49. 10.1016/j.fcr.2004.05.002.

8. Schlemmer MR, Francis DD, Shanahan JF, Schepers JS: Remotely measuring chlorophyll content in corn leaves with differing nitrogen levels and relative water content. Agron J. 2005, 97: 106-112. 10.2134/agronj2005.0106.

9. Arnon DI: Copper Enzymes in Isolated Chloroplasts. Polyphenoloxidase

in Beta Vulgaris. Plant Physiol. 1949, 24: 1-15. 10.1104/pp.24.1.1.

10. Lichtenthaler HK: Chlorophylls and carotenoids: Pigments of photosynthetic biomembranes. Meth Enzymol. 1987, 148: 350-382.

11. Porra RJ, Thompson WA, Kriedemann PE: Determination of accurate extinction coefficients and simultaneous equations for assaying chlorophylls a and b extracted with four different solvents: verification of the concentration of chlorophyll standards by atomic absorption spectroscopy. Biochim Biophys Acta. 1989, 975: 384-394. 10.1016/S0005-2728(89)80347-0.

12. Li Z, Ahn TK, Avenson TJ, Ballottari M, Cruz JA, Kramer DM, Bassi R, Fleming GR, Keasling JD, Niyogi KK: Lutein accumulation in the absence of zeaxanthin restores nonphotochemical quenching in the Arabidopsis thaliana npq1 mutant. Plant Cell. 2009, 21: 1798-1812. 10.1105/tpc.109.066571.

13. Kim EH, Li XP, Razeghifard R, Anderson JM, Niyogi KK, Pogson BJ, Chow WS: The multiple roles of light-harvesting chlorophyll a/b-protein complexes define structure and optimize function of Arabidopsis chloroplasts: a study using two chlorophyll b-less mutants. Biochim Biophys Acta. 2009, 1787: 973-984. 10.1016/j.bbabio.2009.04.009.

14. Venketeswaran S: Studies on the isolation of green pigmented callus tissue of tobacco and its continued maintenance in suspension cultures. Physiol Plant. 1965, 18: 776-789. 10.1111/j.1399-3054.1965.tb06936.x.

15. Benedetti CE, Arruda P: Altering the expression of the chlorophyllase gene ATHCOR1 in transgenic Arabidopsis caused changes in the chlorophyll-to-chlorophyllide ratio. Plant Physiol. 2002, 128: 1255-1263. 10.1104/pp.010813.

16. Kariola T, Brader G, Li J, Palva ET: Chlorophyllase 1, a damage control enzyme, affects the balance between defense pathways in plants. Plant Cell. 2005, 17: 282-294. 10.1105/tpc.104.025817.

17. Bacon MF, Holden M: Changes in chlorophylls resulting from various chemical and physical treatments of leaves and leaf extracts. Phytochemistry. 1967, 6: 193-210. 10.1016/S0031-9422(00)82763-6.

18. Holden M: The breakdown of chlorophyll by chlorophyllase. Biochem J. 1961, 78: 359-

19. Mcfeeters RF, Chichester CO, Whitaker JR: Purification and properties of chlorophyllase from Ailanthus altissima (tree-of-heaven). Plant Physiol. 1971, 47: 609-618. 10.1104/pp.47.5.609.

20. Hörtensteiner S: Chlorophyll degradation during senescence. Annu Rev Plant Biol. 2006, 57: 55-77. 10.1146/annurev.arplant.57.032905.105212.

21. Tsuchiya T, Ohta H, Okawa K, Iwamatsu A, Shimada H, Masuda T, Takamiya K: Cloning of chlorophyllase, the key enzyme in chlorophyll degradation: Finding of a lipase motif and the induction by methyl jasmonate. Proc Natl Acad Sci USA. 1999, 96: 15362-15367. 10.1073/pnas.96.26.15362.

22. Cheong JJ, Choi YD: Methyl jasmonate as a vital substance in plants. Trends Genet. 2003, 19: 409-413. 10.1016/S0168-9525(03)00138-0.

23. Schenk N, Schelbert S, Kanwischer M, Goldschmidt EE, Dörmann P, Hörtensteiner S: The chlorophyllases AtCLH1 and AtCLH2 are not essential for senescence-related chlorophyll breakdown in Arabidopsis thaliana. FEBS Lett. 2007, 581: 5517-5525. 10.1016/j.febslet.2007.10.060.

24. Moran R, Porath D: Chlorophyll determination in intact tissues using n, n-dimethylformamide. Plant Physiol. 1980, 65: 478-479. 10.1104/pp.65.3.478.

25. Rodríguez MT, Gonzélez MP, Linares JM: Degradation of chlorophyll and chlorophyllase activity in senescing barley leaves. J Plant Physiol. 1987, 129: 369-374. 10.1016/S0176-1617(87)80094-9.

26. Inskeep WP, Bloom PR: Extinction coefficients of chlorophyll a and B in n, n-dimethylformamide and 80% acetone. Plant Physiol. 1985, 77: 483-485. 10.1104/pp.77.2.483.

27. Tsuchiya T, Ohta H, Masuda T, Mikami B, Kita N, Shioi Y, Takamiya K: Purification and characterization of two isozymes of chlorophyllase from mature leaves of Chenopodium album. Plant Pell Physiol. 1997, 38: 1026-1031. 10.1093/oxfordjournals.pcp.a029267.

28. Takamiya KI, Tsuchiya T, Ohta H: Degradation pathway(s) of chlorophyll: what has gene cloning revealed?. Trends Plant Sci. 2000, 5: 426-431. 10.1016/S1360-1385(00)01735-0.

29. Schelbert S, Aubry S, Burla B, Agne B, Kessler F, Krupinska K, Hörtensteiner S: Pheophytin pheophorbide hydrolase (pheophytinase) is involved in chlorophyll breakdown during leaf senescence in Arabidopsis. Plant Cell. 2009, 21: 767-785. 10.1105/tpc.108.064089.

30. Wrbitzky R: Liver function in workers exposed to N, N-dimethylformamide during the production of synthetic textiles. Int Arch Occup Environ Health. 1999, 72: 19-25. 10.1007/s004200050329.

31. Zapata M, Rodríguez F, Garrido JL: Separation of chlorophylls and carotenoids from marine phytoplankton, a new HPLC method using a reversed phase C8 column and phridine-containing mobile phases. Mar Ecol Prog Ser. 2000, 195: 29-45.

Chapter 13

A GENOMIC APPROACH TO STUDY ANTHOCYANIN SYNTHESIS AND FLOWER PIGMENTATION IN PASSIONFLOWERS

Lilian Cristina Baldon Aizza and Marcelo Carnier Dornelas

Departamento de Biologia Vegetal. Rua Monteiro Lobato 970, Instituto de Biologia, Universidade Estadual de Campinas, Cidade Universitária Zeferino Vaz, 13083-970 Campinas, SP, Brazil

ABSTRACT

Most of the plant pigments ranging from red to purple colors belong to the anthocyanin group of flavonoids. The flowers of plants belonging to the genus Passiflora (passionflowers) show a wide range of floral adaptations to diverse pollinating agents, including variation in the pigmentation of floral parts ranging from white to red and purple colors. Exploring a database of expressed sequence tags obtained from flower buds of two divergentPassiflora species, we obtained assembled sequences potentially corresponding to 15 different genes of the anthocyanin biosynthesis pathway in these species. The obtained sequences code for putative enzymes are involved in the production of flavonoid precursors, as well as those involved in the formation of particular ("decorated") anthocyanin molecules. We also obtained sequences encoding regulatory factors that control the expression of structural genes and regulate the spatial and temporal accumulation of pigments. The identification of some of the putative Passiflora anthocyanin biosynthesis pathway genes provides novel resources for research on secondary metabolism in passionflowers, especially on the elucidation of the processes involved in floral pigmentation, which will allow future studies on the role of pigmentation in pollinator preferences in a molecular level.

INTRODUCTION

Anthocyanins belong to a diverse group of secondary metabolites of the phenylpropanoid class, the flavonoids, which are found in different plant species. They represent some of the most important natural pigments, which are responsible for the wide range of red to purple colors present in many flowers, fruits, seeds, leaves, and stems. Besides having great economical relevance, flower and fruit pigments play an important ecological role in the animal attraction for pollination and seed dispersal, wich is a spectacular example of coevolution between plants and animals [1–3].

The biosynthetic pathway of anthocyanins has been well characterized biochemically and genetically in species with different floral morphology, pigmentation pattern, and pollination syndromes such as Petunia hybrida [4,5], Matthiola [6], Dianthus [7], Eustoma [8], Gerbera [9], Zea mays [10, 11], Antirrhinum majus [12], andIpomoea [13, 14]. A representation of a general anthocyanin biosynthetic pathway is shown in Figure 1.

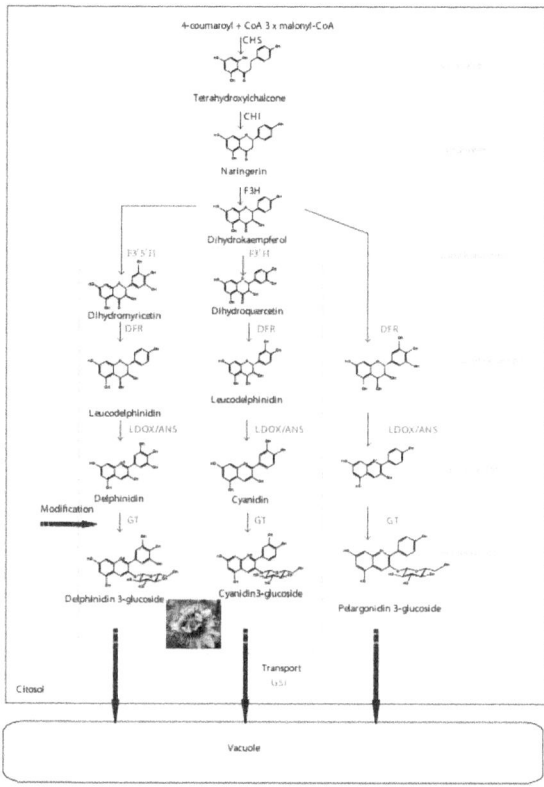

Figure 1: Schematic representation of the anthocyanin biosynthetic pathway (adapted from [16]). Enzymes are indicated in red, and classes of compounds are in green. An-

thocyanidin is further modified with glycosyl, acyl, or methyl groups, resulting in the "decorated" anthocyanin. In this case, UF3GT is responsible for the glycosylation of anthocyanidins. The proposed anthocyanin biosynthetic pathway for Passiflora edulis is highlighted by the colored background. CHS: chalcone sintase; CHI: chalcone isomerase; F3H: flavanone 3-hydroxylase; F3'H: flavanone 3'-hydroxylase; F3'5'H: flavanone 3'5'-hydroxylase; DFR: dihydroflavonol 4-reductase; LDOX/ANS: leuco-anthocyanidin dioxygenase/anthocyanidin synthase; GT: glucosyltransferase; GST: glutathione S-transferase.

Briefly, the pathway is initiated with chalcone synthase (CHS) catalyzing the stepwise condensation of three molecules of acetate residues from malonlyl-CoA with one molecule of 4-coumaroyl-CoA to form the basic structure of flavonoids (tetrahydroxychalcone), which is rapidly isomerized to the colorless naringenin by chalcone isomerase (CHI). Naringenin is then converted to dihydroflavonol by flavanone 3-hydroxylase (F3H). Dihydroflavonol 4-reductase (DFR), which is a specific enzyme for the anthocyanin synthesis, catalyses the production of leucoanthocyanidins from dihydroflavonols, which can be hydroxylated on the 3' or 5' position of the B-ring by flavonoid 3'-hydroxylase (F3'H) to produce dihydroquercetin or by flavonoid 3'5'-hydroxylase (F3'5'H) to form dihydromyricetin. Subsequently, leucoanthocyanidin oxidase/anthocyanidin synthase (LDOX/ANS) is responsible for the formation of the anthocyanidins from the colorless leucoanthocyanidins. GT enzymes (O-glucosyltransferases) represent the final step in anthocyanin biosynthesis: anthocyanidins are converted in differentially "decorated" anthocyanin molecules [15, 16]. Biochemical approaches have demonstrated that all anthocyanin pigments are derived from one of three aglycones: pelargonidin, cyaniding, and delphinidin. The main determinants of the apparent color of these pigments are the hydroxylation and methylation patterns, as well as the number and type of sugars on the beta ring of the flavonoid molecule [1, 3,17–19].

Figure 1 depicts a generalized anthocyanin biosynthesis pathway. At least, two groups of genes are required for anthocyanin biosynthesis: the first group is represented by the structural genes encoding enzymes for the production of the flavonoid precursors, as well as those involved in the formation of particular ("decorated") anthocyanin molecules. The second group includes the genes encoding regulatory factors that control the expression of structural genes which are mainly orenestrated by complexes formed by MYB and basic helix-loop-helix (bHLH) transcription factors that include WDR (WD40 repeats) proteins [2, 4, 15, 16, 20–23].

There are about 600 Passiflora species widely distributed in tropical and subtropical regions. Some Passifloraspecies have economical importance due to the production of fruits (passionfruit) or use as ornamentals. Nevertheless, a large number of Passiflora species are rare and/or endangered, as the environment of their diversity center has been increasingly degraded by human activities [24]. An enormous floral diversity is observed among Passiflora species, including variation in color, size, morphology, and fusion of floral organs. These and other floral characteristics, including evolutionary innovations such as the presence of coronal filaments and an androgynophore, are indicative of the wide range of pollination syndromes found in the genus [24]. Wide passionflowers may be pollinated by insects (bees and wasps), hummingbirds, and bats [24]. The most striking feature of floral variation among passionflowers is the wide range of pigmentation patterns of the corona filaments. Most of the floral pigments in Passiflora are different types of anthocyanin molecules [25, 26]. Among all Passiflora species, P. edulis Deg and P. suberosa L. are of particular interest, because they are modelPassiflora species for which expressed sequences tags (ESTs) were produced within the frame of the "PASSIOMA" Project [27]. P. edulis Deg flowers are pollinated by large bees of genus Xylocopa. These flowers are about 8–12 cm wide, and their coronas contain multiple series of purplish filaments with white tips. The flowers of P. suberosa L. are small (2-3 cm wide) and show two morphologically distinct series of corona filaments: the outer series is greenish, and the inner series is formed by smaller purple filaments. The flowers of P. suberosa are pollinated by wasps [28].

We are particularly interested in the characterization of genes involved in the anthocyanin biosynthetic pathway of these two Passiflora species. With this aim, we searched for putative Passiflora genes responsible for flower pigmentation, using the key proteins known to be involved in the different enzymatic steps of anthocyanin biosynthesis as baits to search for expressed sequences tags (ESTs) in the PASSIOMA database.

MATERIAL AND METHODS

Searching Passiflora ESTs Homologous to Anthocyanin Biosynthetic Genes

The clustered expressed sequence tags (ESTs) from the PASSIOMA Project database [27] were used as a primary source of data for our analyses. These sequences were assembled from ESTs obtained from the sequencing of several P. edulis or P. suberosa cDNA libraries, made from floral buds at different developmental stages (see [27] for details on library construction, sequencing, and database structure). Nucleotide sequences and their respective deduced amino acid sequences from genes known to be involved in anthocyanin biosynthesis (see Figure 1) were obtained from the National Center for

Biotechnology Information (NCBI;http://www.ncbi.nlm.nih.gov/). Searches for putative homolog sequences in the PASSIOMA database were conducted using the tBLASTN module that compares the consensus amino acid sequence with a translated nucleotide sequences database [29]. We generally used Arabidopsis thaliana or Petunia hybrida as query consensus sequences as the anthocyanin biosynthesis pathways in these model species are more thoroughly studied at the molecular level [30–32]. All sequences in the PASSIOMA database that exhibited a significant alignment (-value lower than 10–5) with the query were retrieved from the PASSIOMA database.

The clusterization of all reads identified using a given query sequence was performed using the CAP3 algorithm [33] from the BioEdit software [34]. The novel cluster consensus sequences obtained were reinspected for the occurrence of conserved motives using InterProScan [35] and were compared to NCBI databases using BLAST [29]. Sequences that did not show the main motives present in the query sequence were discarded. Validated sequences were then included in phylogenetic analyses.

Comparison of the Amino Acid Sequences and Phylogenetic Analysis

All amino acid sequences were aligned by CLUSTALX software using default parameters [36]. The obtained alignments were eventually corrected by hand and imported into the molecular evolutionary genetics analysis (MEGA) software [37]. Phylogenetic trees were obtained using parsimony and/or genetic distance calculations (in the later case using pairwise deletion option and with the Poisson correction model). Neighbor-joining [38] and Bootstrap (with 10,000 replicates) trees were also constructed.

RESULTS

The cDNA libraries of the PASSIOMA Project were obtained from mRNA extracted from floral buds at different developmental stages, and it is expected that all EST sequences correspond to genes expressed duringPassiflora flower development [27]. This sequence search detected a total of 75 Passiflora EST sequences, 34 of them corresponding to P. edulis sequences and 41 of them corresponding to sequences derived from P. suberosalibraries. When submitted to the CAP3 algorithm and detailed comparison of their deduced amino acid sequences, the number of valid clusters was reduced to 15, potentially corresponding to 15 different genes. When the validated amino acid sequences obtained from the PASSIOMA database were compared to other plant protein sequences in the public databases, the first BLAST hits generally corresponded to Populus andRicinus sequences. This was expected, as Passiflora and these genera belong to the same order (Malpighiales) and are considered to be closely related [39].

We obtained assembled EST sequences corresponding to genes of the following genes families: CHS, DFR, GT, GST, MYB, and WD40 (see Table 1). Therefore, we used 15 Passiflora assembled sequences from the PASSIOMA database and a selected set of genes from divergent plant species from the public databases to explore their evolutionary relationships. The obtained sequence comparison alignments allowed the construction of phylogenetic trees for each of these families of genes involved in the different enzymatic steps of the anthocyanin pathway.

Table 1: Putative Passiflora homologs of genes encoding elements of the anthocyanin biosynthetic pathway

Enzyme	Passiflora AS*	First BLAST hit	e-value	ID/SM
	PACEPE3010G11.g	ABD24222 CHS Populus alba	$7e^{-72}$	85/90
	PACEPE3014B06.g	ABC86919 CHS Populus alba	$9e^{-67}$	77/84
CHS	PACEPE3007G06.g	XP_002305446 CHS-like Populus trichocarpa	$7e^{-121}$	84/92
	PACEPE3023H10.g	XP_002326830 CHS-like Populus trichocarpa	$3e^{-110}$	82/91
	PACEPS7017D03.g	AAQ62589 CHS3 Glycine Max	$1e^{-98}$	82/88
DFR	PACEPE3003G04.g	XP_002307667 DFR2 Populus trichocarpa	$1e^{-95}$	82/94
GT	PACEPE3030G03.g	XP_002532899 UFGT Ricinus communis	$6e^{-31}$	53/70
	PACEPS7021H07.g	XP_002518725 UFGT Ricinus communis	$8e^{-37}$	56/72
	PACEPE3013H01.g	AF048978 GST Glycine Max	$7e^{-52}$	79/90
GST	PACEPE3007A05.g	XM_002519342 GST theta Ricinus communis	$4e^{-52}$	77/89
	PACEPE3018F08.g	ADB11335 GSTF7 phi Populus trichocarpa	$2e^{-82}$	68/83
	PACEPS4006H06.g	ADB11332 GSTF4 phi Populus trichocarpa	$2e^{-61}$	63/78
	PACEPS7023B03.g	AF243378 GST 23 Glycine max	$1e^{-51}$	80/88
MYB	PACEPS7022E07.g	XP_002530824 R2R3 MYB Ricinus communis	$7e^{-80}$	88/91
WD40	PACEPE3007G07.g	XP_002512788 WD-repeat protein Ricinus communis	$3e^{-124}$	92/96

Abbreviations: CHS: chalcone synthase; DFR: dihydroflavonol 4-reductase; GT: glucosyltransferase and GST: glutathione S-transferase.
Using the BLASTp algorithm [29].
* AS: assembled sequence. Codes refer to the longest cDNA clone. PACEPE: Passiflora edulis; PACEPS: Passiflora suberosa.
ID/SM: identity/similarity (both based on the amino acid sequence) with the first BLAST hit.

The similarities among all genes identified in this study and those reported from other plant species were assembled in Table 1 and ranged from 70% (PACEPE3030G03.g; representing a putative member of the GST, glutathione S-transferase superfamily) to 96% (PACEPE3007G07.g; potentially encoding a WD40 protein).

Some of these gene sequences showed significant similarity to elements required for early or late steps of the pathway; others putatively encode regulatory proteins involved in the control of the spatial and temporal patterns of pigmentation, while others are responsible for intracellular transport of the anthocyanin molecules. The role of each of these genes in the anthocyanin biosynthesis and the probable implications for the understanding of the Passiflora flower pigmentation are presented in the Discussion.

Identification and Phylogenetic Analysis of Passionflower Genes Potentially Involved in Anthocyanin Biosynthesis and Transport

Chalcone Synthases (CHSs)

We have found 5 Passiflora assembled sequences (5 putative genes) encoding enzymes of the CHS family: PACEPE3010G11.g, PACEPE3014B06.g, PACEPE3007G06.g, PACEPE3023H10.g and PACEPS7017D03.g. These sequences are expected to encode proteins with 231, 158, 254, 237, and 222 amino acids, respectively. The deduced CHS proteins showed more than 80% similarity to CHSs of other plant species (Table 1). To determine the phylogenetic relationship of different CHSs, we aligned protein sequences from a diverse range of plant species (moss, ferns, gymnosperms and angiosperms), cyanobacterium (Synechococcus sp.) and Passiflorarepresentatives of the CHS superfamily (Figure 2). The phylogenetic tree was resolved in three clades. These three clades were highly supported with 100% bootstrap values. The Passiflora proteins were consistently positioned into different clades. One of these monophyletic clades (highlighted in Figure 2) contains all the anther-specific CHS-like genes (ASCLs; [40, 41]). The remaining sequences, including three Passifloramembers, were clustered in the other sister clade together with all CHS genes from seed plants.

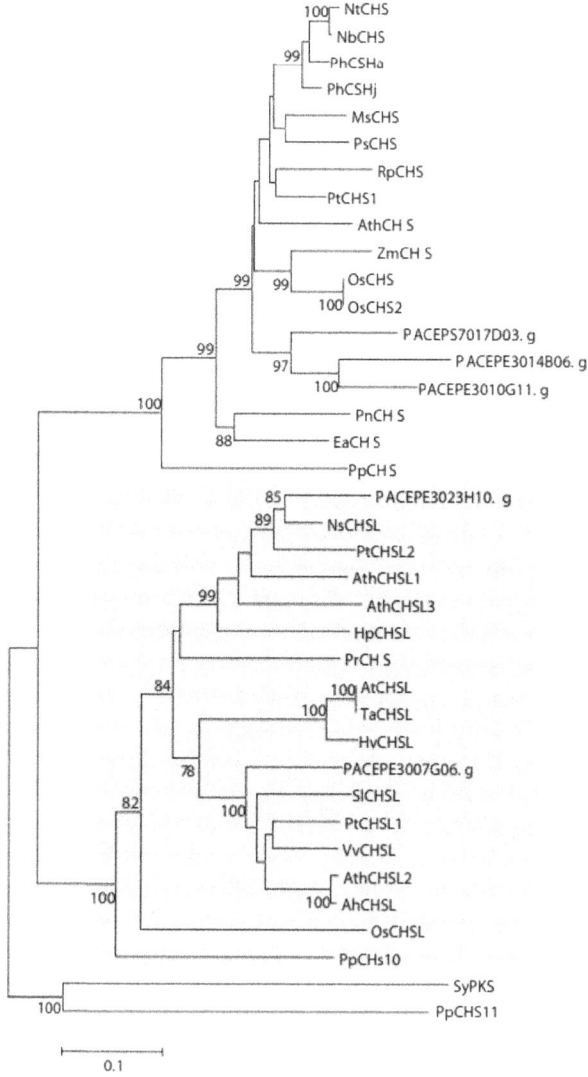

Figure 2: A Neighbor-joining phylogenetic tree of chalcone synthase (CHS) amino acids sequences. The cluster containing all anther-specific CHS-like enzymes is highlighted. Bootstrap values from 1,000 replicates were used to assess the robustness of the trees. Only bootstrap values above 75% are indicated at the nodes. Accession numbers for genes from other species are given in Supplementary data.

Dihydroflavonol 4-Reductases (DFR)

A single Passiflora cDNA sequence of 850 bp encoding a predicted protein of 204 amino acids showed significant -value () and 94% similarity to a Populus

DFR sequence (Table 1). Figure 3 shows an alignment of the deduced amino acid sequence of the Passiflora DFR with some other plant sequences containing an NADP-binding domain, considered the region of substrate preference of DFR enzymes [42, 43]. Additionally, the Passiflora DFR showed an aspartic acid residue at position 134, as it is observed for thePetunia and Populus proteins, whereas Gerbera and some Lotus DFR show an asparagine residue at the same position (Figure 3). We adopted the terminology suggested by Shimada and coworkers [44] to designate the conserved motifs present in the DFR sequence.

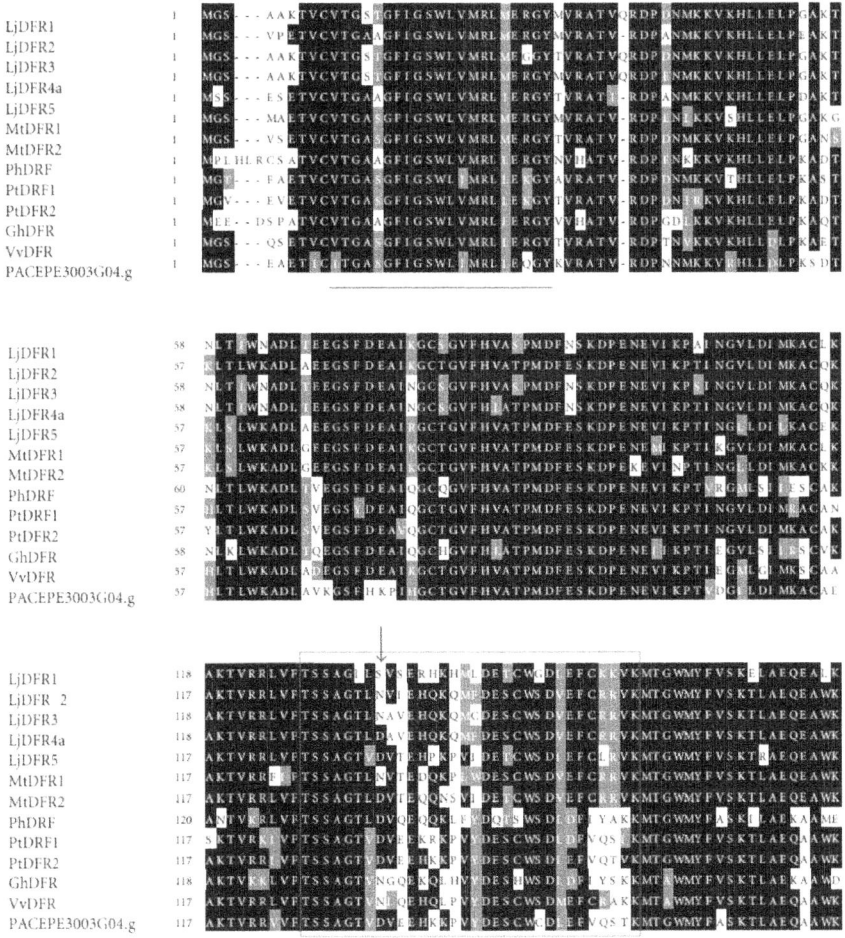

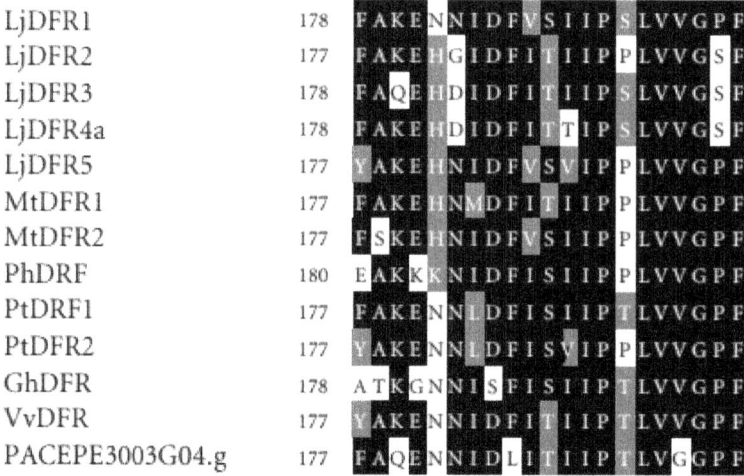

Figure 3: Multiple sequence alignment of the Passiflora sequence with some plant DFR sequences. The identical and similar residues are highlighted on a black and gray background, respectively. NADP-binding domain is underlined. Boxed amino acids have been considered to control the substrate specificity of DFR enzyme [40], and the amino acid residue (indicated by an arrowhead) is especially important for this specificity [41]. The alignment was performed using CLUSTALX and BOXSHADE program.

A neighbor-joining tree was constructed based on the alignment DFR sequences shown in Figure 3. The monocots and eudicots DFRs were positioned separately. While monocot DFR genes formed one clade, the eudicot DFR sequences diverged into two clades. Clearly, Asn-type DFRs are found in a larger number of species. On the other hand, Asp-type DFRs are restricted to some species, including Passiflora and Populus(Figure 4).

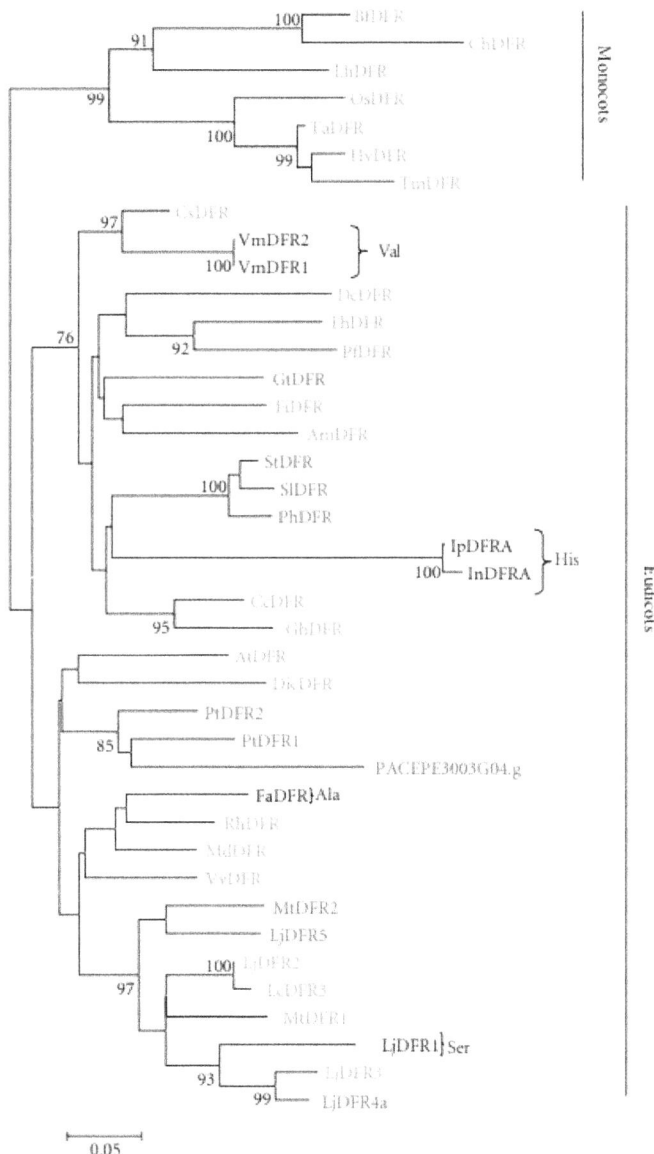

Figure 4: A Neighbor-joining phylogenetic tree of dihydroflavonol 4-reductase (DFR) amino acids sequences. Bootstrap values from 1,000 replicates were used to assess the robustness of the trees. Only bootstrap values above 75% are indicated at the nodes. Asn-type DFRs, Asp-type DFRs, and DFRs of neither Asn nor Asp-type are indicates in blue, red, and black, respectively [42]. Accession numbers for genes from other species are given in Supplementary data which are available online at

Glucosyltransferases (GT)

We identified two Passiflora EST clones, PACEPE3030G03.g and PACEPS7021H02.g, encoding proteins with sequence similarity to Ricinus communis glucosyltransferases (Table 1). The first cDNA sequence contained an ORF specifying a 124 amino acid protein, and the second cDNA encoded a protein of 200 amino acid residues. These putative Passiflora GT proteins were compared with those GT enzymes described by Kovinick and colleagues [45] and retrieved from the NCBI database. The obtained phylogenetic tree resulted in five clades, according to their in vitro substrate specificities [45]. Phylogenetic analysis revealed that the Passiflorasequences were positioned within the Cluster II proteins (Figure 5).

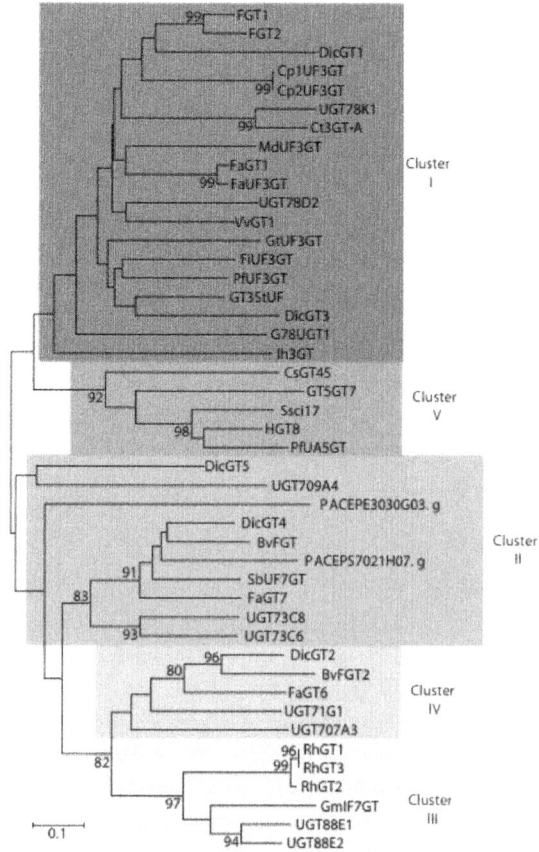

Figure 5: A Neighbor-joining phylogenetic tree of glucosyltransferase (GT) amino acids sequences. Bootstrap values from 1,000 replicates were used to assess the robustness of the trees. Only bootstrap values above 75% are indicated at the nodes. Accession numbers for genes from other species are given in Supplementary data.

Glutathione S-Transferases (GSTs)

We have identified five Passiflora sequences representing putative members of the GST family. Each member was represented by a single EST sequence. Comparison of these deduced GST protein sequences with those in the GenBank database revealed homology with multifunctional GSTs from Populus, Ricinus, and Glycine spp (see Table 1). Phylogenetic relationships among the putative Passiflora GSTs and family members of other plant species were established (Figure 6). Based on sequence similarity, the five Passiflora putative GSTs were grouped into three clades. PACEPE3018F08.g, PACEPS4006H06.g, and PACEPS7023B03.g are type I GSTs, PACEPE3007A05.g is a type II GST, and PACEPE3013H01.g is a type III GST [46].

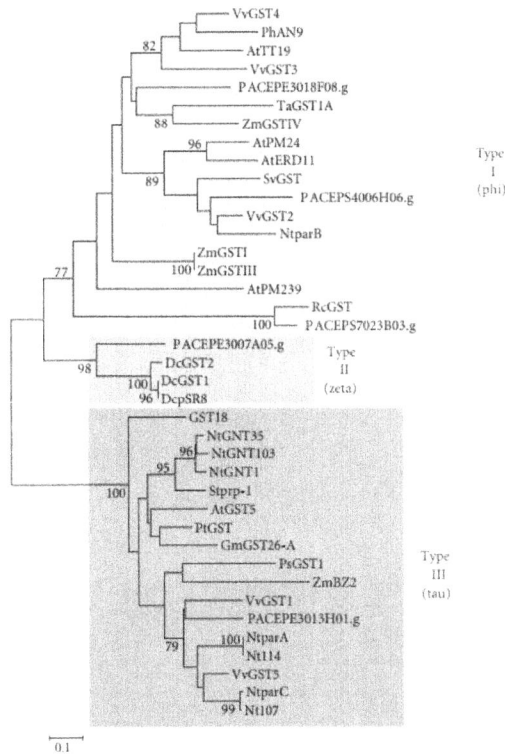

Figure 6: A Neighbor-Joining phylogenetic tree of glutathione S-transferase (GST) amino acids sequences with three types representing phi, tau, and zeta classes. Phi and tau are plant-specific GSTs. Bootstrap values from 1,000 replicates were used to assess the robustness of the trees. Only bootstrap values above 75% are indicated at the nodes. Accession numbers for genes from other species are given in Supplementary data.

We could not find any putative homologs to chalcone isomerases (CHI), flavanone 3-hydroxylases (F3H), and anthocyanidin synthases (ANS; see Figure 1) in the PASSIOMA database. Three EST sequences were identified corresponding to a putative flavonoid 3-O-hydroxylase (F3'H) gene, and one sequence was found that showed significant homology to genes encoding flavonoid 3-5-O-hydroxylases (F3'5'H; data not show). As these sequences were incomplete at their 5' end, they were not considered in our analyses.

Identification and Phylogenetic Analysis of Passionflower Genes Potentially Involved in Spatially and Temporally Patterning Anthocyanin Deposition

Based on the searches in the PASSIOMA database, we identified one potential homolog for an MYB transcription factor of the R2R3 class. The P. suberosa cDNA clone PACEPS7022E07.g encodes a protein of 132 amino acids showing 91% similarity to the Ricinus communis R2R3 MYB. On the other hand, PACEPE3007G07.g is a putative P. edulis WD40 gene of 886 bp encoding 291 amino acid residues showing 96% similarity to an R. communis, WD40 (Table 1).

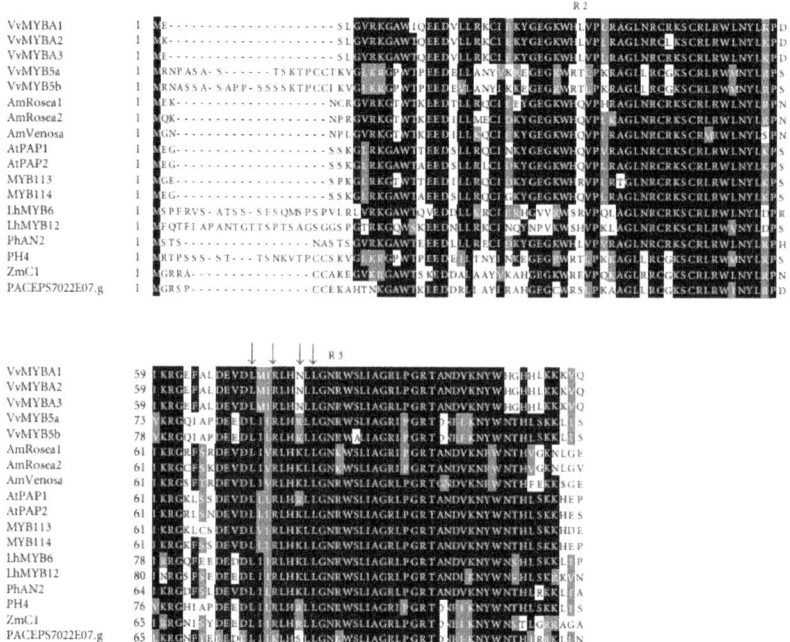

Figure 7: Multiple sequence alignment of the R2R3 MYB domains involved in anthocyanin production including the deduced amino acid sequence of Passiflora suberosa. R2R3 repeats refer to two imperfect repeats of the MYB do-

main. The identical and similar residues are highlighted on a black and gray background, respectively. Red box shows the R/B like bHLH interacting motif in the R3 repeat [45], and arrows indicate four specific residues of maize C1 required for interaction with a bHLH cofactor R [46]. Blue box shows a conserved motif in the R2R3 repeats for eudicots MYB related to the anthocyanin pigments [47]. The alignment was performed using CLUSTALX and BOX-SHADE program.

Figure 7 shows an alignment of the deduced PACEPS7022E07.g protein sequence with 17 other plant anthocyanin-related R2R3-MYB, indicating the presence of a conserved DNA-binding domain, designated as the R2R3 domain. All sequences analyzed also contained a second conserved amino acid motif in the R3 repeat (red box), important for the interaction between MYB and bHLH proteins in Arabidopsis [48]. The four specific residues required for this interaction in maize [49] are also indicated by the arrows in Figure 7. The third conserved motif appears to be ANDV (blue box) in the R3 repeat of all eudicot R2R3-MYB proteins related to anthocyanin biosynthesis.

A phylogenetic tree of selected plant R2R3-MYB transcription factors, including PACEPS7022E07.g, was constructed using the alignment of the conserved R2R3 repeats (Figure 8). The Passiflora sequence was placed within the clade including ZMC1 (Zea mays), PhPH4 (Petunia hybrida), VvMYB5a, and VvMYB5b (Vitis vinifera), which are known to be involved in the regulation of the anthocyanin pathway in these species [49–51].

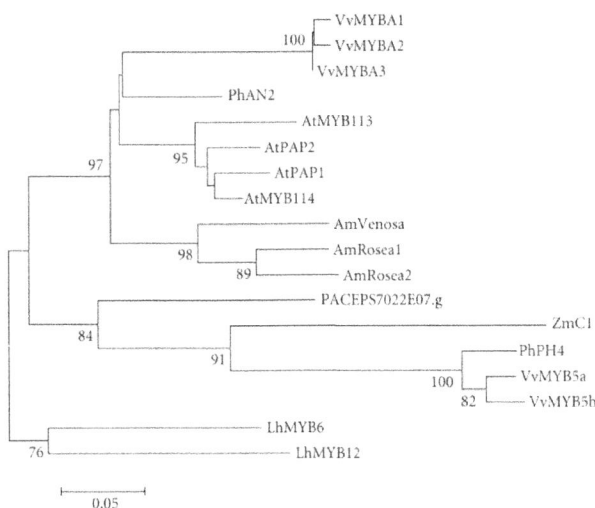

Figure 8: A Neighbor-joining phylogenetic tree of plant R2R3 MYB sequences. Bootstrap values from 1,000 replicates were used to assess the robustness of the trees. Only

bootstrap values above 75% are indicated at the nodes. Accession numbers for genes from other species are given in Supplementary data.

Sequence comparison of selected plant WD40 proteins with the sequence obtained from P. edulis indicated that the four WD repeats are highly conserved among all species analyzed (Figure 9). Phylogenetic analysis of these amino acid sequences confirmed that P. edulis WD40 grouped together with Ricinus communis WD40 and found to be more related to other dicot proteins (Figure 10).

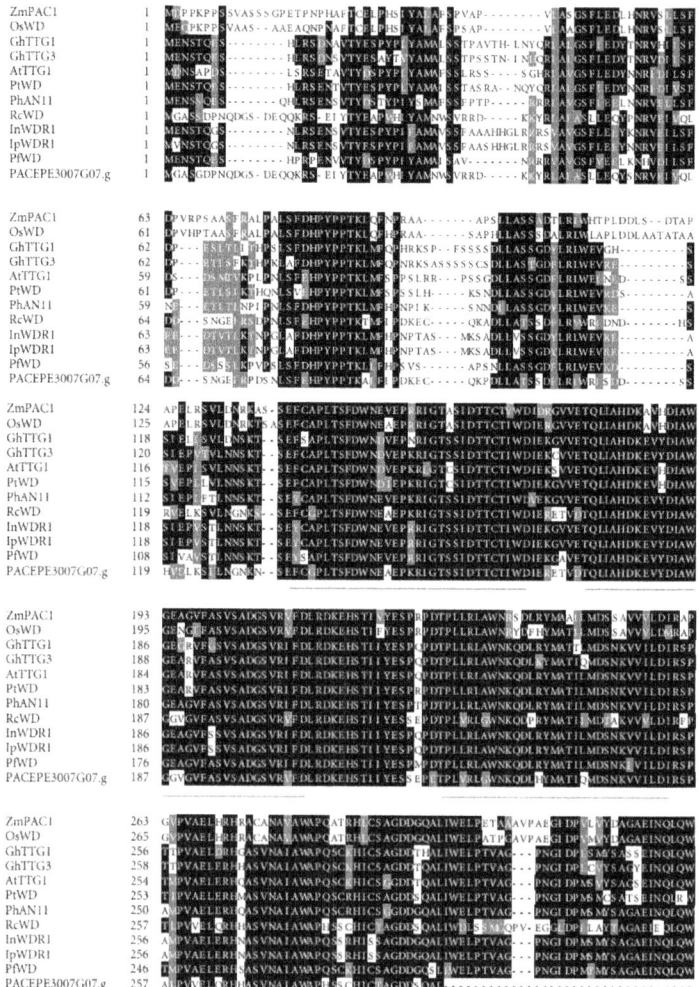

Figure 9: Multiple sequence alignment of the WD40 proteins involved in anthocyanin production, including the deduced amino acid sequence of the Passiflora edulis WD40. The identical and similar residues are highlighted on a black and gray background,

respectively. Four conserved WD repeat domain are underlined in red. The alignment was performed using CLUSTALX and BOXSHADE program.

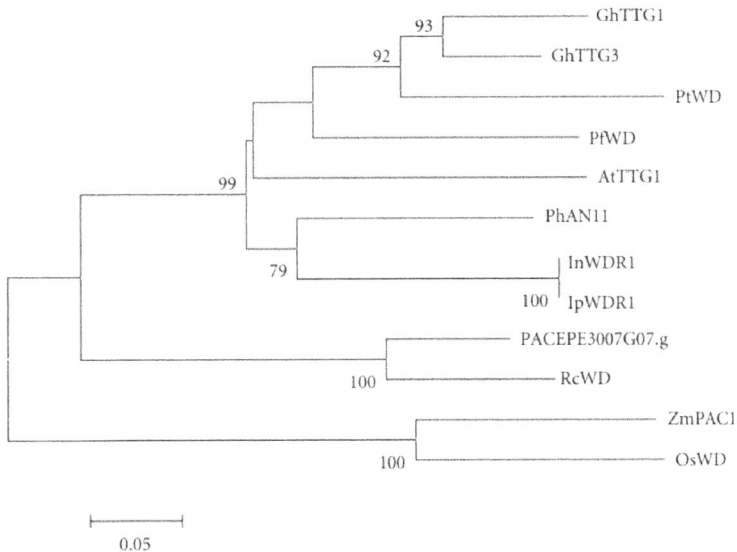

Figure 10: A Neighbor-joining phylogenetic tree of plant WD 40 proteins. Bootstrap values from 1,000 replicates were used to assess the robustness of the trees. Only boot-strap values above 75% are indicated at the nodes. Accession numbers for genes from other species are given in Supplementary data.

No putative homologs to bHLH transcription factors were found in the PASSIOMA database.

DISCUSSION

Flavonoid pathway results in the production of a range of flavonoid compounds, including anthocyanins (Figure 1). CHS is the first enzyme in the phenylpropanoid pathway and is encoded by members of a plant-specific multigene family of polyketide synthases. Nevertheless, genes belonging to the CHS family have been recently described to occur in some microorganisms (Azotobacter vinelandii; [52] and Neurospora crassa; [53]) and, therefore, indicate CHS functions might have evolved previous to the divergence of land plants. Thus, the biological functions of some of the CHS superfamily members are clearly important to plant adaptation. CHS proteins are collectively linked to the biosynthesis of different plant products with diverse functions such as UV protection, defense against pathogens, pigment biosynthesis, and pollen fertility [54, 55].

Sequence analysis indicated that two Passiflora CHS deduced proteins belong to a small distinct group of chalcone synthases that includes angiosperm and gymnosperms homologs to anther-specific chalcone synthase-like genes (ASCLs; highlighted in Figure 2). Furthermore, all ASCLs form a monophyletic clade. Recently, ASCLs transcripts were detected within the tapetum cells during microspore stage in wheat [56]. These genes apparently have important roles in anther development and in pollen fertility [40, 41, 56].

The remaining three Passiflora CHSs were clustered together in a sister clade containing all seed plant CHS genes. Their products are considered key in the biosynthesis of flavonoids. These include CHSA and CHSJgenes, known to be expressed in floral tissues, and involved in floral pigmentation in petunia [30, 31, 57]. Moreover, two nonchalcone genes, divergent from the typical CHSs, formed a separate clade. The SyPKS gene from cyanobacterium encodes an enzyme of the thiolase superfamily [58], whereas the function of thePpCHS11 gene (from Physcomitrella patens) may resemble more the most recent common ancestor of all plant CHSs than do other members of the plant CHS superfamily [55].

We do not have identified putative genes encoding CHI enzymes. Besides the general limitations and drawbacks of the EST-based approach, another possible explanation may be because the rapid isomerization of chalcone to form narigen and the fact that even in the absence of a functional CHI enzyme, chalcone can spontaneously isomerize to form naringenin [15].

DFR is an enzyme catalysing the reduction of three dihydroflavonols: dihydromyricetin (DHM), dihydroquercetin (DHQ), and dihydrokaempferol (DHK) into colorless leucoanthocyanidins. These are further converted to delphinidin, cyaniding, and pelargonidin (Figure 1). The synthesis of three different anthocyanidins is mainly determined by the enzymes activities of two hydroxylases: F3'OH and F3'5'OH. The first converts DHK to DHQ and F3'5'OH converts DHK to DHM [15].

In some plant species, DFR displays distinct substrate specificity in according to the hydroxylation pattern of anthocyanin molecule [30]. A hypothesis to determine substrate specificity was proposed based on the amino acid sequence alignment of Petunia DFR with others plants. The alignment indicated a variable region that controls substrate recognition. Naturally, Petunia hybrida does not produce orange flowers, because the DFR enzyme cannot use dihydrokaempferol as substrate to produce pelargonidin, due to an aspartic acid residue at the 134th position [30, 42], as it was also observed for Passiflora (Figure 3), thus converting dihydroquercetin to leucocyanidin and, more efficiently, the reduction of dihydromyricetin to leucodelphinidin [30, 59]. On the other hand, some Gerbera genotypes have an asparagine residue at this same position and can utilize three dihydroflavonols as substrates of DFR, consequently producing orange to red colored flowers [9, 30]. Thus, the flower color is partly determined by alteration of a single amino acid that changes the substrate specificity of the DFR enzyme.

Almost all anthocyanidins undergo several modifications, which vary across species and involve enzymes of the glucosyltransferase, methyltransferase, and acyltransferase families. The most common is glycosylation of the 3-position of anthocyanidins (represented in Figure 1) to produce stable anthocyanin molecules [15, 30, 31,60]. UDP-glucose:flavonoid 3-O-glucosyltransferase (3GT) belongs to a large multigene glucosyltransferases (GTs) family, representing the final step in anthocyanin biosynthesis.

In this work, we adopted the classification of the GTs into clusters according to Kovinic and colleagues [45]. Cluster I groups includes 3GTs enzymes. Cluster II includes GTs with multiples substrates preferences, generally for chalcones, flavones and flavonols but not anthocyanidins. Enzymes from Cluster III have isoflavone 7-O and anthocyanidin 3,5-O-GT activities. Cluster IV glycosylates flavonol and isoflavonol substrates and Cluster V have anthocyanin 5-O and/ or flavonol 7-O-UGT enzymes [45]. Our results indicated that the obtained Passiflora glucosyltransferase gene sequences were grouped in Cluster II, together with other family members that show a high catalytic specificity for more than one class of flavonoid substrates (Figure 5). DicGT5 (from Dianthus caryophyllus) glycosylates a chalcononaringenin 2'-O-glucosyltransferase [61], whereas the Beta vulgaris GT has a favonoid-7, 4'-O-betanidin-5-O-glucosyltransferase activity [62]. Both GTs have non-anthocyanidin substrate specificity. Despite these results, obviously neither GT substrate specificity, nor in vivo function of the Passiflora GTs can be predicted solely based on amino acid sequence similarities and must be experimentally determined.

Anthocyanin biosynthesis has been demonstrated to occur predominantly in the cytosol, but these pigments are exclusively accumulated in the vacuole of epidermal cells [20]. Transport of pigments to the vacuoles requires a glutathione S-transferase and a specific carrier protein localized in the vacuolar membrane. GSTs are multifunctional proteins encoded by a large familiar present in all cellular organisms. Plants GSTs are classified on the basis of sequence identity into four classes: phi, tau, theta, and zeta [46]. The two small zeta and theta classes include GSTs from animals and plants, while the phi and tau classes are plant-specific. Several studies have confirmed the involvement of GSTs in the vacuolar transport of anthocyanins. PhAN2 (from Petunia), ZmBZ2 (from maize), and AtTT19 (from Arabidopsis) are GST proteins involved in anthocyanin transport [30–32, 63–65].

To characterize their phylogenetic relationships, the deduced amino acid sequences from the Passiflora putative GSTs were compared with other plant GST sequences, including the ones mentioned above. Figure 6 shows that the Passiflora GSTs are included into three different clades: three sequences were positioned in the same clade of PhAN9 and AtTT19 (phi class), whereas one sequence was grouped together with ZmBZ2 (tau class; [66]). Although of these known proteins belong to distinct GST clades, they perform similar functions [63–65].

Interestingly, PACEPE3007A05.g was clustered with carnation (Dianthus caryophyllus) GST type II (zeta class) which is associated to petal senescence in response to ethylene [67, 68].

At the moment, we can classify the Passiflora GSTs into type I (phi), type II (zeta), and type III (tau). At least, four of them might be involved in the anthocyanin pathway and PACEPE3007A05.g might be related to other biological processes related to flower development such as those observed for the carnation GST.

In all analyzed species, the spatial and temporal expression of the structural genes of the anthocyanin biosynthetic pathway is controlled by regulatory genes, which interfere with the intensity and pattern of anthocyanin biosynthesis [15]. MYBs, basic helix-loop-helix (bHLH) transcription factors and WD40 proteins form a transcriptional complex for the activation of the structural genes [4, 12, 20, 47, 69, 70]. MYBs and bHLHs proteins are coded by large multigene families, and those associated with anthocyanin biosynthesis are characterized by a conserved DNA-binding domain consisting of two imperfect repeats (named R2R3), and a specific bHLH domain, respectively. These two gene families have been extensively studied in model plants such as Arabidopsis and maize [48, 49, 71].

A multiple sequence alignment of the R2R3 domains of selected MYB proteins known to be involved in anthocyanin biosynthesis regulation, and the deduced amino acid sequence of PACEPS7022E07.g confirmed the presence of the conserved R2R3-MYB domain in this P. suberosa sequence (Figure 7) as well as that of a second conserved domain in the R3 repeat (red box, Figure 7), which is known to be necessary for the interaction between MYB and bHLH transcription factors [48, 49]. Additionally, a third motif in the R3 repeat (ANDV, blue box in Figure 7) represents a conserved motif shared among all eudicot MYBs involved in the anthocyanin biosynthesis [72].

The phylogenetic tree obtained using the alignment shown in Figure 7 is presented in Figure 8 and indicates LhMYB6 and LhMYB12 clustered outside the eudicot clade. These two genes regulate anthocyanin biosynthesis in the flowers of lily (Lilium hybrid), a monocot [73]. One clade is formed exclusively by eudicot anthocyanin regulators (PhAn2, AtPAP1, AtPAP2, AmROSEA1, and AmROSEA2; [12, 71, 74–77]. Curiously, one regulator of the anthocyanin in maize (a monocot), ZmC1 was positioned in the same clade of other dicot members such as PhPH4 (from Petunia), VvMYB5a, and VvMYB5b (from Vitis), as well as the Passiflora R2R3-MYB sequence. PhPH4 is expressed in the petal epidermis and activates vacuolar acidification in petunia [50]. VvMYB5a and VvMYB5b genes are involved in the regulation of anthocyanin biosynthesis during grape berry development [51].

WD40 proteins are highly conserved and can be found in organisms that do not biosynthesize anthocyanins as algae, fungi, and animals [78, 79]. In plants, these proteins are involved in a plethora of developmental and biochemical functions. As an example, the Arabidopsis TRANSPARENT TESTA GLABRA 1 (TTG1), which is a WD40 protein, is involved in regulating trichome formation, anthocyanin biosynthesis, seed coat pigmentation, and seed coat mucilage production. A common feature of WD40 repeat proteins is that they facilitate protein-protein interactions between the MYB and bHLH proteins [22, 79].

The alignment of the Passiflora WD40 protein sequence with other known WD40s from different plant species revealed the presence of conserved WD40 motifs in the C-terminal region (Figure 9). The phylogenetic tree constructed based on this alignment is shown in Figure 10. The results indicated that the monocot sequences ZmPAC1 and OsWD clustered together, whereas the eudicot WD40s known to function as anthocyanin regulators were grouped into a different clade, with Passiflora WD40 being closely related to the Ricinus communis protein (RcWD, Table 1 and Figure 10). Although WD40 proteins are required to regulate anthocyanins and proanthocyanidin together with MYB and bHLH transcription factors, their potential involvement in other biological processes is enormous, therefore, it is premature to say what functions PACEPE3007G07.g might perform in Passiflora.

The fact that no putative homologs to bHLH transcription factors were found in the PASSIOMA database may reflect the high degree of novelty of most of the libraries of the PASSIOMA project indicating that full gene expression spectra was not completely achieved [27]. Perhaps a more deep sequencing effort would reveal that such homologs are indeed expressed in Passiflora flowers, as these elements are generally essential to MYB-WD40 protein complex stability [30–32].

CONCLUSIONS AND PERSPECTIVES

We took the first steps toward the understanding of the molecular processes involved in the biosynthesis of anthocyanins in Passiflora that could account for the differences in pollinator preferences found in the genus. We identified 15 putative coding sequences derived from two distinct Passiflora species (P. edulis and P. suberosa) expressed in developing flower buds and potentially involved in the anthocyanin biosynthetic pathway. Comparisons of deduced amino acid sequences from the 15 Passiflora cDNAs with selected sequences from other plant species revealed strong similarity with genes that encode key elements involved in the biosynthesis (8 sequences), transcriptional regulation (2 sequences), and transport (5 sequences) of anthocyanin molecules.

Needed research concerning the determination of temporal and spatial expression patterns of all thesePassiflora putative anthocyanin-related genes presented here are already ongoing in our group. We expect that future work on the manipulation of their expression patterns, using transgenic approaches, will help us to unravel important aspects relating anthocyanin biosynthesis, flower pigmentation, and flower pollination in rapidly changing tropical environments.

ACKNOWLEDGMENT

The authors acknowledge FAPESP and CNPq (Brazil) for financial support.

REFERENCES

1.   J. B. Harborne, The Flavonoids: Advances in Research Since 1986, Chapman & Hall/CRC, New York, NY, USA, 1st edition, 1994.

2.   E. Grotewold, "The genetics and biochemistry of floral pigments," Annual Review of Plant Biology, vol. 57, pp. 761–780, 2006.

3.   M. D. Rausher, "Evolutionary transitions in floral color," International Journal of Plant Sciences, vol. 169, no. 1, pp. 7–21, 2008.

4.   F. Quattrocchio, J. Wing, K. van der Woude et al., "Molecular analysis of the anthocyanin2 gene of Petunia and its role in the evolution of flower color," Plant Cell, vol. 11, no. 8, pp. 1433–1444, 1999.

5.   C. Spelt, F. Quattrocchio, J. N. M. Mol, and R. Koes, "Anthocyanin1 of Petunia encodes a basic helix-loop-helix protein that directly activates transcription of structural anthocyanin genes," Plant Cell, vol. 12, no. 9, pp. 1619–1631, 2000.

6.   W. Heller, G. Forkmann, L. Britsch, and H. Grisebach, "Enzymatic

reduction of (+)-dihydroflavonols to flavan-3,4-cis-diols with flower extracts from Matthiola incana and its role in anthocyanin biosynthesis,"Planta, vol. 165, no. 2, pp. 284–287, 1985.

7. K. Stich, T. Eidenberger, F. Wurst, and G. Forkmann, "Enzymatic conversion of dihydroflavonols to flavan-3,4-diols using flower extracts of Dianthus caryophyllus L. (carnation)," Planta, vol. 187, no. 1, pp. 103–108, 1992.

8. K. M. Davies, J. M. Bradley, K. E. Schwinn, K. R. Markham, and E. Podivinsky, "Flavonoid biosynthesis in flower petals of five lines of lisianthus (Eustoma grandiflorum Grise.)," Plant Science, vol. 95, no. 1, pp. 67–77, 1993.

9. Y. Helariutta, P. Elomaa, M. Kotilainen, P. Seppänen, and T. H. Teeri, "Cloning of cDNA coding for dihydroflavonol-4-reductase (DFR) and characterization of dfr expression in the corollas of Gerbera hybrida var. Regina (Compositae)," Plant Molecular Biology, vol. 22, no. 2, pp. 183–193, 1993.

10. E. Grotewold, B. J. Drummond, B. Bowen, and T. Peterson, "The myb-homologous P gene controls phlobaphene pigmentation in maize floral organs by directly activating a flavonoid biosynthetic gene subset," Cell, vol. 76, no. 3, pp. 543–553, 1994.

11. J. M. Hernandez, G. F. Heine, N. G. Irani et al., "Different mechanisms participate in the R-dependent activity of the R2R3 MYB transcription factor C1," The Journal of Biological Chemistry, vol. 279, no. 46, pp. 48205–48213, 2004.

12. K. Schwinn, J. Venail, Y. Shang et al., "A small family of MYB-regulatory genes controls floral pigmentation intensity and patterning in the Genus antirrhinum," Plant Cell, vol. 18, no. 4, pp. 831–851, 2006.

13. Y. Morita, M. Saitoh, A. Hoshino, E. Nitasaka, and S. Iida, "Isolation of cDNAs for R2R3-MYB, bHLH and WDR transcriptional regulators and identification of c and ca mutations conferring white flowers in the Japanese morning glory," Plant and Cell Physiology, vol. 47, no. 4, pp. 457–470, 2006.

14. K. I. Park, N. Ishikawa, Y. Morita, J. D. Choi, A. Hoshino, and S. Iida, "A bHLH regulatory gene in the common morning glory, Ipomoea purpurea, controls anthocyanin biosynthesis in flowers, proanthocyanidin and phytomelanin pigmentation in seeds, and seed trichome formation," Plant Journal, vol. 49, no. 4, pp. 641–654, 2007. ·

15. T. A. Holton and E. C. Cornish, "Genetics and biochemistry of anthocyanin

biosynthesis," Plant Cell, vol. 7, no. 7, pp. 1071–1083, 1995.

16. R. Koes, W. Verweij, and F. Quattrocchio, "Flavonoids: a colorful model for the regulation and evolution of biochemical pathways," Trends in Plant Science, vol. 10, no. 5, pp. 236–242, 2005.

17. K. Davies, "Plant pigments and their manipulation," Annual Plant Reviews, vol. 14, p. 368, 2004.

18. H. S. Lee and V. Hong, "Chromatographic analysis of anthocyanins," Journal of Chromatography, vol. 624, no. 1-2, pp. 221–234, 1992.

19. J. B. Harborne, T. J. Mabry, and H. Mabry, The Flavonoids, Academic Press, New York, NY, USA, 1975.

20. J. Mol, E. Grofewold, and R. Koes, "How genes paint flowers and seeds," Trends in Plant Science, vol. 3, no. 6, pp. 212–217, 1998.

21. B. Winkel-Shirley, "Flavonoid biosynthesis. A colorful model for genetics, biochemistry, cell biology, and biotechnology," Plant Physiology, vol. 126, no. 2, pp. 485–493, 2001.

22. N. A. Ramsay and B. J. Glover, "MYB-bHLH-WD40 protein complex and the evolution of cellular diversity," Trends in Plant Science, vol. 10, no. 2, pp. 63–70, 2005.

23. L. Lepiniec, I. Debeaujon, J. M. Routaboul et al., "Genetics and biochemistry of seed flavonoids," Annual Review of Plant Biology, vol. 57, pp. 405–430, 2006.

24. T. Ulmer and J. M. MacDougal, Passiflora, Passion Flowers of the World, Timber Press, Cambridge, UK, 2004.

25. M. M. Halim and R. P. Collins, "Anthocyanins of Passiflora quadrangularis," Bulletin of the Torrey Botanical Club, vol. 97, no. 5, pp. 247–248, 1970.

26. L. KidØy, A. M. Nygård, Ø. M. Andersen, A. T. Pedersen, D. W. Aksnes, and B. T. Kiremire, "Anthocyanins in fruits of Passiflora edulis and P. suberosa," Journal of Food Composition and Analysis, vol. 10, no. 1, pp. 49–54, 1997.

27. M. C. Dornelas, S. M. Tsai, and A. P. M. Rodriguez, "Expressed sequence tags of genes involved in the flowering process of Passiflora spp.," in Floriculture, Ornamental and Plant Biotechnology, J. A. Teixeira da Silva, Ed., pp. 483–488, Global Science Books, London, UK, 2006.

28. I. G. Varassin, J. R. Trigo, and M. Sazima, "The role of nectar production, flower pigments and odour in the pollination of four species of Passiflora (Passifloraceae) in south-eastern Brazil," Botanical Journal of the Linnean Society, vol. 136, no. 2, pp. 139–152, 2001.

29. S. F. Altschul, W. Gish, W. Miller, E. W. Myers, and D. J. Lipman, "Basic local alignment search tool,"Journal of Molecular Biology, vol. 215, no. 3, pp. 403–410, 1990.

30. T. Nakatsuka, Y. Abe, Y. Kakizaki, S. Yamamura, and M. Nishihara, "Production of red-flowered plants by genetic engineering of multiple flavonoid biosynthetic genes," Plant Cell Reports, vol. 26, no. 11, pp. 1951–1959, 2007.

31. M. Hanumappa, G. Choi, S. Ryu, and G. Choi, "Modulation of flower colour by rationally designed dominant-negative chalcone synthase," Journal of Experimental Botany, vol. 58, no. 10, pp. 2471–2478, 2007.

32. S. Martens, A. Preuß, and U. Matern, "Multifunctional flavonoid dioxygenases: flavonol and anthocyanin biosynthesis in Arabidopsis thaliana L.," Phytochemistry, vol. 71, no. 10, pp. 1040–1049, 2010.

33. X. Huang and A. Madan, "CAP3: a DNA sequence assembly program," Genome Research, vol. 9, no. 9, pp. 868–877, 1999.

34. T. A. Hall, "BioEdit: a user-friendly biological sequence alignment editor and analysis program for Windows 95/98/NT," Nucleic Acids Symposium Series, vol. 41, no. 41, pp. 95–98, 1999.

35. S. Hunter, R. Apweiler, T. K. Attwood et al., "InterPro: the integrative protein signature database," Nucleic Acids Research, vol. 37, no. 1, pp. D211–D215, 2009.

36. J. D. Thompson, T. J. Gibson, F. Plewniak, F. Jeanmougin, and D. G. Higgins, "The CLUSTAL X windows interface: flexible strategies for multiple sequence alignment aided by quality analysis tools," Nucleic Acids Research, vol. 25, no. 24, pp. 4876–4882, 1997.

37. S. Kumar, K. Tamura, and M. Nei, "MEGA3: integrated software for molecular evolutionary genetics analysis and sequence alignment," Briefings in Bioinformatics, vol. 5, no. 2, pp. 150–163, 2004.

38. N. Saitou and M. Nei, "The neighbor-joining method: a new method for reconstructing phylogenetic trees," Molecular Biology and Evolution, vol. 4, no. 4, pp. 406–425, 1987.

39. M. J. Moore, P. S. Soltis, C. D. Bell, J. G. Burleigh, and D. E. Soltis, "Phylogenetic analysis of 83 plastid genes further resolves the early diversification of eudicots," Proceedings of the National Academy of Sciences of the United States of America, vol. 107, no. 10, pp. 4623–4628, 2010.

40. A. Ageez, Y. Kazama, R. Sugiyama, and S. Kawano, "Male-fertility genes expressed in male flower buds ofSilene latifolia include homologs

of anther-specific genes," Genes and Genetic Systems, vol. 80, no. 6, pp. 403–413, 2005.

41. C. Jiang, C. K. Schommer, S. Y. Kim, and D. Y. Suh, "Cloning and characterization of chalcone synthase from the moss, Physcomitrella patens," Phytochemistry, vol. 67, no. 23, pp. 2531–2540, 2006.

42. M. Beld, C. Martin, H. Huits, A. R. Stuitje, and A. G. M. Gerats, "Flavonoid synthesis in Petunia hybrida: partial characterization of dihydroflavonol-4-reductase genes," Plant Molecular Biology, vol. 13, no. 5, pp. 491–502, 1989.

43. E. T. Johnson, S. Ryu, H. Yi, B. Shin, H. Cheong, and G. Choi, "Alteration of a single amino acid changes the substrate specificity of dihydroflavonol 4-reductase," Plant Journal, vol. 25, no. 3, pp. 325–333, 2001.

44. N. Shimada, R. Sasaki, S. Sato et al., "A comprehensive analysis of six dihydroflavonol 4-reductases encoded by a gene cluster of the Lotus japonicus genome," Journal of Experimental Botany, vol. 56, no. 419, pp. 2573–2585, 2005.

45. N. Kovinich, A. Saleem, J. T. Arnason, and B. Miki, "Functional characterization of a UDP-glucose:flavonoid 3-O-glucosyltransferase from the seed coat of black soybean (Glycine max (L.) Merr.)," Phytochemistry, vol. 71, no. 11-12, pp. 1253–1263, 2010.

46. D. P. Dixon, A. Lapthorn, and R. Edwards, "Plant glutathione transferases," Genome Biology, vol. 3, no. 3, pp. 1–10, 2002.

47. Y. Borovsky, M. Oren-Shamir, R. Ovadia, W. De Jong, and I. Paran, "The A locus that controls anthocyanin accumulation in pepper encodes a MYB transcription factor homologous to Anthocyanin2 of Petunia," Theoretical and Applied Genetics, vol. 109, no. 1, pp. 23–29, 2004.

48. I. M. Zimmermann, M. A. Heim, B. Weisshaar, and J. F. Uhrig, "Comprehensive identification of Arabidopsis thaliana MYB transcription factors interacting with R/B-like BHLH proteins," Plant Journal, vol. 40, no. 1, pp. 22–34, 2004.

49. E. Grotewold, M. B. Sainz, L. Tagliani, J. M. Hernandez, B. Bowen, and V. L. Chandler, "Identification of the residues in the Myb domain of maize C1 that specify the interaction with the bHLH cofactor R," Proceedings of the National Academy of Sciences of the United States of America, vol. 97, no. 25, pp. 13579–13584, 2000.

50. F. Quattrocchio, W. Verweij, A. Kroon, C. Spelt, J. Mol, and R. Koes, "PH4 of petunia is an R2R3 MYB protein that activates vacuolar acidification through interactions with basic-helix-loop-helix transcription factors of the anthocyanin pathway," Plant Cell, vol. 18, no. 5, pp. 1274–1291,

2006.

51. L. Deluc, J. Bogs, A. R. Walker et al., "The transcription factor VvMYB5b contributes to the regulation of anthocyanin and proanthocyanidin biosynthesis in developing grape berries," Plant Physiology, vol. 147, no. 4, pp. 2041–2053, 2008.

52. N. Funa, H. Ozawa, A. Hirata, and S. Horinouchi, "Phenolic lipid synthesis by type III polyketide synthases is essential for cyst formation in Azotobacter vinelandii," Proceedings of the National Academy of Sciences of the United States of America, vol. 103, no. 16, pp. 6356–6361, 2006.

53. N. Funa, T. Awakawa, and S. Horinouchi, "Pentaketide resorcylic acid synthesis by type III polyketide synthase from Neurospora crassa," The Journal of Biological Chemistry, vol. 282, no. 19, pp. 14476–14481, 2007.

54. M. L. Durbin, B. McCaig, and M. T. Clegg, "Molecular evolution of the chalcone synthase multigene family in the morning glory genome," Plant Molecular Biology, vol. 42, no. 1, pp. 79–92, 2000.

55. P. K. H. Koduri, G. S. Gordon, E. I. Barker, C. C. Colpitts, N. W. Ashton, and D. Y. Suh, "Genome-wide analysis of the chalcone synthase superfamily genes of Physcomitrella patens," Plant Molecular Biology, vol. 72, no. 3, pp. 247–263, 2010.

56. S. Wu, S. J. B. O›Leary, S. Gleddie, F. Eudes, A. Laroche, and L. S. Robert, "A chalcone synthase-like gene is highly expressed in the tapetum of both wheat (Triticum aestivum L.) and triticale (x Triticosecale Wittmack)," Plant Cell Reports, vol. 27, no. 9, pp. 1441–1449, 2008.

57. R. E. Koes, C. E. Spelt, P. J. M. van den Elzen, and J. N. M. Mol, "Cloning and molecular characterization of the chalcone synthase multigene family of Petunia hybrida," Gene, vol. 81, no. 2, pp. 245–257, 1989.

58. C. Jiang, S. Y. Kim, and D. Y. Suh, "Divergent evolution of the thiolase superfamily and chalcone synthase family," Molecular Phylogenetics and Evolution, vol. 49, no. 3, pp. 691–701, 2008.

59. G. Forkmann and B. Ruhnau, "Distinct substrate specificity of dihydroflavonol-4-reductase from flowers of Petunia hybrida," Zeitschrift für Naturforschung—Section C: Biosciences, vol. 42, pp. 1146–1148, 1987.

60. S. Rudd, "Expressed sequence tags: alternative or complement to whole genome sequences?" Trends in Plant Science, vol. 8, no. 7, pp. 321–329, 2003.

61. J. Ogata, Y. Itoh, M. Ishida, H. Yoshida, and Y. Ozeki, "Cloning and heterologous expression of cDNAs encoding flavonoid glucosyltransferases from Dianthus caryophyllus," Plant Biotechnology, vol. 21, no. 5, pp. 367–375, 2004.

62. J. Isayenkova, V. Wray, M. Nimtz, D. Strack, and T. Vogt, "Cloning and functional characterisation of two regioselective flavonoid glucosyltransferases from Beta vulgaris," Phytochemistry, vol. 67, no. 15, pp. 1598–1612, 2006.

63. M. R. Alfenito, E. Souer, C. D. Goodman et al., "Functional complementation of anthocyanin sequestration in the vacuole by widely divergent glutathione S-transferases," Plant Cell, vol. 10, no. 7, pp. 1135–1149, 1998.

64. K. A. Marrs, M. R. Alfenito, A. M. Lloyd, and V. Walbot, "A glutathione S-transferase involved in vacuolar transfer encoded by the maize gene Bronze-2," Nature, vol. 375, no. 6530, pp. 397–400, 1995.

65. S. Kitamura, N. Shikazono, and A. Tanaka, "TRANSPARENT TESTA 19 is involved in the accumulation of both anthocyanins and proanthocyanidins in Arabidopsis," Plant Journal, vol. 37, no. 1, pp. 104–114, 2004.

66. L. Loyall, K. Uchida, S. Braun, M. Furuya, and H. Frohnmeyer, "Glutathione and a UV light-induced glutathione S-transferase are involved in signaling to chalcone synthase in cell cultures," Plant Cell, vol. 12, no. 10, pp. 1939–1950, 2000.

67. R. C. Meyer, P. B. Goldsbrough, and W. R. Woodson, "An ethylene-responsive flower senescence-related gene from carnation encodes a protein homologous to glutathione s-transferases," Plant Molecular Biology, vol. 17, no. 2, pp. 277–281, 1991.

68. H. Itzhaki, J. M. Maxson, and W. R. Woodson, "An ethylene-responsive enhancer element is involved in the senescence- related expression of the carnation glutathione-S-transferase (GST1) gene," Proceedings of the National Academy of Sciences of the United States of America, vol. 91, no. 19, pp. 8925–8929, 1994.

69. P. Elomaa, A. Uimari, M. Mehto, V. A. Albert, R. A. E. Laitinen, and T. H. Teeri, "Activation of anthocyanin biosynthesis in Gerbera hybrida (Asteraceae) suggests conserved protein-protein and protein-promoter interactions between the anciently diverged monocots and eudicots," Plant Physiology, vol. 133, no. 4, pp. 1831–1842, 2003.

70. H. Mathews, S. K. Clendennen, C. G. Caldwell et al., "Activation tagging in tomato identifies a transcriptional regulator of anthocyanin

biosynthesis, modification, and transport," Plant Cell, vol. 15, no. 8, pp. 1689–1703, 2003.

71. R. Stracke, M. Werber, and B. Weisshaar, "The R2R3-MYB gene family in Arabidopsis thaliana," Current Opinion in Plant Biology, vol. 4, no. 5, pp. 447–456, 2001.

72. K. Lin-Wang, K. Bolitho, and K. Grafton, "An R2R3 MYB transcription factor associated with regulation of the anthocyanin biosynthetic pathway in Rosaceae," BMC Plant Biology, vol. 10, no. 1, pp. 50–67, 2010.

73. M. Yamagishi, Y. Shimoyamada, T. Nakatsuka, and K. Masuda, "Two R2R3-MYB genes, homologs of petunia AN2, regulate anthocyanin biosyntheses in flower tepals, tepal spots and leaves of asiatic hybrid Lily," Plant and Cell Physiology, vol. 51, no. 3, pp. 463–474, 2010.

74. C. Jiang, X. Gu, and T. Peterson, "Identification of conserved gene structures and carboxy-terminal motifs in the Myb gene family of Arabidopsis and Oryza sativa L. ssp. indica," Genome Biology, vol. 5, no. 7, p. R46, 2004.

75. S. Kobayashi, M. Ishimaru, K. Hiraoka, and C. Honda, "Myb-related genes of the Kyoho grape (Vitis labruscana) regulate anthocyanin biosynthesis," Planta, vol. 215, no. 6, pp. 924–933, 2002.

76. A. R. Walker, E. Lee, J. Bogs, D. A. J. McDavid, M. R. Thomas, and S. P. Robinson, "White grapes arose through the mutation of two similar and adjacent regulatory genes," Plant Journal, vol. 49, no. 5, pp. 772–785, 2007.

77. A. Gonzalez, M. Zhao, J. M. Leavitt, and A. M. Lloyd, "Regulation of the anthocyanin biosynthetic pathway by the TTG1/bHLH/Myb transcriptional complex in Arabidopsis seedlings," Plant Journal, vol. 53, no. 5, pp. 814–827, 2008.

78. N. de Vetten, F. Quattrocchio, J. Mol, and R. Koes, "The an11 locus controlling flower pigmentation in petunia encodes a novel WD-repeat protein conserved in yeast, plants, and animals," Genes and Development, vol. 11, no. 11, pp. 1422–1434, 1997.

79. J. Brueggemann, B. Weisshaar, and M. Sagasser, "A WD40-repeat gene from Malus × domestica is a functional homologue of Arabidopsis thaliana TRANSPARENT TESTA GLABRA1," Plant Cell Reports, vol. 29, no. 3, pp. 285–294, 2010.

CITATION

CHAPTER 1

F. Pacheco, H. Oliveira Silveira, A. Alvarenga, I. Almeida Alvarenga, J. Pereira Pinto and J. Sousa Lira, "Gas Exchange and Production of Photosynthetic Pigments of Piper aduncum L. Grown at Different Irradiances," American Journal of Plant Sciences, Vol. 4 No. 12C, 2013, pp. 114-121. doi: 10.4236/ajps.2013.412A3014.

CHAPTER 2

Huang J, Wei C, Zhang Y, Blackburn GA, Wang X, Wei C, et al. (2015) Meta-Analysis of the Detection of Plant Pigment Concentrations Using Hyperspectral Remotely Sensed Data. PLoS ONE 10(9): e0137029. doi:10.1371/journal.pone.0137029

CHAPTER 3

Herrera, A. (2015) The Biological Pigments in Plants Physiology. Agricultural Sciences, 6, 1262-1271. doi: 10.4236/as.2015.610121.

CHAPTER 4

Azamjon B. Soliev, Kakushi Hosokawa, and Keiichi Enomoto, "Bioactive Pigments from Marine Bacteria: Applications and Physiological Roles," Evidence-Based Complementary and Alternative Medicine, vol. 2011, Article ID 670349, 17 pages, 2011. doi:10.1155/2011/670349.

CHAPTER 5

Aiman Yusoff, N. T. R. N. Kumara, Andery Lim, Piyasiri Ekanayake, and Kushan U. Tennakoon, "Impacts of Temperature on the Stability of Tropical Plant Pigments as Sensitizers for Dye Sensitized Solar Cells," Journal of Biophysics, vol. 2014, Article ID 739514, 8 pages, 2014. doi:10.1155/2014/739514.

CHAPTER 6

Giorgio Manenti & Giuliano Tedesco (1977) Ultrastructure and Pigments of Ivy (Hedera Helix L.) Varieties with Green and Variegated Leaves, Caryologia, 30:2, 163-176, DOI: 10.1080/00087114.1977.10796689.

CHAPTER 7

Kristin R. Abney, Dean A. Kopsell, Carl E. Sams, Svetlana Zivanovic, and David E. Kopsell, "UV-B Radiation Impacts Shoot Tissue Pigment Composition in Allium fistulosum L. Cultigens," The Scientific World Journal, vol. 2013, Article ID 513867, 10 pages, 2013. doi:10.1155/2013/513867

CHAPTER 8

Reena Kushwaha, Pankaj Srivastava, and Lal Bahadur, "Natural Pigments from Plants Used as Sensitizers for TiO2 Based Dye-Sensitized Solar Cells," Journal of Energy, vol. 2013, Article ID 654953, 8 pages, 2013. doi:10.1155/2013/654953

CHAPTER 9

Alítcia Moraes Kleinowski; Isabel Rodrigues Brandão; Andersom Milech Einhardt; Márcia Vaz Ribeiro; José Antonio Peters; Eugenia Jacira Bolacel Braga, Pigment production and growth of Alternanthera plants cultured in vitro in the presence of tyrosine, http://dx.doi.org/10.1590/S1516-89132013005000012.

CHAPTER 10

Harris et al.: Betalain production is possible in anthocyanin-producing plant species given the presence of DOPAdioxygenase and L-DOPA. BMC Plant Biology 2012 12:34. doi:10.1186/1471-2229-12-34.

CHAPTER 12

Xueyun Hu, Ayumi Tanaka and Ryouichi Tanaka, Simple extraction methods that prevent the artifactual conversion of chlorophyll to chlorophyllide during pigment isolation from leaf samples, DOI: 10.1186/1746-4811-9-19

CHAPTER 13

Lilian Cristina Baldon Aizza and Marcelo Carnier Dornelas, "A Genomic Approach to Study Anthocyanin Synthesis and Flower Pigmentation in Passionflowers," Journal of Nucleic Acids, vol. 2011, Article ID 371517, 17 pages, 2011. doi:10.4061/2011/371517

INDEX